Modeling Indoor Air Pollution

Modeling Indoor Air Pollution

Darrell W Pepper
University of Nevada, Las Vegas, USA

David Carrington
Los Alamos National Laboratory, USA

ICP
Imperial College Press

Published by

Imperial College Press
57 Shelton Street
Covent Garden
London WC2H 9HE

Distributed by

World Scientific Publishing Co. Pte. Ltd.
5 Toh Tuck Link, Singapore 596224
USA office: 27 Warren Street, Suite 401-402, Hackensack, NJ 07601
UK office: 57 Shelton Street, Covent Garden, London WC2H 9HE

British Library Cataloguing-in-Publication Data
A catalogue record for this book is available from the British Library.

MODELING INDOOR AIR POLLUTION

Copyright © 2009 by Imperial College Press

All rights reserved. This book, or parts thereof, may not be reproduced in any form or by any means, electronic or mechanical, including photocopying, recording or any information storage and retrieval system now known or to be invented, without written permission from the Publisher.

For photocopying of material in this volume, please pay a copying fee through the Copyright Clearance Center, Inc., 222 Rosewood Drive, Danvers, MA 01923, USA. In this case permission to photocopy is not required from the publisher.

ISBN-13 978-1-84816-324-9
ISBN-10 1-84816-324-X

Printed in Singapore.

To our families, friends, and colleagues

Acknowledgments

This book is the result of many years of developing and applying various models to simulate the dispersion of contaminants. Some of the information presented stems from numerous interactions and collaborations with students and colleagues. Much of the material presented in this book can be found in other texts. Some of these texts briefly describe a few of the more simple numerical techniques, but generally don't provide source codes or describe the latest numerical techniques. We have attempted to integrate fundamental information and the basic formulations regarding indoor air quality and ventilation from the more popular textbooks and monographs.

We wish to especially acknowledge Dr. Xiuling Wang, who diligently converted many of our old FORTRAN codes into MATLAB files, and also developed the COMSOL example files. Also we thank Ms. Kathryn Nelson who developed the website for the book and indoor air quality computer codes.

We are grateful to Mrs. Jeannie Pepper, who typed the manuscript and made sure that we followed proper procedures and format. Her professionalism and patience in typing and proofreading the chapters are greatly appreciated.

Finally, we wish to acknowledge the assistance and support of Miss Elizabeth Bennett, our Editor with ICP. This book was written using Microsoft Word and MathType, with conversion to Adobe pdf. The figures were made using TECPLOT, Microsoft Vizio, and PhotoShop.

Preface

Indoor air pollution is becoming a serious problem. This is especially pertinent in situations where sick building syndrome, under design air flow, and contaminant dispersion occur. Recent interest in homeland security and the aftermath of terrorist activities have created a desire among many governmental agencies to more fully understand interior pollutant dispersion and risk assessment. It is the intention of this book to acquaint the reader with enough information to begin using various modeling tools for assessing indoor air pollution. There are many levels of models, ranging in sophistication from simple analytical expressions to elegant, 3-D schemes for solving the Navier–Stokes equations for fluid flow and species transport. The level of modeling effort resides ultimately with the user, and the desired level of accuracy. While 3-D numerical schemes based on finite difference, finite volume, or finite element techniques provide elegant solutions, they also require a great deal of understanding, patience, and computational resources. Analytical solutions, while fast and simple, may be orders of magnitude off in comparison to actual values. This book presents these most common of numerical and analytical tools that can be used for modeling indoor air pollution and the types of problems where a particular model is best suited.

Chapter 1 presents an overview of indoor air pollution, types of ventilation systems, exposure, and general modeling techniques. Chapter 2 discusses the governing mathematical equations that serve as the basis for modeling air pollutant and flow patterns. In Chapter 3 a general discussion of contaminant sources routinely associated with indoor air quality assessment studies is given, with a presentation on particulates and evaporation of droplets. Assessment criteria are described in Chapter 4, including what to consider in exposure levels as well as economical issues associated with design. Chapter 5 introduces the fundamental analytical tools, along with advection and the classic box model approach, for performing simple model simulation, including their limitations. The dynamics of particle motion, including particle drag and flow in inlets and flanges, are given in Chapter 6. Chapter 7 describes the fundamental numerical approaches commonly used in CFD-type simulations, which are based on finite difference, finite volume, and finite element techniques. Additional discussions are given in Chapter 8 on more advanced methods that include boundary element, particle-in-cell, and meshless methods, which is relatively new. Several modeling examples are included. In Chapter 9, an extensive description of turbulence modeling is presented with consideration to both finite volume and finite element techniques. A time-dependent, two-equation closure model is presented in detail using the finite element method with adaptive meshing; results are shown for example problems. Issues regarding homeland security and the potential threats attributed to terrorist activities are discussed in Chapter 10, including an example scenario.

The examples and computer techniques discussed in the book are available on the web. The website is: www.iaqcodes.com. We have elected to write the majority of the codes in MATLAB. The website lists locations where you can also find FORTRAN and C/C++ versions of some of the example codes. We have found that most engineering graduates today as well as science and engineering students are familiar with MATLAB, and prefer using it as their primary coding environment. We have also used COMSOL to run the example problems. COMSOL, with headquarters in Sweden, is a very versatile multiphysics finite element package used throughout the world; the package permits easy interface and flexibility in setting up problems, along with MATLAB scripting. Many universities and companies are now using COMSOL. In addition, an adaptive finite-element based model that can be used for indoor air pollution is also available from the authors. This model utilizes h-adaptation (mesh refinement) to accurately simulate the dispersion of contaminant within any shaped interior.

Darrell W. Pepper
David B. Carrington
2008

Contents

Chapter 1

Introduction

The study of indoor air pollution has evolved into a unique discipline requiring knowledge in several areas. One must be adept at understanding fundamental principles of fluid mechanics, species transport, heat transfer, and systems engineering. Today, buildings have become complex entities with considerable electronic control features embedded within the structures. Of particular concern are issues involving contaminants that routinely enter or lie dormant within building interiors, and their effects upon human health. Articles can be commonly found in newspapers printed throughout the world describing groups of people becoming sick while staying in a hotel, cruising on a ship, or traveling in planes or buses.

Today, efforts to define and describe pollutant transport within buildings and interiors has become complex. Modeling pollutant transport within indoor environments now requires knowledge of computational tools and techniques that were utilized only in research laboratories a few years ago. Knowledge of fundamental principles of ventilation and building systems, including HVAC, must now be coupled with computational fluid dynamics techniques in order to accurately assess human health and predict contaminant exposure. We begin with a brief background in understanding exactly what is meant by indoor air pollution.

1.1. What is Indoor Air Pollution

The study of indoor air pollution (IAP) involves dealing with the emission, accumulation, and assessment of pollutants generally attributed to poor ventilation and air exchange. Of particular concern are issues involving air quality and human comfort within buildings. Toxic fumes and airborne diseases are known to produce undesirable odors, eye and nose irritations, sickness, and occasionally death. Other products such as tobacco smoke and carbon monoxide can also have serious health effects on people exposed to a poorly ventilated environment; studies indicate that indirect or passive smoking can also lead to lung cancer. Recommendations for outdoor airflow rates to dilute indoor polluted air vary considerably.

1.2. Ventilation Systems

Ventilation systems are designed to either prevent contaminants from entering a room or remove contaminants from interior sources within the room. Since ventilation systems are integral to the study of indoor air pollution, it is prudent to at least identify them.

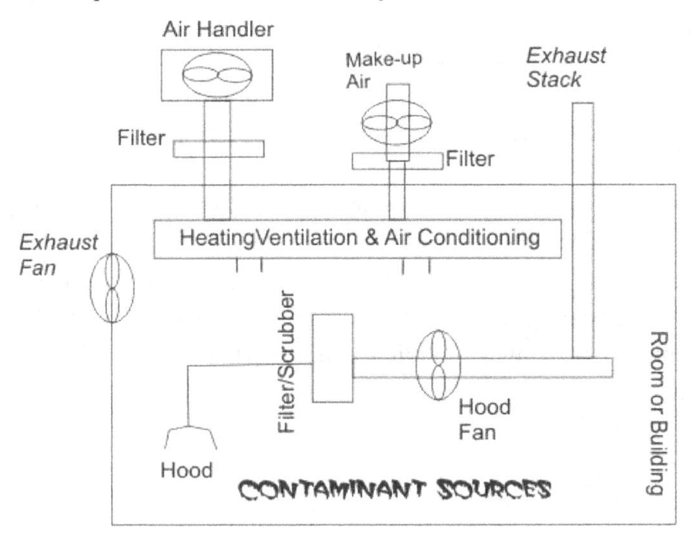

Fig. 1.1 Schematic of a typical ventilation system.

A ventilation system consists of several key components: (1) the contaminant source, (2) an exhaust hood, (3) an air mover, (4) ducts and fittings, (5) makeup air, (6) exhaust air, (7) a pollutant removal device, (8) a discharge stack, and (9) air recirculation. Variations of these components are typically found in most ventilation systems designed to deal with indoor air quality and pollutant removal. Figure 1.1 shows a schematic overview of a general ventilation system.

In particular, the contaminant source typically consists of particulates, gases, and vapors generated by various activities. An exhaust hood is used to contain contaminants emitted from a source, e.g., hoods are used to cover grills in kitchens, an air mover, or fan, is used to draw air into a hood ducts and fittings make up the piping network connecting the hood to the fan, makeup air is air that is brought into the room from the outside – this air is usually temperature and humidity controlled, exhaust air is the air discharged from the room, a pollutant removal device is a specific piece of equipment used to remove excess contaminant from the room (when environmental standards are exceeded), a discharge stack is a stack that exhausts air into the atmosphere, and air recirculation is air that is returned into the room (clean air).

These components are fairly common in rooms containing ventilation systems, especially industrial settings that deal with dirty environments. More detail describing these components and their proper selection can be found in the ASHRAE Handbook (1981) and the textbook by Heinshohn (1991).

1.3. Exposure Risks

The assessment of risk attributed to exposure from hazardous materials is a formal field of study. A great deal of effort was spent in developing risk limits during the early years of the nuclear industry, i.e., in the design and operation of nuclear reactors. A significant amount of mathematical development and theory exists on the subject (see Brain and Beck, 1985).

Assessing risk requires information dealing with the types and amounts of hazardous material and the percent discharged to the environment. It is essential that one have a good grasp of the materials and processes being undertaken before an accurate assessment of risk can be made. For example, there are over 56,000 manufactured or imported substances used in industrial operations (defined by the EPA in response to the Toxic Substances Control Act). The National Institute for Occupational Safety and Health (NIOSH) also lists a registry of toxic effects of chemical substances (RTECS). Likewise, the Occupational Safety and Health Administration (OSHA) maintains a list of toxic and hazardous materials. These registries are updated every few years and can be obtained from respective agency websites.

Risk is generally depicted in terms of events per year (usually a small number) and uncertainty (%). Exposure limits are usually depicted in parts per million or billion, denoted as PPM or PPB, or can be expressed in terms of milligrams per cubic meter (mg/m^3). For example, the risk of getting cancer due to smoking cigarettes (1 pack/day) is 3.6×10^{-3} (annual risk) or a factor of 3 (order of magnitude) in percent. The permissible exposure limit for acetone, for example, is 750 PPM; respirable dust from working with marble is around 5 mg/m^3. Table 1.1 shows a list of some common materials and activities and their permissible exposure limits.

Table 1.1 Permissible Exposure Limits of Several Materials and Activities

Material or Activity	**Annual Event**	%	**PPM**	**mg/m^3**
smoking	3.6×10^{-3}	10^{-3}		
chloroform in drinking water	6×10^{-7}	10^{-7}		
acetone			750	
chlorine			0.5	
fluorine			0.1	
ozone			0.1	
mercury vapor				0.05
marble dust (respirable)				5
grain dust (oat, wheat, barley)				10
wood dust				5

While one can envision various techniques used to establish risk, there is a simple technique to obtain a human exposure dose (Ames *et al.*, 1987). This Human Exposure Dose index is related to the Rodent

Potency Dose, or HERP, and relates the carcinogenicity of certain chemical agents to animal cancer tests. While one cannot use animal cancer tests to exactly predict human risk, the index does provide a good guide for establishing priorities and potential carcinogenic hazards. The HERP is defined as

$$\text{HERP} = \text{daily lifetime human dose (mg/kg) X rodent TD}_{50}\text{ (mg/kg)}, \quad (1.1)$$

where TD_{50} are values taken from a data base for 975 chemicals (Ames *et al.*, 1987). Table 1.2 lists several HERP values commonly encountered by humans.

Table 1.2 Risk Based on HERP Index (from Ames *et al.*, 1987)

Daily Human Exposure	**Dose (μg/70-kg person)**	**HERP (%)**
Chlorinated tap water	Chloroform	0.001
Swimming pool	Chloroform	0.008
Conventional home	Formaldehyde	0.6
Mobile home air	Formaldehyde	2.1
Beer (12 oz)	Ethyl alcohol	2.8
High exposure farm worker	Ethylene dibromide	140.0

1.4 Numerical Modeling of Indoor Air Flow

In recent years there has been extensive activity in the development and use of Computational Fluid Dynamics (CFD) software and special programs for room air movement and contaminant transport applications. These investigations range from the prediction of air jet diffusion, air velocity and temperature distribution in rooms, spread of contamination in enclosures, to fire and smoke spread inside buildings. In most cases the predicted results have been promising when compared to available experimental data. However, numerical modeling of ventilation and associated interior contaminant transport is still at an early stage of development and confidence level. A considerable amount of research and development work is still needed, particularly in the areas of more efficient computational schemes, irregular and adaptive grids, turbulence modeling and wall functions.

One of the earliest attempts to numerically simulate airflow in rooms was conducted by Nielsen (1974) using the stream function-vorticity approach for the dependent variables, along with a two-equation (k-ε) model for turbulence based on the numerical procedure developed by Gosman *et al.* (1969). The computations produced realistic room flows, but was limited to 2-D. Numerous papers have appeared over the years utilizing the stream function-vorticity approach for simulating 2-D flows within enclosures; however, the approach is *practically* limited to 2-D flows, and does not permit one to easily incorporate turbulence and 3-D effects inherent in actual ventilated enclosures. Efforts were later undertaken by Hjertager and Magnussen (1977) using the finite volume approach and the SIMPLE algorithm developed by Patankar and Spalding (1972) to solve the 3-D primitive equations of motion with the k-ε two-equation model for turbulence. They modeled the flow from an air jet exhausting into a rectangular room with two ceiling exits. While the point of jet separation from the ceiling was well predicted, the predicted velocity of the jet near the lower region of the room was higher than the measured value.

Gosman *et al.* (1980) extended their two-dimensional finite volume model to solve isothermal flows within 3-D enclosures with small ventilation openings. They achieved good correlations of velocity profiles and jet velocity decay with measurements. Sakamoto and Matsuo (1980) similarly predicted 3-D isothermal flow in a room using the marker and cell (MAC) technique (Harlow and Welch, 1965) and two turbulence models: the k-ε approach and the large eddy simulation (LES) technique (Deardorff, 1970). Results compared favorably with measured velocity profiles; they recommended that the k-ε approach for turbulence be used for room flow predictions over the LES model because it is simpler to use and requires less computing time for comparable accuracy. A computer program called CAFE, developed by Moult and Dean (1980), was used to solve the 3-D velocity components, temperature, concentration, and k-ε turbulence parameters for flow in industrial enclosures and clean rooms. Results were in good agreement with measurements in regions where velocities were large.

Murakami *et al.* (1987) investigated the three-dimensional airflow and contamination dispersion in six (rectangular) types of ceiling supply clean rooms both numerically and experimentally for isothermal flow. They used the MAC method coupled with a central difference approach for the velocity components, and a second-order upwind scheme for k, ε, and concentration, to solve the transient transport equations. Results showed good agreement between prediction and measurement, as well as some interesting flow phenomena regarding the spread of a jet exhaust as it reached the floor. Awbi (1989) numerically solved 2-D air flow and temperature distributions within rooms with diffusers and various vent locations in an effort to simulate 3-D effects; the 2-D non-isothermal predictions compared well to measured vertical velocity and temperature profiles in the room. An early historical discussion and descriptions of numerical methods for solving 2-D and 3-D ventilation and contaminant transport is given by Awbi (1991). A collection of chapters dealing with various issues regarding the modeling of indoor air quality and exposure was published by the ASTM (edited by Nagda, 1993). An overview of indoor climate and air quality issues is discussed by Hoppe and Martinac (1998). A detailed discussion of fire dynamics within enclosures, including modeling, is given by Karlsson and Quintiere (2000). More recent descriptions of modeling efforts can be found in such journals at *Numerical Heat Transfer*, the *ASHRAE Transactions*, *Indoor Air*, and other related technical journals. Today, one can simply log onto Google and do a search on indoor air quality to find numerous articles dealing with the many facets of IAQ.

1.5 Comments

The study of indoor air pollution and ways in which to assess and evaluate contaminant transport and exposure can quickly become overwhelming. There are numerous techniques and schemes now being used to examine IAQ issues, and new developments underway in many research facilities and universities.

While a solid background in engineering or science with familiarity in basic numerical methods is a plus, it is not critical that one be well trained or experienced in the intricacies or details of such fields. Much of the information and numerical schemes addressed in this text can be quickly digested and tried. Experience and confidence in dealing with indoor contaminant problems comes from repeated use and application of some of the tools addressed in this text.

Chapter 2

Fluid Flow Fundamentals

The underlying physics associated with indoor air pollution and ventilation resides in the governing equations for fluid motion, heat transfer, and species transport. These equations stem from the time dependent form of the general partial differential transport equation. While the equations have been known for over 150 years, they are nonlinear in their most complex (i.e., full physics) forms. Reducing the equations to permit analytical or empirical solutions has been a challenge for many decades. Only since the advent of the digital computer have the full set of equations become tractable, even if coarsely approximated. We begin with the full form of the set of governing equations, then examine simplifications that can be made which often can be used to provide quick, reasonable estimates – in lieu of employing more advanced numerical schemes to yield approximations.

2.1 Conservation Equations

The partial differential equations that describe the flow of fluid, heat, and concentration are all based on the conservation of mass, momentum, thermal energy, and species concentration. The dependent variables are the velocity components, temperature, concentration, and some turbulence variables to account for turbulent flow. These governing equations are generally written in the following form:

Conservation of Mass

$$\frac{\partial \rho}{\partial t}+\frac{\partial \rho u}{\partial x}+\frac{\partial \rho v}{\partial y}+\frac{\partial \rho w}{\partial z}=0, \tag{2.1}$$

Conservation of Momentum

x-direction

$$\rho(\frac{\partial u}{\partial t}+u\frac{\partial u}{\partial x}+v\frac{\partial u}{\partial y}+w\frac{\partial u}{\partial z})=-\frac{\partial p}{\partial x}+\frac{\partial \sigma_{xx}}{\partial x}+\frac{\partial \sigma_{xy}}{\partial y}+\frac{\partial \sigma_{xz}}{\partial z}+f_x, \tag{2.2}$$

y-direction

$$\rho(\frac{\partial v}{\partial t}+u\frac{\partial v}{\partial x}+v\frac{\partial v}{\partial y}+w\frac{\partial v}{\partial z})=-\frac{\partial p}{\partial y}+\frac{\partial \sigma_{yx}}{\partial x}+\frac{\partial \sigma_{yy}}{\partial y}+\frac{\partial \sigma_{yz}}{\partial z}+f_y, \tag{2.3}$$

z-direction

$$\rho(\frac{\partial w}{\partial t}+u\frac{\partial w}{\partial x}+v\frac{\partial w}{\partial y}+w\frac{\partial w}{\partial z})=-\frac{\partial p}{\partial z}+\frac{\partial \sigma_{zx}}{\partial x}+\frac{\partial \sigma_{zy}}{\partial y}+\frac{\partial \sigma_{zx}}{\partial z}+f_z, \tag{2.4}$$

Conservation of Energy

$$\rho c_p(\frac{\partial T}{\partial t}+u\frac{\partial T}{\partial x}+v\frac{\partial T}{\partial y}+w\frac{\partial T}{\partial z})=\frac{\partial q_x}{\partial x}+\frac{\partial q_y}{\partial y}+\frac{\partial q_z}{\partial z}+Q, \tag{2.5}$$

Species Concentration

$$\begin{aligned}&\frac{\partial C}{\partial t}+u\frac{\partial C}{\partial x}+v\frac{\partial C}{\partial y}+w\frac{\partial C}{\partial z}=\\&\frac{\partial}{\partial x}(D_{xx}\frac{\partial C}{\partial x})+\frac{\partial}{\partial y}(D_{yy}\frac{\partial C}{\partial y})+\frac{\partial}{\partial z}(D_{zz}\frac{\partial C}{\partial z})+S\end{aligned}, \tag{2.6}$$

where ρ is density, u, v, and w are horizontal, lateral and vertical velocities, respectively, p is pressure, T is temperature, $f_{x,y,z}$ are velocity body force terms, Q and S are source/sink terms, and D_{xx}, D_{yy}, and D_{zz} are the species concentration diffusion coefficients. The normal and tangential viscous stress terms are defined as

$$\sigma_{xx} = \frac{2\mu}{3}\left(2\frac{\partial u}{\partial x} + \frac{\partial v}{\partial y} + \frac{\partial w}{\partial z}\right) \quad \sigma_{yy} = \frac{2\mu}{3}\left(2\frac{\partial v}{\partial y} + \frac{\partial u}{\partial x} + \frac{\partial w}{\partial z}\right)$$

$$\sigma_{zz} = \frac{2\mu}{3}\left(2\frac{\partial w}{\partial z} + \frac{\partial u}{\partial x} + \frac{\partial v}{\partial y}\right) \qquad \sigma_{xy} = \sigma_{yx} = \mu\left(\frac{\partial u}{\partial y} + \frac{\partial v}{\partial x}\right), \tag{2.7}$$

$$\sigma_{xz} = \sigma_{zx} = \mu\left(\frac{\partial u}{\partial z} + \frac{\partial w}{\partial x}\right) \qquad \sigma_{yz} = \sigma_{zy} = \mu\left(\frac{\partial v}{\partial z} + \frac{\partial w}{\partial y}\right)$$

with

$$q_x = \kappa\left(\frac{\partial T}{\partial x}\right) \quad q_y = \kappa\left(\frac{\partial T}{\partial y}\right) \quad q_z = \kappa\left(\frac{\partial T}{\partial z}\right), \tag{2.8}$$

where μ is dynamic viscosity and κ is thermal conductivity.

2.2 Ideal Fluids

As one can readily see from the complexity of the PDEs described in Eqs. 2.1–2.8 for general viscous fluid motion, obtaining solutions to these formidable equations are difficult, generally requiring a numerical approach – computational fluid dynamics (CFD). There are instances when one can make simple assumptions regarding overall fluid motion, and the solutions are fairly accurate. These assumptions are based on the premise of the flow being *ideal*, or that the flow is (1) incompressible, (2) inviscid, and (3) irrotational. If the flow under question can be considered to be ideal, analytical solutions may be used to obtain values for the components of flow, pressure, temperature, and even concentrations. -

If the flow is incompressible, the density is constant. This helps in eliminating the effects of compressibility and density variation. An inviscid flow is one in which the viscosity is zero – hence there are no effects attributed to molecular or turbulent diffusion, i.e., no mixing. If these two criteria are valid, then the governing equations reduce to the simpler, steady state conditions,

Conservation of Mass

$$\frac{\partial u}{\partial x}+\frac{\partial v}{\partial y}+\frac{\partial w}{\partial z}=0, \tag{2.9}$$

Conservation of Momentum

x-direction

$$u\frac{\partial u}{\partial x}+v\frac{\partial u}{\partial y}+w\frac{\partial u}{\partial z}=-\frac{1}{\rho}\frac{\partial p}{\partial x}, \tag{2.10}$$

y-direction

$$u\frac{\partial v}{\partial x}+v\frac{\partial v}{\partial y}+w\frac{\partial v}{\partial z}=-\frac{1}{\rho}\frac{\partial p}{\partial y}, \tag{2.11}$$

z-direction

$$u\frac{\partial w}{\partial x}+v\frac{\partial w}{\partial y}+w\frac{\partial w}{\partial z}=-\frac{1}{\rho}\frac{\partial p}{\partial z}+g, \tag{2.12}$$

Conservation of Energy

$$\rho c_p(u\frac{\partial T}{\partial x}+v\frac{\partial T}{\partial y}+w\frac{\partial T}{\partial z})=\frac{\partial q_x}{\partial x}+\frac{\partial q_y}{\partial y}+\frac{\partial q_z}{\partial z}+Q, \tag{2.13}$$

Species Concentration

$$\begin{aligned} & u\frac{\partial C}{\partial x}+v\frac{\partial C}{\partial y}+w\frac{\partial C}{\partial z}= \\ & \frac{\partial}{\partial x}(D_{xx}\frac{\partial C}{\partial x})+\frac{\partial}{\partial y}(D_{yy}\frac{\partial C}{\partial y})+\frac{\partial}{\partial z}(D_{zz}\frac{\partial C}{\partial z})+S. \end{aligned} \tag{2.14}$$

A further simplification can be made if the flow is irrotational. Irrotational flow is one in which there is no recirculation or rotation, i.e., the absence of vorticity. This implies a predominance of flow direction with no lateral components. Hence, the velocity components can be grouped into a single value, U, and the momentum equations reduce to Bernoulli's equation x-direction

$$\frac{\partial}{\partial x}\left(\frac{p}{\rho}+\frac{u^2}{2}+gz\right)=0, \tag{2.15a}$$

y-direction

$$\frac{\partial}{\partial y}\left(\frac{p}{\rho}+\frac{v^2}{2}+gz\right)=0, \tag{2.15b}$$

z-direction

$$\frac{\partial}{\partial y}\left(\frac{p}{\rho}+\frac{w^2}{2}+gz\right)=0. \tag{2.15c}$$

Equation (2.15a–c) is typically written in vector form,

$$\nabla\left(\frac{p}{\rho}+\frac{V^2}{2}+gz\right)=0,$$

where ∇ is the gradient operator and V is vector velocity. The quantity $(p/\rho+V^2/2+gz)$ is constant everywhere, and the flow is irrotational, steady, incompressible, and frictionless, i.e., the flow is *ideal*.

There are numerous solutions to cases involving ideal flow. This is usually achieved by introducing the scalar potential functions,

$$u=\frac{\partial\phi}{\partial x};\quad v=\frac{\partial\phi}{\partial y};\quad w=\frac{\partial\phi}{\partial z}, \tag{2.16}$$

where ϕ is the scalar potential function. Substituting these expressions into the continuity equation, one obtains the Laplacian

$$\frac{\partial^2\phi}{\partial x^2}+\frac{\partial^2\phi}{\partial y^2}+\frac{\partial^2\phi}{\partial z^2}=0. \tag{2.17}$$

Similarly, a scalar value for the stream function can be introduced for two-dimensional flow where

$$u=-\frac{\partial\psi}{\partial y};\quad v=\frac{\partial\psi}{\partial x}, \tag{2.18}$$

and a Laplacian equation written as

$$\frac{\partial^2 \psi}{\partial x^2} + \frac{\partial^2 \psi}{\partial y^2} = 0. \tag{2.19}$$

Table 2.1 lists velocities and derivatives of the potential functions for two-dimensional planar, axisymmetric cylindrical, and spherical coordinates.

Table 2.1 Velocities as a Function of ϕ or ψ

Coordinate System	ϕ	ψ
2-D Planar	$u = \frac{\partial \phi}{\partial x}; \quad v = \frac{\partial \phi}{\partial y}$	$u = -\frac{\partial \psi}{\partial y}; \quad v = \frac{\partial \psi}{\partial x}$
Axisymmetric	$u = \frac{\partial \phi}{\partial r}; \quad v = -\frac{1}{r}\frac{\partial \phi}{\partial \theta}$	$u = -\frac{1}{r}\frac{\partial \psi}{\partial \theta}; \quad v = \frac{\partial \psi}{\partial r}$
3-D Spherical	$u = \frac{\partial \phi}{\partial r}; \quad v = -\frac{1}{r}\frac{\partial \phi}{\partial \theta}$	$u = -\frac{1}{r^2 \sin\theta}\frac{\partial \psi}{\partial \theta}$ $v = \frac{1}{r \sin\theta}\frac{\partial \psi}{\partial r}$

Utilizing these two variables, potential flow solutions can be obtained for many different geometries, including inlets and flanges.

Analytical solutions to Laplace's equation are harmonic functions, i.e., since the equation is linear and homogeneous, the combination of several solutions to subsets of the problem is also the solution to the overall problem. Hence, a flow field produced by two independent flow fields, each of which can be treated as ideal, can be combined (*superposition principle*) to yield the overall solution. This is a well-known mathematical maneuver used extensively by aerodynamicists when designing flows over wings and bodies. For example, if ϕ_1 and ϕ_2 are two independent solutions, then the horizontal velocity, u, can be obtained for the entire problem using the relation

$$u = \frac{\partial \phi}{\partial x} = \frac{\partial(\phi_1 + \phi_2)}{\partial x} = \frac{\partial \phi_1}{\partial x} + \frac{\partial \phi_2}{\partial x} = u_1 + u_2. \tag{2.20}$$

1.1. What is Indoor Air Pollution

The study of indoor air pollution (IAP) involves dealing with the emission, accumulation, and assessment of pollutants generally attributed to poor ventilation and air exchange. Of particular concern are issues involving air quality and human comfort within buildings. Toxic fumes and airborne diseases are known to produce undesirable odors, eye and nose irritations, sickness, and occasionally death. Other products such as tobacco smoke and carbon monoxide can also have serious health effects on people exposed to a poorly ventilated environment; studies indicate that indirect or passive smoking can also lead to lung cancer. Recommendations for outdoor airflow rates to dilute indoor polluted air vary considerably.

1.2. Ventilation Systems

Ventilation systems are designed to either prevent contaminants from entering a room or remove contaminants from interior sources within the room. Since ventilation systems are integral to the study of indoor air pollution, it is prudent to at least identify them.

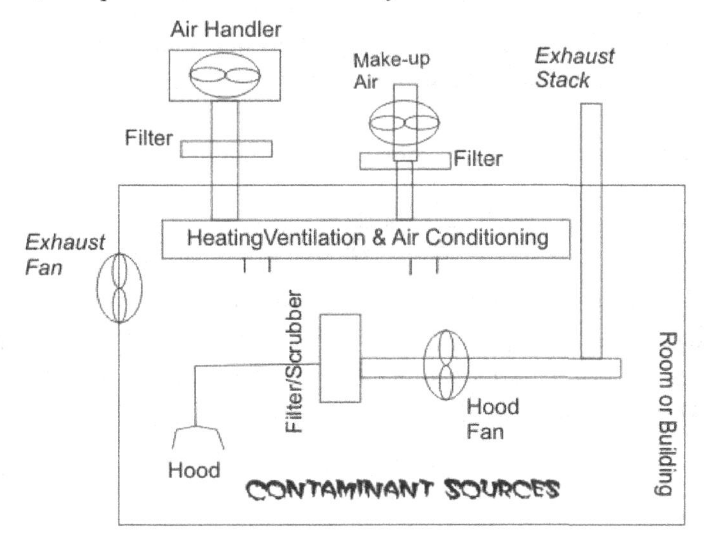

Fig. 1.1 Schematic of a typical ventilation system.

2.2.1 Conformal mapping

If a flow is conservative, that is, if it can be defined by a potential (harmonic) function, then the method of conformal mapping can be employed to solve for ideal flow. Conformal mapping has roots in complex analysis; the mathematics of complex calculus.

Conformal mapping allows for the use a known solution for a given domain to determine a solution in a geometrical domain of interest. The method of conformal mapping provides a quick and general idea of the ideal flow in the domain of interest. Many solutions exist in simple domains that can be mapped to domains of interest. In fact, the previously presented potential functions are of this variety. In this section we investigate how this mapping process works, from the geometric domain, represented by a complex coordinate

$$z = x + iy \ , \tag{2.21}$$

to some analytic function of z given by

$$w = \phi + i\psi \ . \tag{2.22}$$

Depending on how the construction of $w = f(z)$ is performed, certain geometric similarities exist between the z plane and the w plane. Those transformation, which preserve angles of intersection are conformal. The functions ϕ and ψ are precisely the potentials and streamlines we seek in the geometric domain z; these define the ideal flow in the geometric domain z.

If $\psi(x, y) = c$, where c is some constant, and if

$$\frac{\partial^2 \psi}{\partial x^2} + \frac{\partial^2 \psi}{\partial y^2} = 0 , \tag{2.23}$$

then $\psi(x, y)$ represents streamlines under the assumptions of ideal flow. It is the imaginary part of $w = f(x, y)$. By selecting appropriate mappings, w to z, various geometries and flow situations with sources and sinks can be developed.

Example 2.2.1.1 Flow near an inside corner: Solving for flow in a corner demonstrates the usefulness and basic procedures of conformal mapping for finding solutions to ideal flow. Represented in the z plane are all positive x and y points, as shown in Fig. 2.1. This is the corner flow domain and will be the range of the conformal map from the w plane. Picking the function

$$w = z^2 = x^2 - y^2 + i2xy\,, \tag{2.24}$$

maps this corner z plane onto the upper half of the w=f(ϕ,ψ) plane.

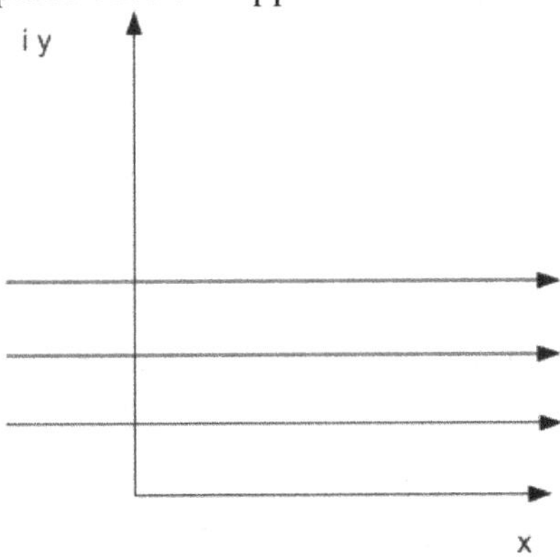

Fig. 2.1 Constant flow in the w plane.

Flow in upper half of the w plane moving right to left is given by the complex potential function

$$f(w) = Cw = C(\phi + i\psi)\,, \tag{2.25}$$

where C is some positive real value and now

$$w = x + iy\,. \tag{2.26}$$

Then the potential equation for flow is given by

$$\phi(x, y) = Cx\,, \tag{2.27}$$

and the stream function is

$$\psi(x, y) = Cy\,. \tag{2.28}$$

Under the described mapping the potential function in the z plane

$$f = Cz^2 = C(x^2 - y^2) + i2xy\,, \tag{2.29}$$

with the stream function in the z plane being

$$\psi = 2Cxy, \tag{2.30}$$

as plotted in Fig. 2.2.

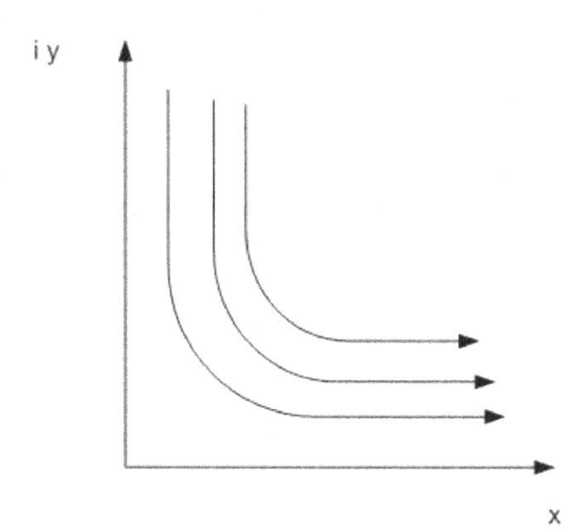

Fig. 2.2 Flow in a corner in the z plane under the mapping.

There are numerous catalogs of various transformations available in the literature. This includes from the simple to more complex doubly connected domains.

Example 2.2.1.2 Flow around an outside corner: Flow around corner can be solved in the *w* plane with the mapping

$$w = z^{2/3}, \tag{2.31}$$

which yields the potential function

$$f(w) = C(\phi + i\psi) = Cz^{2/3} = C(x + iy)^{2/3}. \tag{2.32}$$

The stream function is given by the imaginary part of *w*,

$$\psi = \mathrm{Im}\left(C\sqrt[3]{(x^2 - y^2 + i2xy)}\right). \tag{2.33}$$

By reversing the map, that is, making

$$z = w^{a/\pi}, \tag{2.34}$$

the flow around the corner is more easily ascertained. By evaluating this function further via its complex polar values and noting that

$$z = x + iy = r(\cos\theta + i\sin\theta) = re^{i\theta}, \tag{2.35}$$

then

$$w = r^{a/\pi} e^{i\theta a/\pi}. \tag{2.36}$$

Here the upper half of the z plane $0 < \theta < \pi$ is mapped onto the section $0 < \theta < a$ in the w plane. For the function $w = f(z) = z$ having some component $y = C$ as constant flow lines parallel to the x axis of the z plane, has under the mapping stream lines satisfying the imaginary part of $w^{\pi/a} = C$ in the w plane. This is stated mathematically as

$$\mathrm{Im}(w^{\pi/a}) = C, \tag{2.37}$$

which describes stream lines around an outside corner as shown in Fig. 2.3 (for a/π = 3/2) .

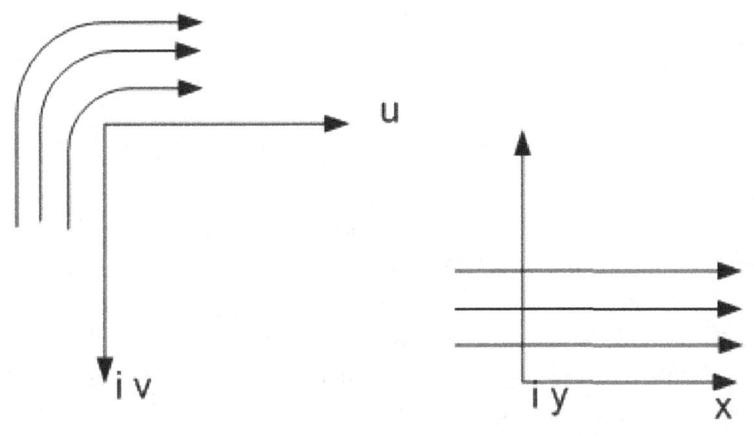

Fig. 2.3 Mapping potential function and streamlines for flow around a corner.

2.2.2 Schwarz–Christoffel transform

One of the primary methods for handling polygonal domains is the Schwarz–Christoffel mapping. This mapping relates a polygon's vertices in the w plane to the real axis in the z plane. The polygon's interior region is mapped to the upper half plane in z.

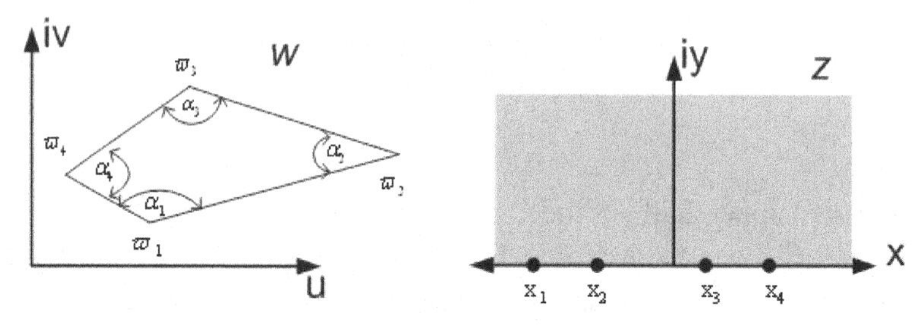

Fig. 2.4 Schwarz–Christoffel transformation – mapping polygon to the real axis.

Given vertices of the polygon in the w plane, w_i with corresponding interior angles α_i as shown in Fig. 2.4, then

$$\frac{dw}{dz} = A(z-x_1)^{\alpha_1/\pi-1}(z-x_2)^{\alpha_2/\pi-1}....(z-x_n)^{\alpha_n/\pi-1} + B, \quad (2.38)$$

for any complex constants A and B. So, after integrating the mapping is found to be

$$w = A\int(z-x_1)^{\alpha_1/\pi-1}(z-x_2)^{\alpha_2/\pi-1}....(z-x_n)^{\alpha_n/\pi-1}dz + B \cdot \quad (2.39)$$

Illustrating the Schwarz–Christoffel mapping method for solving ideal flow from a duct is the next example.

Example 2.2.2.1 Flow into a duct: By mapping a cut in the *w* plane (this is a half line extending to -∞) which is some distance away from the real axis, parallel to the real axis, as shown in Fig. 2.5, to the real axis of the *z* plane via the Schwarz–Christoffel method results in

$$\frac{dw}{dz} = \frac{\gamma(z+1)}{z} = \gamma\left(1+\frac{1}{z}\right). \quad (2.40)$$

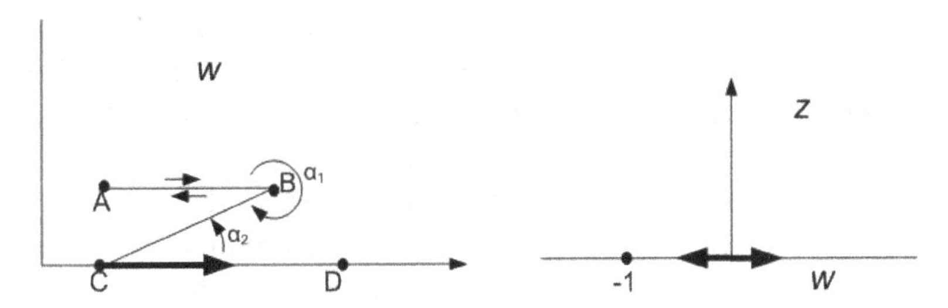

Fig. 2.5 Plane *w* with a cut (AB) and point ABCD being mapped onto the real axis in the *z* plane using the Schwarz–Christoffel transform.

Here the domain is bounded below by the polygonal line ABCD as shown, also described in Levinson and Redheffer (1970). The points A, C, D are considered to be at ∞. The angle between B and C is 2π when C is at $-\infty$ and the angle between C and D is zero. The mapping given by Eq. (2.40) places A, B, C, and D in the *w* plane onto, $-\infty$, -1. 0, and ∞ (respectively) on the *z* plane real axis for some $\gamma > 0$ also shown in Fig. 2.4. Integrating the differential Eq. (2.40), results in at least one solution

$$w = \gamma(z + \ln(z)) = \gamma(\phi + i\psi)\,. \tag{2.41}$$

Only the lines given by Im (ln(*z*)) = C are associated with the streamlines, ψ (x,y) for flow. These constant lines in *z* are mapped onto *w* by

$$\begin{aligned} w &= z + \ln(z) = re^{i\theta} + \ln(re^{i\theta}) = \\ &re^{i\theta} + \ln(r) + \ln(e^{i\theta}) = re^{i\theta} + \ln(r) + i\theta \end{aligned}, \tag{2.42}$$

and are shown in Fig. 2.6.

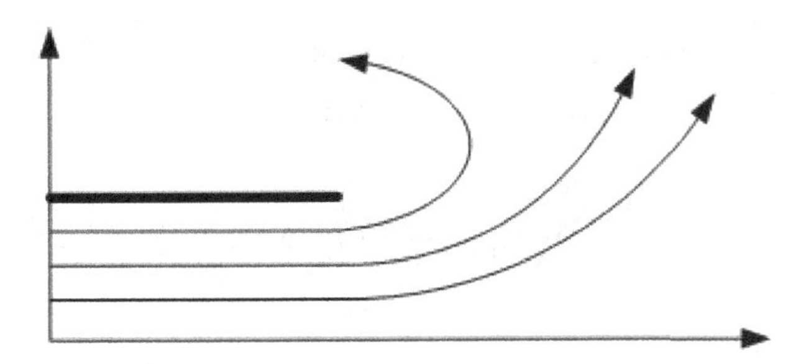

Fig. 2.6 Streamlines for flow from duct in upper half of *w* plane.

By reflecting this solution about the real axis, flow in a duct can be established, where the flow originates at the origin along the ordinate axis of the w plane. This is accomplished by taking a slit in the lower half-plane of w and using the transformation as described. That is, the upper half of w can be reflected about the x axis, so the same flow is formed in the lower half of the w under a similar mapping, as shown in Fig. 2.7.

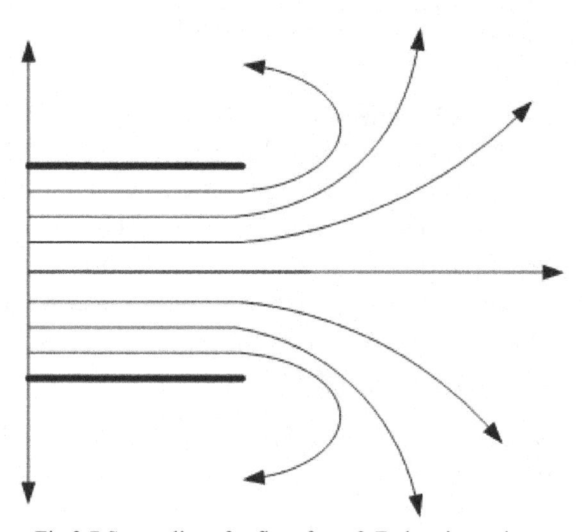

Fig 2.7 Streamlines for flow from 2-D duct in w plane.

Also, simply noting that by Schwarz's reflection principle,

$$F_2(z) = \overline{F_1(\tilde{z})}, \tag{2.43}$$

where $z = x - iy$, the complex conjugate of z,

$$F_1(z) = u_1(x, y) + iv_1(x, y), \tag{2.44}$$

and the conjugate of

$$F_1(\tilde{z}) = u_1(x, -y) + iv_1(x, -y). \tag{2.45}$$

Therefore,

$$\overline{F_1(\tilde{z})} = u_1(x, -y) - iv_1(x, -y) = F_2(z). \tag{2.46}$$

As mentioned, solutions may be additive as shown by the above reflection for the duct flow example, but this is not always true. Analysis of the flow should be made before applying the principle of reflection.

2.2.3 Numerical mapping

Many methods exist to create mappings, and there are many numerical methods to evaluate mappings. For even more complex transformations, numerical conformal mapping can be applied on local elements of the w plane. Some recent texts describe the evaluation process in more detail; many either have software with them or describe software which is available as open source or add-ons to existing packages. These software include Mathematica®, Maple®, SC Toolbox for Matlab® described by Driscoll (1996), and also Conform developed by Ivanov and Trubetskov (1994).

Investigating the subject of complex analysis and conformal mapping has rich reward for help in analyzing ideal flow. Because the Schwarz–Christoffel mapping lends itself to numerical evaluation, it is widely employed in the software mention in this section. The numerical evaluation of the mappings has greatly extended the usefulness of the conformal mapping technique for evaluating ideal flow. We give a few of the results of conformal mappings here, the potential functions and their associated real and imaginary parts for flow in ducts and around corners. Later, in Chapter 6, we use the results of mappings to discuss particle trajectories near ducts, and the associated flow near sinks and sources. In any regard, precise representation of fluid flow is not part of ideal flow assumptions. In Chapter 7 and 8 we describe in more detail numerical solutions to the complete Reynolds averaged Navier–Stokes equations.

2.2.4 Superposition for stream functions

There are four basic two-dimensional flows associated with irrotational fluid mechanics: (1) uniform flow, (2) line source (or sink), (3) line vortex, and (4) the doublet. The stream function relations are shown below with their corresponding patterns in Fig. 2.8 (Chow, 1983).

Uniform flow

$$\psi = U(y\cos\alpha - x\sin\alpha), \tag{2.47a}$$

Line source

$$\psi = \frac{Q}{2\pi}\tan^{-1}\frac{y-y_0}{x-x_0} = \frac{Q}{2\pi}\theta', \tag{2.47b}$$

Line vortex

$$\psi = \frac{\Gamma}{2\pi}\ln\left[(x-x_0)^2+(y-y_0)^2\right]^{1/2} = \frac{\Gamma}{2\pi}\ln r', \tag{2.47c}$$

Doublet

$$\psi = -\frac{k}{2\pi}\frac{y-y_0}{(x-x_0)^2+(y-y_0)^2} = -\frac{k}{2\pi}\frac{\sin\theta'}{r'}, \tag{2.47d}$$

where U denotes a uniform flow speed, Γ is circulation, and k is the strength of the doublet.

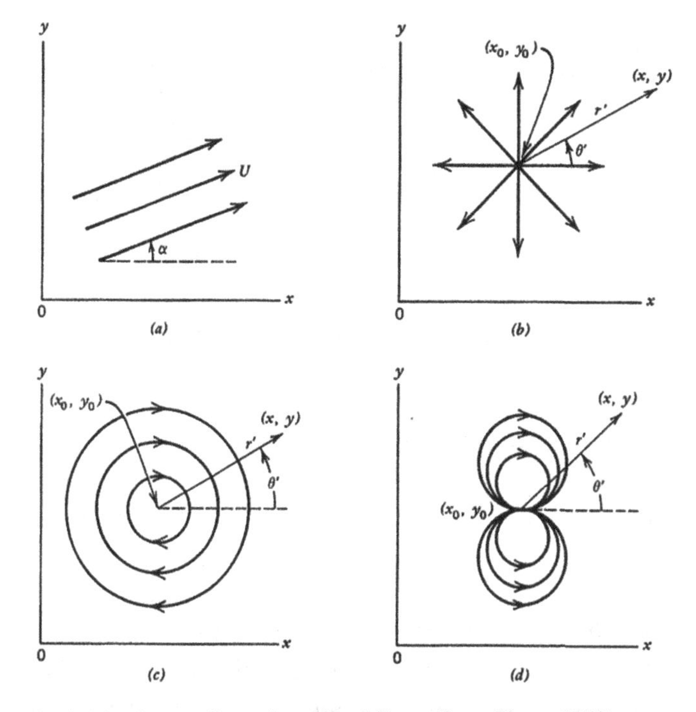

Fig. 2.8 Basic two-dimensional ideal flows (from Chow, 1983).

Example 2.2.4.1 Flow over a semi-infinite body: Assume a polar coordinate system with a source, Q, located at the origin. The resulting flow pattern if this flow is superimposed on a uniform stream with velocity U_{oo} moving from left to right is shown in Fig. 2.9.

The resulting streamline shapes obtained by adding the solutions to the uniform stream and source, shown at the right of Fig. 2.9, create a potential flow stream that splits at the stagnation point, denoted by point B. For further details, see Anderson (2001).

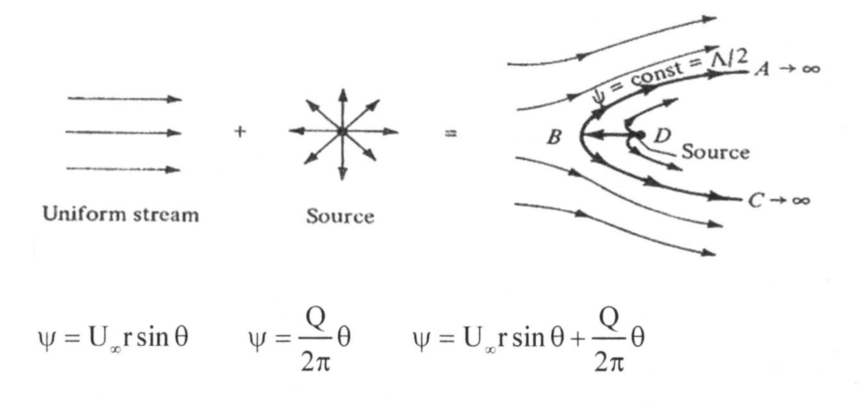

Fig. 2.9 Flow over a semi-infinite body.

2.3 Turbulence

A CFD code must be capable of modeling both laminar and turbulent fluid motion. Current approaches to modeling turbulence are based on either "first" or "second-order" closure models in which the governing equations are closed by equations for various turbulence correlation terms (kinetic energy, shear stress, etc.). Examples of such closure schemes are discussed in detail by Wilcox (2006). Results show that advanced turbulence closure schemes, incorporating more physics and less empiricism, provide the generality for modeling wider classes of problems and more accurately account for the irregular nature of turbulent flow.

An effective viscosity is usually employed to simplify solution of the turbulent equations. This concept allows the turbulent stress terms to be conveniently combined with the molecular viscosity (laminar flow) into an overall viscosity term for numerical solution. Two of the earliest and most frequently used approaches to model the effective viscosity (and effective diffusion coefficients) are the Prandtl mixing length model (0th order) and the k-ε two-equation model (1st order).

Although the mixing length hypothesis has been successfully applied to solving numerous turbulent flow problems, it has little application in complex flows due to the difficulty in specifying an appropriate length. The method is essentially unsuitable for situations in which recirculation occurs.

In an attempt to more accurately model turbulence within complex regions, especially when recirculation is present, a two-equation turbulence model was first proposed by Jones and Launder (1972). The most common two-equation model is one based on solution of the turbulent kinetic energy, k, and its dissipation rate, ε. This model is known as the k-ε scheme, and is popular because of its applicability to a wide range of flow problems (as well as low computational demand over more complex models). The k-ε model has been applied to numerous flow problems with good predictive accuracy, and is still the preferred choice for simulating flows where there is the potential for recirculation and/or swirl. However, recent advances in the use of k-w and algebraic closure schemes appear promising, and more users are now including simulations using these approaches (see Wilcox, 2006).

In the k-ε model, the equation for k (which is derived from the general Navier–Stokes equations) is written as

$$\begin{aligned}
&\frac{\partial \rho k}{\partial t}+\frac{\partial \rho u k}{\partial x}+\frac{\partial \rho v k}{\partial y}+\frac{\partial \rho w k}{\partial z}= \\
&\frac{\partial}{\partial x}(\Gamma_k\frac{\partial k}{\partial x})+\frac{\partial}{\partial y}(\Gamma_k\frac{\partial k}{\partial y})+\frac{\partial}{\partial z}(\Gamma_k\frac{\partial k}{\partial z})+ \\
&\mu_\tau(2[(\frac{\partial u}{\partial x})^2+(\frac{\partial v}{\partial y})^2+(\frac{\partial w}{\partial z})^2] \qquad , \qquad (2.48) \\
&+(\frac{\partial u}{\partial y}+\frac{\partial v}{\partial x})^2+(\frac{\partial u}{\partial z}+\frac{\partial w}{\partial x})^2+(\frac{\partial w}{\partial y}+\frac{\partial v}{\partial z})^2) \\
&-C_\mu\rho\frac{k^{1.5}}{L}+\beta g\frac{\mu_\tau \partial T}{\sigma_t \partial y},
\end{aligned}$$

where $\mu_k = \mu_e/\mu_k$ with $\mu_k \sim 1$, μ_t is the turbulent Prandtl number (0.5 to 0.9) and C_μ is a constant ~ 0.09. The last term represents the effect of buoyancy.

The transport equation for ε is as follows:

$$
\begin{aligned}
&\frac{\partial \rho\varepsilon}{\partial t}+\frac{\partial \rho u\varepsilon}{\partial x}+\frac{\partial \rho v\varepsilon}{\partial y}+\frac{\partial \rho w\varepsilon}{\partial z}= \\
&\frac{\partial}{\partial x}(\Gamma_\varepsilon\frac{\partial \varepsilon}{\partial x})+\frac{\partial}{\partial y}(\Gamma_\varepsilon\frac{\partial \varepsilon}{\partial y})+\frac{\partial}{\partial z}(\Gamma_\varepsilon\frac{\partial \varepsilon}{\partial z})+ \\
&C_1\frac{\varepsilon}{k}\mu_\tau(2[(\frac{\partial u}{\partial x})^2+(\frac{\partial v}{\partial y})^2+(\frac{\partial w}{\partial z})^2]+ \\
&(\frac{\partial u}{\partial y}+\frac{\partial v}{\partial x})^2+(\frac{\partial u}{\partial z}+\frac{\partial w}{\partial x})^2+(\frac{\partial w}{\partial y}+\frac{\partial v}{\partial z})^2) \\
&-C_2\rho\frac{\varepsilon^2}{k}+C_1\beta g\frac{\varepsilon}{k}\Gamma_t\frac{\partial T}{\partial y}.
\end{aligned}
\tag{2.49}
$$

Here, $\Gamma_t = \mu_e/\mu_\tau$ where μ_τ is a constant equal to 1.22, $C_1 = 1.44$, and $C_2 = 1.92$. The equation for concentration species can likewise be written in similar fashion, i.e.,

$$
\begin{aligned}
&\frac{\partial \rho C}{\partial t}+\frac{\partial \rho uC}{\partial x}+\frac{\partial \rho vC}{\partial y}+\frac{\partial \rho wC}{\partial z}= \\
&\frac{\partial}{\partial x}(D_{xx}\frac{\partial C}{\partial x})+\frac{\partial}{\partial y}(D_{yy}\frac{\partial C}{\partial y})+\frac{\partial}{\partial z}(D_{zz}\frac{\partial C}{\partial z}) \\
&+\frac{\partial}{\partial x}(-\rho u'c')+\frac{\partial}{\partial y}(-\rho v'c')+\frac{\partial}{\partial z}(-\rho w'c')+S,
\end{aligned}
\tag{2.50}
$$

where c' is the deviation from the mean. The terms -ρu'c', -ρv'c', and -ρw'c' are the turbulent diffusion fluxes.

Attempts to simplify the Reynolds stress transport equations are usually made by approximating the advection and diffusion terms into algebraic expressions; such models are referred to as algebraic stress models (ASM). This technique reduces the computational time required to obtain a solution of the transport equations. However, these models have not found wide-scale application in fluid flow problems due to their complexity, and the fact that they still require a large amount of computing time and do not always produce better predictions than the k-ε model.

In large eddy simulation (LES) models, large-scale turbulence fluctuations are solved directly by appropriate transport equations and only the small-scale fluctuations contribute to ε. The nonlinear interaction between the large-scale and small-scale turbulence motion is approximated through a subgrid-scale turbulent viscosity model. Success of this type of turbulence modeling lies with the computational grid being fine enough to lie within the inertial subrange (Kolmogorov scale) where energy cascade takes place and the dissipation rate, ε, has a constant value. The LES method has the ability to freeze the flow at any moment in time; if mean flow quantities are required, the calculations must be conducted over a very long time scale. The application of LES has been relatively limited to isothermal flows in channels and over a cube. However, considerably more work is needed before the method can be applied to a wider range of flow problems.

The accuracy of the solution of the discretized turbulence equations depends on the accuracy of specifying the physical quantities at the boundaries of the flow domain, and on the methods of linking these relations to the bulk flow. Close to a solid boundary, the local Reynolds number is extremely small and turbulent fluctuations are damped out by the proximity of the surface – laminar shear becomes a locally dominant force as a result of the steep velocity gradient. Because of the damping effect of the wall, the transport equations for the turbulence quantities do not apply close to the wall. One way of dealing with this problem is to add extra source terms to the transport equations for k and ε, and use an extremely fine grid close to the surface so that the first few points are within the laminar sublayer. This technique is effective, but requires a vast number of grid points (especially in three-dimensions).

An alternative, and more popular, approach is to use Couette flow analysis and apply algebraic relations (logarithmic laws or wall functions) close to the surface. This approach does not require an ultra-fine grid near the surface. At a point close to the wall, the momentum equation is reduced to a one-dimensional form with gradients in the direction normal to the surface.

Boundary conditions at vent inlets are usually set to fully developed profiles, unless specified directly by the user from experimental data. Likewise, at exits, the transverse velocity components are normally set to zero and the longitudinal exit velocity calculated from mass balance. Exit values for k and ε are usually not required because the Reynolds number at the exit is typically large; likewise, the gradients normal to the flow direction of the dependent variables may also be set to zero at the exit plane. A particularly nice feature when using finite element methods is the ability to set the traction terms (i.e., the RHS of the governing equations) equal to zero at the exit. This is the true mathematical formulation for proper specification of the outflow boundary conditions, and does not require *a priori* judgment by the user (Gresho *et al.*, 1984) when using finite volume or finite difference schemes.

2.4 Species Transport

It is well known than contamination produced in a ventilated room can quickly spread over the whole zone, especially in a mixing ventilation system with a large rate of entrainment and a circulatory motion created by jets. Normally, the transport equation for concentration is solved either in time-average form or time-dependent form after a converged solution has been achieved for the other transport equations (velocity, temperature, and k-ε turbulence parameters).

When low concentration levels exist in a room environment (~100 ppm), the difference in density between the contaminant and air is usually ignored. This practice is fairly common in both research and industrial applications with regards to either gas or small particulate transport. Nielsen (1981) used this approach to model 2-D concentration distributions within enclosures to investigate the importance of room aspect ratios on concentration distribution; a decrease in height of the room air supply slot produced a decrease in concentration in the enclosure. *Higher room concentrations were found to exist when the contamination source was placed in a relatively stagnant region in the room.* Murakami *et al.* (1983) obtained similar conclusions from their three-dimensional simulations, and were later confirmed by Davidson (1989) using a 3-D, k-ε turbulence model.

The spread of smoke within an L-shaped (rectangular) shopping mall was investigated by Markatos and Cox (1984) using the PHOENICS finite volume code. Both steady-state and transient spread of smoke from a fire was modeled, and results compared with experimental measurements. Agreement between measurement and prediction was generally satisfactory with small differences in the velocity profiles near the top of the doorway openings, and in the temperature profiles at the center of the doorways (where cold air entering from the lower region meets the hot smoke leaving the upper region).

Several commercially available CFD codes are being used for room ventilation and contaminant dispersal. The code CFX, developed by AERE Harwell, is a variant of the SIMPLE technique (Patankar, 1980), and resembles the PHOENICS and FLUENT codes. This code incorporates finite volumes and unstructured meshes to account for irregular surfaces, and was used to simulate the fire that occurred in Kings Cross Station in London several years ago. Likewise, the FLOVENT code, which is similar to FLUENT and CFX, allows one to perform 2-D and small-scale 3-D problems on high performance PCs. Unfortunately, the code does not allow one to handle irregular geometries – curved surfaces must be approximated by orthogonal grids (this effect leads to the stair-step appearance for irregular boundaries and can degrade the ability of a code to accurately resolve boundary layer effects and turbulence near surfaces).

We use both FLUENT and COMSOL 3.4 as our choices for commercial packages to model indoor air contaminant dispersion. FLUENT is relatively easy to use, and has a wide following of CFD users throughout the world. COMSOL is a very user friendly and fairly inexpensive package that permits one to model a wide range of multiphysics problems, including chemically based problems, using a moderately powered PC.

2.5 Comments

The set of nonlinear, partial differential equations that fully describe the flow and transport of air and contaminants is formidable. These equations are among the most difficult to solve. Simple assumptions allow one to reduce the set of equations to forms that can quickly give solutions describing the general dispersive effects of the spread of contaminants. Such simple solutions are ideal for designers and responders where a quick assessment is needed. However, to fully capture all the physics and intricate recirculatory nature of fluid motion and contaminant dispersion found in real world situations, the complete equations must be solved. These equations can only be solved using numerical techniques – no analytical solutions exist. While numerous empirical expressions can be found in the literature for various classes of flows, especially when dealing with turbulence, great care must be exercised when using such expressions beyond their ranges of applicability.

Chapter 3

Contaminant Sources

Contaminants consist of gases, solids, or liquids (or combinations) and come in many types and forms. Some of the more common contaminants typically attributed to indoor air quality include smoke and odors attributed to perfumes, tobacco, and the cooking of food. The variety of contaminants are as plentiful as their source locations and origins. We begin with a brief description of the types of contaminants, followed by a discussion of the concentration equation and its various terms and units.

3.1 Types of Contaminants

Contaminant in buildings generally consists of either particles or gases. Particles can either be in the form of solids or liquids. Gases are generally gaseous or exist as a vapor, both of which obey the perfect gas law. Indoor contamination is generally due to humans and animals, including contaminant releases from furnishings and processes within interior spaces, and by intrusion of contaminants from outside air. Another form of contaminants is mold (fungal material). Humans and animals (mammals) exhale CO_2; this can become very troublesome in confined spaces (such as submarines) or heavily occupied interiors since it serves as an indicator of poor ventilation. The other major culprit to human health is carbon monoxide (CO), which is highly toxic. CO results from incomplete combustion of hydrocarbon fuels and tobacco smoking.

The Glossary from *Fundamentals of Industrial Hygiene* (Plog, 1988) gives the following definitions for specific airborne contaminants:

Dusts: Solid particles typically created from crushing, handling, detonation, and impact of organic or inorganic materials; particles do not diffuse in air but settle under the influence of gravity.

Gas: Material state of matter with very low density and viscosity that respond to changes in temperature and pressure; diffuses and uniformly distributes itself throughout any enclosure.

Vapors: Gaseous form of substances normally in solid or liquid state at room temperature and pressure; vapors diffuse and mix with the environment – evaporation is the changing of a liquid into a vapor state.

Aerosols: Liquid droplets or solid particles that are dispersed in air with diameters generally in the range of 0.01–100 μm; aerosols generally remain suspended in air for some time.

Fume: Particulate created from the evaporation of solid materials and dispersed into the air; fumes are usually less than 1 μm in diameter.

Mists: Suspended liquid droplets generated from condensation as a gas transforms to a liquid state or by a liquid dispersing into the air due to foaming, splashing, or atomizing; mist forms when a finely divided liquid becomes suspended in air.

Smoke: Particles (suspension of aerosols in air) created from combustion or sublimation and consists of droplets as well as dry particles, e.g., tobacco produces a wet smoke composed of tarry droplets; carbon or soot particles are generally less than 0.1 μ in size and result from incomplete combustion of carbon-based materials.

The sources of building contamination and the multitude of contaminants are numerous. Many of the indoor pollution problems stem from construction activities of operations within a facility. Such contaminants include volatile organic compounds (VOCs), pesticides, biological contaminants promoted by moisture, asbestos, radon, lead, and PCBs. A variety of units are used for concentration, which is discussed next in section 3.2.

3.2 Units

It is important to know the common forms of units used to describe concentration. A concentration is essentially a quantity of material per unit volume, unit mass, or unit moles. However, one must be careful when referring to a concentration within the air, within water, or soil (which is a multiphase media). Chemists typically refer to the number of moles per unit volume to define a concentration. A mole is defined as 1 mole = 6.023 x 10^{23} molecules (from Avogadro's Number). Chemical engineers commonly use moles per volume of water, mass per mass of solid, or moles per mole of gas, depending on the medium. For example, assume there is 2 g/m^3 of CO_2 dissolved in water, where concentration in water is usually given in terms of mass per unit volume of moles per unit volume. The molecular weight of CO_2 is 44 g/mole. The concentration in moles/m^3 would be

$$\frac{2g/m^3CO_2}{44g/mole} = 0.0455 \text{ moles/m}^3.$$

Concentration in air is usually given in units of partial pressure at one atmosphere of total pressure. Since the pressure of a gas is proportional to the number of molecules in a given volume,

$$\frac{\text{partial pressure}}{\text{total pressure}} = \frac{\text{molecules of compound}}{\text{total molecules}} = \frac{\text{moles of compound}}{\text{total moles}}.$$

The most common abbreviations found in the literature are units based in parts per million, billion, or trillion:

In water:

ppm	parts per million by weight	mg/L or g/m^3
ppb	parts per billion by weight	μg/L or mg/m^3
ppt	parts per trillion by weight	ng/L or μg/m^3

In air:

ppm	parts per million by volume	mL/m^3 or 10^{-6} atm/atm
ppb	parts per billion by volume	$\mu g/m^3$ or 10^{-9} atm/atm
ppt	parts per trillion by volume	ng/m^3 or 10^{-12} atm/atm

The Environmental Protection Agency (EPA) recommends a maximum level of 1.8 g/m^3 (or 1000 ppm) for continuous CO_2 exposure. On the other hand, CO levels near 15 ppm are harmful to humans, with cumulative effects.

Assume there is 2 ppm of methane in 1 atm of air. What is the partial pressure of methane? Using the ideal gas law to convert ppm to atmospheres of methane/atmosphere of total pressure,

$$\frac{P_{meth}V}{P_{air}V} = \frac{n_{meth}RT}{n_{air}RT}$$

$$\frac{P_{meth}}{P_{air}} = \frac{n_{meth}}{n_{air}} = 2x10^{-6},$$

where the concentration by volume is equal to the concentration by moles. Thus, the methane partial pressure is 2 x 10^{-6} atm.

3.3 Materials

A major portion of indoor air contaminants come from building materials and equipment. VOCs resulting from the manufacturing and installation processes typically migrate into the air. The majority of VOCs can be classified into the following categories (from Hays *et al.*, 1995):

***Adhesives, sealants, and architectural coatings*:** these types of coatings are installed wet and dry or cure on the premises; the solvents used in the formulation of these materials directly relate to the VOCs emitted. The resins used in the base of adhesives are either natural or synthetic and range from low to high emission rates; sealants consist of putties, caulking compounds, rubber, acrylic latexes, and silicones while architectural coatings include paints, stains, sealers, and varnishes.

Particleboard and plywood: particleboard is a composite produce made from wood chips or residues that are bonded together with adhesives and typically come from milling or woodworking waste. Plywood consists of several thin layers or plies of wood that are bonded by adhesive and are generally classified as either softwood or hardwood; the IAQ effects of softwood and hardwood vary with the adhesive (PF and UF resins).

Carpet, resilient flooring, and wall covering: these types of materials bring VOC-emitting composition into the building interior along with the use of adhesives to attach the material to various surfaces. Carpets typically consist of fibers of either wool or synthetics. Resilient flooring is generally either tile or sheet (vinyl or rubber). Wall coverings are made from paper, fabric, and vinyl.

Insulation, acoustical ceiling tile, and furnishings: these types of materials include a variety of paints, adhesives, backing, fabrics, and fibrous materials all of which combine to contribute VOCs. Insulation is commonly thermal oriented, but acoustical and fireproofing also are used; these usually exist in the form of batt and rigid foam consisting of fiberglass or mineral wool. Furnishings include such items as prefabricated movable partitions, workstations, desks, chairs, couches, photocopiers, computers, etc.

Table 3.1 is a partial list of materials and some of the chemicals emitted from their surfaces, along with emission rates when known. When building materials have a high-surface-area-to-room-volume ratio, it is important to quantify the emissions and their rates to avoid harmful effects to occupants.

Table 3.1 Partial List of Building Materials and their Emissions (from Hays *et al.*, 1995)

Material	Chemical emitted	Emission rate
Adhesives	Alcohols	
	Amines	
	Benzene	
	Toluene	
Sealants	Alcohols	
	Amines	
	Benzene	
	Xylenes	
Architectural coatings	paints – C4–benzene	
	paints – Toluene	
	stains/varnishes – Amines	
	stains/varnishes – Benzene	
Particleboard	Amines	
	Formaldehyde	0.2-2 $mg/m^2/h$
	n-Hexane	15-26 $\mu g/m^2/h$
Carpeting	4-Phenylcyclohexene	0.1 $mg/m^2/h$(latex backed)
	Styrene	
Resilient Flooring	Amines	
	Alkanes	
	Linoleum – Trichloroethylene	3.6 $\mu g/m^2/h$
Wall Coverings	Amines	
	Xylenes	
Insulation	foam – Acetone	ND-0.02 $mg/m^2/h$
	Chloroform	ND-0.002 $mg/m^2/h$
Furnishings	Upholstery – Formaldehyde	

3.4 Typical Operations

The most common carriers of pollutants are ventilation systems and the human body (general work activity and socialization). The ventilation system serves as an ideal transport mechanism for dispersing particulates and gaseous compounds throughout a building. Similarly, the human body acts as a repository from transporting all forms of pollutants within a room as well as to other humans.

Operations commonly found in many industrial and office environments include such processes as maintenance and housekeeping, which permit dust or particulate buildup that leads to indoor air contamination. Likewise, office equipment, including such devices as wet and dry copying machines, computers, laser printers, and color inkjet printers, emit VOCs during operation. Pest control, construction activities in occupied buildings, moisture leaks, and many industrial activities including chemical spills, grinding, pouring, and sprays lead to indoor contamination. Operations involving food preparation and consumption are particularly sensitive to emissions and unsanitary conditions that lead to indoor air quality problems. Even the natural process of evaporation and diffusion of volatile liquids stored in rooms are common contributors to overall air quality.

3.5 The Diffusion Equation

The concentration equation mathematically describes the transport and diffusion of a chemical species. This equation, first introduced in Eq. (2.6), is typically written in the differential form as

$$\begin{aligned}&\frac{\partial C}{\partial t}+u\frac{\partial C}{\partial x}+v\frac{\partial C}{\partial y}+w\frac{\partial C}{\partial z}=\\&\frac{\partial}{\partial x}\left(D_{xx}\frac{\partial C}{\partial x}\right)+\frac{\partial}{\partial y}\left(D_{yy}\frac{\partial C}{\partial y}\right)+\frac{\partial}{\partial z}\left(D_{zz}\frac{\partial C}{\partial z}\right)+S\end{aligned}, \tag{3.1}$$

where the advective transport is described by the gradient flux terms on the left-hand side of the equation and the diffusion terms by the second derivatives on the right-hand side of the equation. S is the source or sink term (also known as a body force term in solid and fluid mechanics). The derivative of concentration with respect to time denotes the transient nature of the transport. The units attributed to this equation are

$$\frac{\partial C}{\partial t} + u\frac{\partial C}{\partial x} + v\frac{\partial C}{\partial y} + w\frac{\partial C}{\partial z} =$$

$$(g/m^3/s) \quad (m/s)(g/m^4) \quad (m/s)(g/m^4) \quad (m/s)(g/m^4)$$

$$\frac{\partial}{\partial x}\left(D_{xx}\frac{\partial C}{\partial x}\right) + \frac{\partial}{\partial y}\left(D_{yy}\frac{\partial C}{\partial y}\right) + \frac{\partial}{\partial z}\left(D_{zz}\frac{\partial C}{\partial z}\right) + S \quad .(3.2)$$

$$(m^{-1})(m^3/s)(g/m^4) \quad (m^{-1})(m^3/s)(g/m^4) \quad (m^{-1})(m^3/s)(g/m^4) \quad (g/m^3/s)$$

There are many books that delve into the derivation of this equation and conservation of mass principles. Fick's law is used to establish the diffusive flux rate and stems from the analogy to heat transfer. Notice that the diffusion depends on the diffusion coefficients, D_{xx}, D_{yy}, and D_{zz}, and the gradient of concentration with distance. The diffusion coefficients generally vary from around $10^{-5} m^2/s$ (or $0.1\ cm^2/s$) in gases to $10^{-9}\ m^2/s$ (or $10^{-5}\ cm^2/s$) in liquids. There are also various relations attributed to the form of S, depending on the type of chemical reaction. These can be simple zeroth order (S = constant, g/m^3-s), first order ($S = k_1C$, $k_1 = S^{-1}$), second order ($S = k_2C^2$, $k_2 = m^3/g$-s), or other forms.

3.6 Diffusion in Air

Contamination enters the air by either intermittent (i.e., puff) or continuous source emission. A puff is an instantaneous release, or burst, of material of short duration. A continuous emission occurs when a source of pollutant is emitted over a long time, leading to a discernable plume emanating from the source location. The transport physics attributed to both occurrences obey the conservation of mass, as previously described by Eq. 2.1. Much has been written on the atmospheric dispersion of puffs and plumes of contaminants (see Pasquill and Smith, 1985), especially if one can reduce the PDE form of Eq. 3.1 to a more manageable form that can be solved analytically. These analytical solutions are based on the use of Gaussian assumptions, i.e., statistical representations of the probability of concentration being found at specific locations. However, use of such reduced equation sets requires information from the user that may not be known. This is discussed in more detail in the chapter on Gaussian and analytical solutions. For convenience, we introduce the relations here:

Puff:

$$C(x,y,z)=\left(\frac{Q}{(2\pi)^{3/2}\sigma_x(t)\sigma_y(t)\sigma_z(t)}\right) \exp-\left[\frac{(x-Ut)^2}{2\sigma_x^2}+\frac{y^2}{2\sigma_y^2}+\frac{z^2}{2\sigma_z^2}\right], \tag{3.3}$$

Plume:

$$C(x,y,z)=\left(\frac{Q}{2\pi\sigma_y\sigma_z U}\right)\exp\left(\frac{-y^2}{2\sigma_y^2}\right) \left[\exp\left(\frac{-(z-H)^2}{2\sigma_z^2}\right)+\exp\left(\frac{-(z+H)^2}{2\sigma_z^2}\right)\right], \tag{3.4}$$

where Q is the source term, U is the principal velocity (or speed) of the air, x,y,z are spatial distances (from either the puff center or the plume source), H is the height of the release, and σx, σy, and σz are the standard deviations, or diffusion coefficients (which are found using empirical relations developed by Pasquill and Gifford – the Pasquill–Gifford curves (see Pasquill and Smith, 1985). Solutions for C from Eqs. 3.3–3.4 produce Gaussian probability values which yield circular distributions that can be plotted for specific deviations from the center of the puff or plume – these are usually calculated out to $\pm 3\sigma$ standard deviations.

The diffusion coefficient of particles of diameter D_p can be estimated from the relation (see Fuchs, 1964)

$$D = \frac{\kappa TC}{3\pi\mu D_p}, \qquad (3.5)$$

$$\sigma = \sqrt{2Dt}$$

where κ is the Boltzmann constant, μ is the molecular viscosity of the carrier gas, C is a constant, D_p is the droplet diameter, t is time, and σ is the standard deviation, or diffusion coefficient. Table 3.2 shows particle size versus diffusion coefficient in air at STP.

Table 3.2 Particle Diffusion Coefficients in Air (STP) (from *Industrial Ventilation*, R. J. Heinsohn, J. Wiley & Sons, New York, 1991, pg. 180)

D_p (μm)	D (cm²/s)
0.01	1.35×10^{-4}
0.05	6.82×10^{-6}
0.10	2.21×10^{-6}
0.50	2.74×10^{-7}
1.00	1.27×10^{-7}
5.00	2.38×10^{-8}
10.00	1.38×10^{-8}

The Chapman–Enskog equation (see Chapman and Cowling, 1970; Gulliver, 2007; Cussler, 1997) is commonly used to establish diffusion coefficients for compounds; Wilke–Lee (1955) made an adjustment to the original relation to account for diffusivities of lower molecular weight compounds. There are numerous source books that give the molecular diffusion coefficients for a variety of gases. One of the most commonly used sources is the CRC Handbook of Physics and Chemistry, which can be found in nearly every library. An equation developed by Chen and Othmer (see Vargaftik, 1975) can also be used to obtain the binary gas diffusion coefficient (D_{12} in cm²/s)

$$D_{12} = \left(\frac{0.43}{K}\right)\left(\frac{T}{100}\right)^{1.81}\sqrt{\left(\frac{1}{M_1}\right)+\left(\frac{1}{M_2}\right)}$$

$$K = EP\left(\frac{T_{c1}T_{c2}}{10000}\right)^{0.1405}, \quad E = \left[\left(\frac{v_{c1}}{100}\right)^{0.4}+\left(\frac{v_{c2}}{100}\right)^{0.4}\right]^2, \tag{3.6}$$

where v_c and T are the critical volume (cm^3/g-mol) and temperature (°K), M_1 and M_2 are the molecular weights, and pressure P is in atmospheres. To obtain the diffusion coefficient at temperatures and pressures other than STP,

$$D(P,T) = D(STP)\left(\frac{1}{P}\right)\left(\frac{T}{298}\right)^{1.81}. \tag{3.7}$$

Table 3.3 lists some common diffusion coefficients in air for several chemical compounds. Appendix 1 lists many gas pairs and their diffusion coefficients.

Table 3.3 Diffusion Coefficients for Several Contaminants in Air

Substance	M	**D** (10^{-5} m^2/s)
Acetone	56	0.83
Ammonia	17	2.2
Benzene	78	0.77
Chloroform	119	0.87
Hexane	86	0.8
Methane	16	2.2
Sulfur dioxide	64	1.3
Toluene	92	0.71

3.7 Evaporation of Droplets

Drops of liquids are formed from a myriad of industrial and everyday operations. Droplets are basically formed as a result of spraying, aerating, or atomizing. In addition, gas rising through a liquid may ultimately collapse at the liquid's surface and produce liquid droplets.

Every drop has a liquid–air interface; it is this interface through which the liquid of the drop, or the liquid contaminant within the drop, evaporates. The physics associated with droplet formation and evaporation are well known, and can be found in detail in various textbooks dealing with cloud physics – a well-known reference is the work by Pruppacher and Klett (1978).

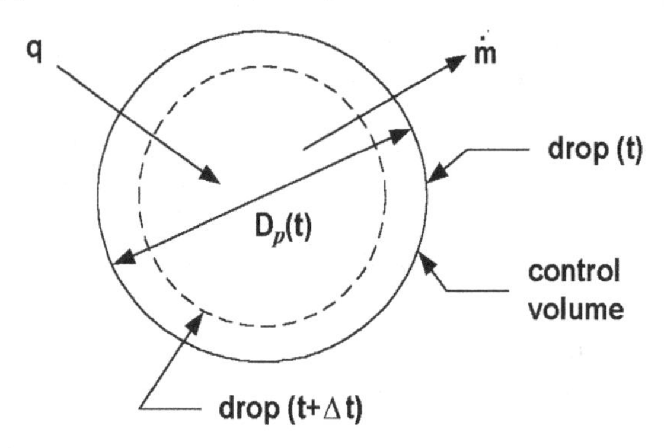

Fig. 3.1 Schematic of an evaporating drop.

The physical processes associated with droplet evaporation can best be illustrated using the Fig. 3.1. Vapor escapes from the surface of the drop due to the vapor pressure of the saturated liquid being greater than the partial pressure of the vapor in the far field. The drop diameter, D_p, decreases as the liquid evaporates which in turn affects the rate of evaporation. The evaporating liquid removes energy from the drop and lowers the drop temperature below the ambient air temperature; this process lowers the drop pressure at the drop–air interface. Since evaporation lowers the drop temperature below the air temperature, energy is transferred to the drop by convection from the surrounding air. The mass and heat transfer are strongly coupled and thus control the rate of drop evaporation.

The set of differential equations that describe the evaporation rate, temperature, and diameter of a drop are fairly well established. These three simultaneous differential equations are typically written in the form

$$\dot{m} = C_1 \pi M D_p \left(\frac{D_{ac}}{R_u}\right) Sc^{b1} Re^{a1} \left[\left(\frac{P_{r,i}}{T_p}\right) - \left(\frac{P_{v,o}}{T_o}\right)\right]$$

$$\dot{m} = -\frac{\rho \pi D^2}{2} \frac{dD_p}{dt} \quad , \tag{3.8}$$

$$\frac{d(T_o - T_p)}{dt} = -\left(\frac{6h_L}{\rho D_p c_v}\right)(T_o - T_p) + \frac{6\dot{m}h_{fg}}{\rho \pi D_p^3 c_v}$$

where Sc is the Schmidt number, Re is the Reynolds number, R_u is the universal gas constant, h_L is the average heat transfer coefficient, c_V is the specific heat at constant volume, and h_{fg} is the enthalpy of vaporization. The equation set in Eq. 3.6 is best solved numerically. If the drop temperature at the liquid–air interface is known, the diameter of the drop can be calculated as a function of time by equating the first two relations. To compute the drop temperature, a simple energy balance as shown in Fig. 3.1 gives the expression for q,

$$q = \frac{mc_v dT_p}{dt} + u_f \frac{dm}{mt} + \dot{m}h_g(T) \quad \text{where} \quad m = \frac{\rho \pi D_p^3}{6} \, , \tag{3.9}$$

where u_f is the internal energy of the saturated liquid and h_g is the enthalpy of the saturated vapor. For droplets with diameters less than 100 μm, the heat transfer within the drop is so rapid that the temperature within the drop can be considered to be uniform. For drops larger than 100 μm, the equation set (3.6) must be solved. The evaporation rate can be calculated from the first relation in Eq. (3.6). Table 3.4 lists several drop sizes, mass, the mass flow rate (or evaporation rate), and temperature difference between ambient and drop temperatures.

Table 3.4 Particle diameter, mass, m, and temperature difference (from *Industrial Ventilation*, R. J. Heinsohn, J. Wiley & Sons, New York, 1991, pg. 216)

D_p (μm)	m (kg)	$\dot{m}$ (kg/s)	T_o-T_p (°C)
5	6.54×10^{-14}	1.08×10^{-10}	0.45
10	5.23×10^{-13}	2.16×10^{-10}	0.63
50	6.54×10^{-11}	1.18×10^{-9}	6.58
100	5.23×10^{-10}	2.74×10^{-9}	14.48

For particle sizes less than 10 μm, the particle temperature is essentially the same as the ambient temperature.

3.8 Resuspension of Particulate

Resuspension refers to the entrainment of a particulate into the air stream. The amount that is entrained into the air stream can be estimated using resuspension factors, resuspension rates, and fractional releases. Factional releases provide an initial amount of contaminant injected into the fluid media.

Resuspension factors are defined as

$$\chi = \frac{C(m^3)}{C(m^2)}, \tag{3.10}$$

where the quotient χ, is airborne concentration per cubic meter of air divided by the surface concentration per square meter. Resuspension factors are not very useful for estimating quantities of particulate being entrained over time (changing or depleting surface concentration). They do however, supply an effective method to evaluate the amount injected into the airflow by an activity at any one time, provided the surface concentration is known. Resuspension rates or mass fractions rates are defined as the fraction of contaminant released over time.

For low flow rates, resuspension coefficients must be specified. Approximations to resuspension rates or factors (mass flux into the domain) are based on the activity occurring and are listed below in this section. For disturbances from turbulent mixing, analytical calculation as developed by Martin *et al.* (1983) may be sufficient. An injection rate based on empirical evidence is desired.

Particulate entrainment is accomplished when attached particles move. A stream velocity large enough to accomplish this is defined as the threshold speed or threshold friction velocity '$u_{+threshold}$'. Once particles are moving the adhesive forces are much weaker, and the particles are available for entrainment. The forces responsible for breaking the attachment are a function of shear stresses acting on the particle, particulate impingement from already suspended material, and adhesive forces between the surface and the particulate. Martin *et al.* (1983) determine resuspension analytically, giving the resuspension rate as a function of friction velocity,

$$u_* = \sqrt{\frac{\tau_{wall}}{\rho_{fluid}}}, \tag{3.11}$$

where τ_{wall} is the shear stress at the wall.

Threshold friction speed is determined from a semi-empirical relation as

$$A = (0.108 + 0.0323/B - 0.00173/B^2)(1 + 0.055/\rho_{fluid}\, g\, d_{part}^2)^{1/2}, \tag{3.12a}$$

where

$$A = u_{*thershold} / \left[(\rho_{part} - \rho_{fluid})\, g\, d_p / \rho_{part}\right]^{1/2} \text{ and}$$
$$B = u_{*thershold}\, d_{part}\, \rho_{fluid} / \mu_{fluid}.$$

The equation is used for the range of $0.22 \le B \le 10$. For $B \le 0.22$

$$A = 0.266(1 + 0.055/\rho_{fluid}\, g\, d_{part}^2)^{1/2}(1. + 2.123/B)^{-1/2}, \tag{3.12b}$$

is used. Since '$u_{+threshold}$' appears in both terms of the equality, iteration is required to obtain a solution.

Suspension occurs for particles having physical diameters smaller than 52 μm when the threshold velocity is reached. Particle suspension is assumed to occur when the terminal settling velocity 'v_s' is equal to the friction velocity and the friction velocity is greater than the threshold velocity.

The amount of material suspended is given by

$$q_s = q_{horizontal}(c_{vert} / u_*^3 c_{hort}) \left[\left(\frac{u_*}{u_{*threshold}}\right)^{Mass\%Suspended/3} - 1\right], \tag{3.13}$$

where

$$q_{horizontal} = 2.61 \frac{\rho_{fluid}}{g} (u_* + u_{*threshold})^2 (u_* - u_{*threshold})$$

and

$$c_{vert} = 2x10^{-10} \text{ and } c_{hort} = 1x10^{-6}.$$

There are limitations on the use of this equation since the empirical constants were found by using light soil particles laying in flat thick beds without obstructions to disturb air flow. However, the equation form is proper, only needing experimental results for determining constants.

Threshold velocities for dense substances such as lead are calculated by Martin *et al.* (1983) to have a minimum value at about 0.3 m/sec for a 49 μm diameter particle. Smaller particles have much greater threshold friction velocities. Determination of the friction velocities on the walls in the laminar sublayer will allow for the incorporation of this resuspension equation provided the species is lying in a thick bed. Application of the above equation to other circumstances will require empirical data.

3.9 Coagulation of Particulate

Another source (and sink for particles) is by coagulation of smaller particles into larger particles as they collide. The time rate of change of concentration from agglomeration for particles with different sizes is given by (Reist, 1993)

$$\frac{dC}{dt} = \frac{-K_o}{2} C^2, \tag{3.14}$$

where

$$K_o = K_{12} = 2\pi (d_{part_1} + d_{part_2})(D_1 + D_2).$$

Over a relative short period of time small particles will coagulate by diffusion into larger particles. For a monodispersed particulate (1 = 2) of initial concentration of 1000 / cm^3, the time for half the particles to coagulate is 55 hours. The time for the particle size to double for this case is 16 days. For an initial concentration of 100,000 / cm^3 the coagulation half-life is 33 minutes and the size doubling time is 4 hours (Hinds, 1982). The time dependent relationship does not include source and sink terms that would also be affecting equations of concentration.

A deposition velocity by diffusion for particles with a micron aerodynamic diameter is insufficient to remove many of the small particles. However, time for coagulation is of the order of the air exchange rate. Therefore, any small particles would have a propensity to agglomerate to a size large enough for settling velocities to possibly be an effective scavenger. Typically, there will be some concentration of particles in the ambient air referred to as Total Suspended Particulate or TSP.

3.10 Comments

There are many sources of contaminants, most found in everyday environments. The emission of carbon dioxide comes principally from humans and mammals. Measurement of CO_2 can be used to assess the effectiveness of ventilation and air exchange. Carbon monoxide, attributed to incomplete combustion common to vehicle exhausts, can become deadly in confined spaces.

Contaminants are typically in the forms of gases (vapors) or particulates (particles). The unit most commonly associated with contaminant concentration is g/m^3. The diffusion coefficients associated with the concentration transport equation are usually expressed in m^2/s (or cm^2/s). The diffusion coefficient determines the rate of the spread of the compound as it diffuses into a medium (typically air).

Chapter 4

Assessment Criteria

There are various criteria that can be used for assessment studies associated with a contaminant dispersing within a room or building. Such factors are typically predefined by the assessor. One of the most important factors is exposure to a contaminant, i.e., how long someone is exposed to the source and the resulting health effects. A low-level exposure of a carcinogen over a long period of time can be just as deadly as a high-level exposure within a short time – the end result is the same. This is certainly the case when dealing with radioactive material, as discovered from the Chernobyl catastrophe in 1986. Another factor to consider is economics, especially when considering the cost of remediation versus total rebuild.

4.1 Exposure

Prevention or remediation of indoor air pollution requires expertise in optimizing geometrical configurations, knowledge of HVAC systems, perceived or expected contaminants and source locations, and economics. Much of the design concept involves ways in which to optimize benefits or balancing the advantages and disadvantages of various configurations and equipment. The fact that a room or building will conceivably become contaminated is generally an accepted fact – to what extent indoor air pollution will become critical is not really known until it happens.

Most companies have a somewhat formal process when developing assessment criteria and procedures – much of this relies on company administrative policies and the experience of the person conducting the assessment. In addition, consultants with specific areas of expertise can play a major role in orchestrating the overall evaluation and assessment of equipment, materials, and potential exposures. In general, one must take into account the activities and processes being undertaken in a room or building, the movement of people, and the anticipated costs associated with using the best versus barely acceptable.

There are numerous agencies and organizations that have attempted to establish exposure limits to various chemicals and materials. These standards are typically referred to as threshold limit values (TLV), permissible exposure limits (PEL), and maximum acceptable concentrations (MAC). The American Conference of Governmental and Industrial Hygienists use TLV; the Occupational Safety and Health Administration (OSHA) publishes PEL values; the American National Standards Institute use MAC. While all three are generally compatible, PEL values are backed by law – it is usually prudent for the engineer or scientist to always check with OSHA for the PEL values. Table 4.1 shows a partial list of substances and the OSHA established PEL.

Table 4.1 Partial List of OSHA Permissible Exposure Limits

Substance	**PEL*** (ppm)
Acetic acid	10
Benzene	10
Chloroform	2
Formaldehyde	3
Ozone	0.1
Turpentine	100

*TWA values

There are several limits that are commonly used in evaluating exposure. The first of these is the time-weighted average of concentration. This is the amount of concentration that is exposed to workers during a normal, 8-hr day, 5 days per week with causing adverse effects. The time-weighted values are calculated from the expressions

$$\mathrm{TWA}(8-\mathrm{hr}) = \left(\frac{1}{8}\right)\int_0^8 c(t)dt$$
$$\mathrm{TWA}(40-\mathrm{hr}) = \left(\frac{1}{40}\right)\int_0^{40} c(t)dt \quad (4.1)$$

Short-term exposure limit is the maximum concentration to which workers can be exposed continuously up to 15 minutes without suffering from side effects. This relation is normally written as

$$\mathrm{STEL} = \left(\frac{1}{15}\right)\int_0^{15} c(t)dt\,. \quad (4.2)$$

Exposure hazards for a mixture of gaseous contaminants are defined by OSHA using an exposure parameter

$$\mathrm{En} = \sum_i \left(\frac{c}{L}\right)_i, \quad (4.3)$$

where c_i is the measured concentration and L_i is the PEL in comparable units of concentration. If En > 1, exposure is considered to be beyond acceptable limits.

An interesting contaminant that gets greatly overlooked is noise. Longitudinal pressure waves ranging from 20–20,000 Hz are known as sound waves. Hearing can be impaired when individuals are exposed to sound or noise above certain amplitudes and lengths of time. Sound power (W) is related to sound intensity by the relation

$$W = \left(4\pi r^2\right)\left(\frac{P^2}{\rho a}\right), \quad (4.4)$$

where a is the speed of sound, ρ is density, r is distance from the source, and P is pressure. Sound intensity (I) is usually used for the expression ($P^2/\rho a$). A sound-intensity level (L_I) can be defined as

$$L_I = 10\log_{10}\left(\frac{I}{I_o}\right), \quad (4.5)$$

where $I_o = 10^{-12} W/m^2$ and corresponds to the intensity at reference pressure ($P_o = 2 \times 10^{-5}$ N/m^2). Sound pressure (L_P) can be calculated at locations from a piece of equipment or process generating noise using the simple formula

$$L_p = L_W - 20\log_{10} r(m) + 10\log_{10} Q - 11, \quad (4.6)$$

where L_W is the sound power and Q is the directivity factor defined as the ratio of the sound power of a small omnidirectional hypothetical source to the sound power of an actual source at the same sound pressure level. Units used to express sound pressure level, sound intensity level, and sound power are decibels (dB).

4.2 Economics

Economics are certainly a factor that must be considered when dealing with issues involving design and remediation of indoor air pollution. The two major costs are Total Capital Cost (TCC) and Total Revenue Requirements (TRR). TCC is essentially the initial costs consisting of money spent to design, build, and install various systems and equipment. TRR are monies spent that must be factored in to the TCC and the revenue needed to provide annual operating costs. Total Indirect Costs (TIC) are monies needed to pay for overhead, i.e., construction expenses, contractors fees, loan interest, building rental, etc. Total Direct Costs (TDC) consists of the TCC plus TIC. The TIC is usually a fraction of the TDC. The equation is simply

$$TCC = TDC + TIC = TDC(1 + ICF), \quad (4.7)$$

where ICF is the Indirect Cost Factor. The TRR is composed of total variable costs (TVC) plus total fixed costs (TFC), or

$$TRR = TVC + TFC. \quad (4.8)$$

Capital recovery cost (CRC) and the fixed cost factor (FCF) are calculated as follows:

$$\text{CRC} = \text{TCC x FCF}$$

$$\text{FCF} = \frac{\left[\text{i}(1+\text{i})^{\text{t}}\right]}{\left[(1+\text{i})^{\text{t}} - 1\right]}, \tag{4.9}$$

where i is the annual interest rate and t is the capital recovery period (years). Table 4.2 outlines the various costs and economic factors that should be considered.

Table 4.2 Cost factors for designing and building ventilation systems (from *Industrial Ventilation*, R. J. Heinsohn, J. Wiley & Sons, New York, 1991, pg. 154).

1. TCC	
TDC	TIC in % of TDC
Equipment	Construction expense (10–15%)
Labor	Contingencies (5–30%)
Materials	Contractors fees (4–6%)
Structures	Engineering (4–6%)
Consulting fees	Interest during construction (10–25%)
	Start-up costs (10–15%)
	Working capital (2–4%)
	Total ICF (45–100%)
2. TRR	
TVC	TFC in % of TCC
Administration	Capital recovery cost (11–23%)
Electric, gas, water	Taxes (3–7%)
Maintenance labor	Insurance (1–3%)
Maintenance material	Interim replacement (1–7%)
Operating labor	Tax credits (0–5%)
Supervision	
Raw materials	Total FCF (16–40%)

4.3 Comments

The assessment of indoor air quality is a fairly broad topic. Defining assessment criteria generally depends upon those factors deemed important to the person conducting the evaluation and the requirement guidelines. Clearly exposure to a contaminant is crucial to any assessment study. This is especially evident when dealing with radioactive contaminants (commonly referred to as radioactive "shine," e.g., a plume passing overhead will "shine" gamma, beta, and alpha radiation to a person on the ground).

It is also important to factor in economics. This is important when considering a remediation approach or cleanup activity. Current problems dealing with asbestos in buildings is an example where the economics can be the leading factor in whether to retrofit a building or raze it to the ground.

You can perform your own simple contaminant assessment study in a room. Pop a bag of popcorn (best in a microwave), then leave the bag sealed and take it into a room full of people. Open the bag and ask the participants to raise their hands when they smell the popcorn. You will find that the smell of popcorn rapidly permeates throughout the room. Only a few parts per million are needed to initially smell the popcorn. Over time, the smell becomes strong to nearly everyone in the room. Someone with an overabundance of aftershave or perfume is also effective in demonstrating the rapid dispersion of an odor (and exposure) in a room.

Chapter 5

Simple Modeling Techniques

Simple models typically consist of either first order approximations which may crudely define the problem domain or elegant, sophisticated analytical solutions for ideal conditions. Rarely do such models exist which can provide intricate details at minute levels within an interior. However, the use of simple modeling tools can quickly provide great insight and an overall grasp of the problem. Such models are useful in establishing at least an order of magnitude assessment, and in some instances may be sufficient for determining IAQ values.

There are a variety of analytical tools and simple model configurations that can be useful to a designer in predetermining contaminant levels within an interior. We will start with a description of the simplest of these models assuming rather ideal conditions.

5.1 Analytical Tools

There are generally two concepts used when developing simple models for indoor air quality calculations: (1) well-mixed and (2) partially mixed ventilation models. In a well-mixed model, the concentration is spatially uniform within the enclosure; in a partially mixed model, the concentration is nonuniform generally due to poor mixing. In some situations, it is convenient and relatively safe to assume well-mixed conditions – this type of assumption leads to the use of simple analytical models.

Unfortunately, most real world situations involve dealing with partially mixed hypothesis. Analytical procedures are available for these situations as well, but are somewhat limited; a mixing factor (m) is generally employed to modify the equations for a well-mixed model to account for the nonuniform distribution of concentration. It is usually preferable for these types of problems to employ CFD techniques and numerical models for dispersive transport.

Assume an enclosed space exists in which the concentration is considered to be spatially uniform, as shown in Fig. 5.1.

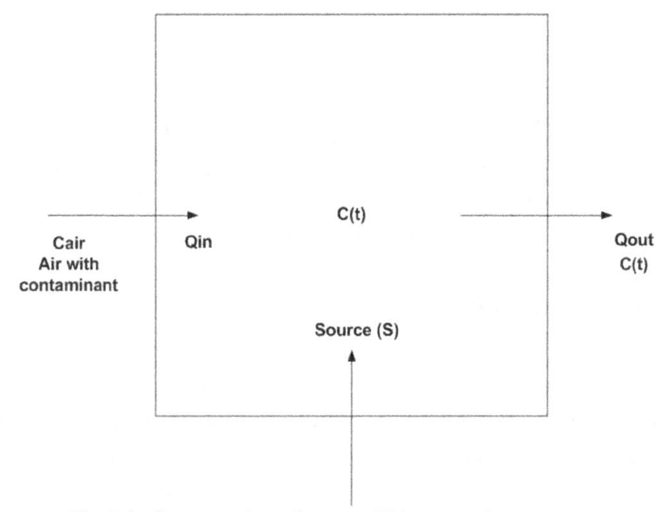

Fig. 5.1 Conservation of mass within an enclosure.

The mass concentration at t = 0 is C_o. A source begins to generate contaminants at a constant rate (S). Outside air with contaminant C_{air} is added to the enclosure at a constant volumetric flow rate Q – contaminated air is removed from the space at the same rate. Applying the equation for conservation of mass, the governing equations for the contaminant concentration entering and leaving the enclosure can be written as

$$\frac{\partial}{\partial t}\int_{cv} C d\forall + \int_{cs} CVdA + S = 0, \tag{5.1}$$

where C is the concentration, ρ is density, V is velocity, $\forall$ is volume, cv denotes the control volume, and cs is the control surface. Letting Q ≡ AV (flow rate), Eq. (5.1) can be written as

$$\forall\frac{dC}{dt}+\sum AVC_{out}-\sum AVC_{in}=S, \tag{5.2}$$

which can be simplified to the relation

$$\forall\frac{dC}{dt}=QC_{air}-QC+S. \tag{5.3}$$

Since the flow of air into and out of the enclosure is in equilibrium, it is easy to obtain an expression for C. If we now integrate over time,

$$\int_{C_0}^{C(t)}\frac{dC}{[QC_{air}+S-QC]}=\frac{1}{\forall}\int_0^t dt. \tag{5.4}$$

solution of Eq. 5.4 becomes

$$\frac{(C_s-C(t))}{(C_s-C_o)}=\exp\left(-\frac{Qt}{\forall}\right), \tag{5.5}$$

where C_s is the steady-state concentration (letting $C_s = C_{air} + S/Q$) obtained by setting the LHS of Eq. 5.3 equal to zero. Assuming both initial and ambient concentrations are zero, one obtains the reduced form of Eq. 5.5

$$\frac{C(t)}{C_s}=1-\exp(-Nt), \tag{5.6}$$

where N = Q/V and is known as the number of room air changes per minute. To illustrate, Eq. (5.6) would predict a concentration that would be 64% of its steady-state value after 15 minutes for a ventilation room rate of 4 changes per hour (N = 4/hr).

Example 5.1.1 Assume a contaminant (1000 mg/m^3) enters an office that is 64 m^3 in volume. The contaminant fills the room. An exhaust fan is used to remove air at 25 m^3/min from the room and outside air enters the room at the same rate. Also assume the outside air is contaminated at 1 mg/m^3.

What is the length of time before you can enter the office if 5 mg/m^3 is the minimum threshold exposure limit?

Solution:

We set $C_s = C_{air} = 1$ mg/m^3, $C_o = 1000$ mg/m^3, $C(t) = 5$ mg/m^3, $Q = 25$ m^3/min, and $V = 64$ m^3; $S = 0$. Thus,

$$\frac{(5-1)}{(1000-1)} = \exp\left(-\frac{25t}{64}\right), \quad t = 14.13 \text{ min.}$$

For a time-varying source or ventilation flow rate, Eq. 5.3 can be rewritten as

$$\frac{dC}{dt} = -\frac{QC}{\mathcal{V}} + \frac{(S + QC_{air})}{\mathcal{V}}. \tag{5.7}$$

Equation 5.7 must be solved numerically. Using a simple difference scheme and averaging C between unknown and known values – $(C^{n+1} + C^n)/2$ – on the right hand side, Eq. 5.7 can be rewritten as

$$\frac{C^{n+1} - C^n}{\Delta t} = -\frac{Q}{\mathcal{V}}\left(\frac{C^{n+1} + C^n}{2}\right) + \frac{S + QC_{air}}{\mathcal{V}}, \tag{5.8}$$

or

$$C^{n+1} = \frac{\left(\frac{1}{\Delta t} - \frac{Q}{2\mathcal{V}}\right)C^n + \left(\frac{S + QC_{air}}{\mathcal{V}}\right)\Delta t}{\frac{1}{\Delta t} + \frac{Q}{2\mathcal{V}}}, \tag{5.9}$$

where n is the iteration and Δt is the time step. The solution begins with $t = 0$ and $n = 0$ with C_o as the initial concentration.

To account for wall losses, i.e., removal of contaminant by solid surfaces, Eq. 5.7 can be modified to include the adsorption rate (k_{ad}) of contaminant on walls

$$\mathcal{V}\frac{dC}{dt} = QC_{air} + S - C(Q + A_{surface}k_{ad}). \tag{5.10}$$

The surface area of the room (A) must now be considered to account for the contaminant sticking to the walls (for gases or vapors, this is called adsorption; for particles, this is referred to as deposition). This is very evident when one cleans a room after someone has been smoking in the room – various surfaces absorb tobacco smoke and then desorption occurs when the smoking ceases. Equation 5.8 can be integrated to give

$$\frac{(C_s - C(t))}{(C_s - C_o)} = \exp\left(-\frac{(Q + A_{surface}k_{ad})t}{V}\right), \quad (5.11)$$

assuming constant values for Q, S, and k_{ad}. For the case when Q = S = 0 (starting with a contaminated room), Eq. 5.11 can be modified to

$$\frac{C(t)}{C_{max}} = \exp\left(-\frac{Atk_{ad}}{V}\right), \quad (5.12)$$

where C_{max} represents the maximum concentration at the beginning of the integration. Figure 5.2 shows the importance of adsorption of tobacco smoke (see Repace and Lowery, 1980; Heinsohn, 1991) and the effects of mixing in a room. In this instance, a single cigarette was burned and then extinguished in a 22 m^3 room and the total mass of suspended particle matter measured during the entire period.

Well-mixed conditions were produced by fans; the natural mixing occurred as a result of natural air currents. Notice that the concentration in the well-mixed experiment fell rapidly, as expected. The slope of the curve allows one to estimate the removal of contaminants by adsorption on solid surfaces. The rate of adsorption was found to be equivalent to an exhaust ventilation rate of 1.4 m^3/min (50 CFM). In this case, desorption acts as a source term in Eq. 5.10.

If only a fraction (f) of the return flow into an enclosed space is fresh air, Eq. 5.10 can be modified to the following form

$$V\frac{dC}{dt} = -CQ[1-(1-f)(1-\eta)] + S + QfC_{air}(1-\eta), \quad (5.13)$$

where f is the makeup air fraction (makeup of fresh air/input air – Q_m/Q) and η is the efficiency of the air cleaning device. Integrating Eq. 5.13,

$$\frac{(C_s - C(t))}{(C_s - C_o)} = \exp\left[-\left(\frac{Qt}{V}\right)\{1-(1-\eta)(1-f)\}\right]. \quad (5.14)$$

For variable source or volumetric flow rates, Eq. 5.14 must also be solved numerically.

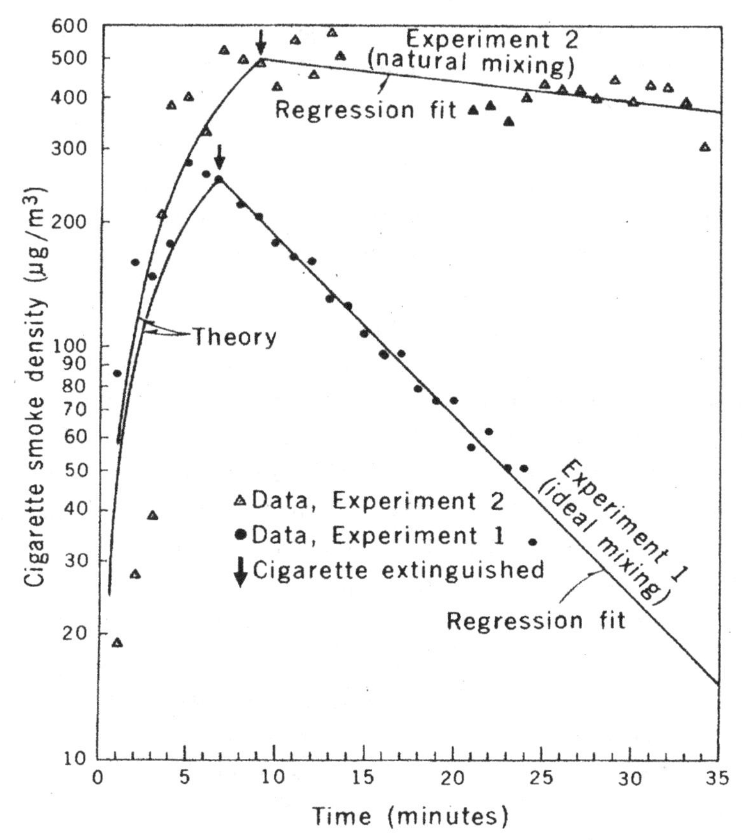

Fig. 5.2 Smoke concentration within a room with and without internal mixing (from J. L. Repace and A. H. Lowrey, *Science*, Vol. 208, May 2, 1980, pg. 467).

For partially mixed conditions, the concentration varies both spatially and temporally. This condition is normally found in most industrial applications. The technique employed here is to introduce a mixing factor (m_f) to account for the spatial variations in concentration. Equation 5.11 now becomes

$$\forall \frac{dC}{dt} = S + (m_f Q C_{air}) - (m_f Q C) - (m_f C \eta Q_r), \qquad (5.15)$$

where Q_r is the volumetric flow rate of recirculated air. Integration of Eq. 5.15 gives

$$\frac{(C_s - C(t))}{(C_s - C_o)} = \exp\left[-\frac{m_f t(Q + Q_r \eta)}{V\!\!\!-}\right]. \quad (5.16)$$

For this particular type of situation, setting $m_f = 1$ indicates a well-mixed model while $m_f < 1$ implies nonuniform mixing and spatial variations in concentration, i.e., $m_f = 0.5$ is used for a perforated ceiling, $m_f = 0.166$ is for natural draft and ceiling exhaust fans, $m_f = 0.10$ is used for infiltration and natural drafts.

The source emission rate, or source strength (S), is usually not known and must be determined from experiment. A source can be released in a clean room and measurements made of the rise in concentration. The governing equation is

$$V\!\!\!-\frac{dC}{dt} = S - C(Q_s + A_{surface} k_{ad}), \quad (5.17)$$

where Q_s represents the volumetric flow rate through the sampling device. Immediately after the source is activated and while the concentration is small, Eq. 5.17 reduces to the simple form

$$V\!\!\!-\frac{dC}{dt} = S, \quad (5.18)$$

and the source strength can be found from the slope of concentration versus time. A more accurate means of determining S is to measure two concentrations, C_1 and C_2, at two times t_1 and t_2, and obtain S from the integration of Eq. 5.17, i.e.,

$$S = -(A_{surface} k_{ad} + Q_s)$$
$$\left[C_1 \exp\left(\frac{-(K_1 k_{ad} + Q_s)(t_2 - t_1)}{V\!\!\!-}\right) - C_2\right] / G \quad (5.19)$$
$$G = 1 - \exp\left[\frac{-(A_{surface} k_{ad} + Q_s)(t_2 - t_1)}{V\!\!\!-}\right],$$

Example 5.1.2 Exhaust hood simulation: An exhaust hood is installed within a few feet of a makeup air inlet in a room. Ethyl alcohol is evaporated in the hood. What is the steady-state concentration of ethyl alcohol in the room and the amount of time before one begins to smell alcohol? Assume the threshold odor limit for ethyl alcohol is 40 mg/m^3 and the following criteria apply:

V = volume of operating room (50 m^3)
$A_{surface}$ = total area of adsorbing surfaces in operating room (85 m^2)
k_{ad} = adsorption rate constant (0.001 m/s)
S = rate at which ethyl alcohol is vaporized inside operating room (1 g/min)
C_o = initial alcohol concentration inside operating room (10 mg/m^3)
C_{air} = concentration of ethyl alcohol entering makeup air duct (100 mg/m^3)
Q_e, Q_r, Q_a, Q_s = volumetric flow rate of exhausted air, recirculated air, makeup air, and supply air (Q_s = 20 m^3/min)
$f = Q_a/Q_s$ = make up air fraction (0.9)
η_1, η_2 = efficiencies of activated charcoal filter (0.5)

Solution:

The governing equation to be solved is of the form

$$V\frac{dC}{dt} = S + Q_s C_s - CQ_e - CA_{surface}k_{ad}\,. \tag{5.20}$$

A mass balance for the air results in the expression

$$\begin{aligned} Q_s &= Q_a + Q_r \\ Q_e &= Q_s. \end{aligned} \tag{5.21}$$

At the fan inlet, the mass balance for alcohol is

$$C_{air}Q_a(1-\eta_2) + CQ_r(1-\eta_1) = C_s Q_s\,. \tag{5.22}$$

Using the definition of f,

$$C_s = C_{air}f(1-\eta_2) + C_r(1-\eta_1)(1-f)\,. \tag{5.23}$$

The differential equation to be solved is of the form

$$\frac{dC}{dt} = K_1 C + K_2, \tag{5.24}$$

where K_1 and K_2 are evaluated as

$$K_1 = \frac{\{-Q_s - A_s k_{ad} + Q_s(1-f)(1-\eta_1)\}}{\forall} =$$

$$-\frac{25\text{m}^3/\text{min}}{50\text{m}^3} = -0.5\,\text{min}^{-1} \tag{5.25}$$

$$K_2 = \frac{[S + fQ_s C_{air}(1-\eta_2)]}{\forall} = 0.0218\text{g}/\text{m}^3\,\text{min}.$$

Setting dC/dt = 0 gives the steady-state concentration,

$$C_s = -\frac{K_2}{K_1} = 45.6\text{mg}/\text{m}^3. \tag{5.26}$$

Since K_1 and K_2 are known constants, the time can be calculated from the integral expression

$$\int_{10}^{40} \frac{dC}{[K_1 C + K_2]} = \int_0^t dt; \qquad t = 4.47\,\text{min}. \tag{5.27}$$

5.2 Advection Model

Many times a source exists that is moving within a confined space. Examples of such situations are automobiles or trains that are traveling through tunnels, or a smoker walking from one room to another. In this instance, a simple control volume approach can be used to establish the governing equation for concentration. In many instances, makeup air consisting of fresh air is used to provide local ventilation, e.g., for tunnels less than 600 m in length.

A nice example of contaminant from an automobile traveling within a tunnel is discussed in Heinsohn (1991). In similar fashion, a schematic of air and contaminant transport from a train traveling within a tunnel is shown in Fig. 5.3. An elemental volume denoted by Adx exists within a tunnel with uniform makeup air and exhausts. The conservation of mass for air within the volume gives the following expression

$$\frac{dU}{dx} = q_m - q_e, \tag{5.28}$$

where $q_m = Q_m/LA$ and $q_e = Q_e/LA$. If q_m and q_e are constant, the air velocity in the tunnel at any location x is

$$\frac{U(x)}{U_o} = 1 + \frac{x(q_m - q_e)}{U_o}, \tag{5.29}$$

where U_o denotes air entering the tunnel. If $q_m > q_e$, then U(x) increases linearly with x; if $q_e > q_m$, U(x) decreases. The conservation of mass for contaminant transport can be written as

$$\frac{dUC}{dx} = s + q_m C_m - Cq_e - kC, \tag{5.30a}$$

or

$$C\frac{dU}{dx} + U\frac{dC}{dx} = s + q_m C_m - Cq_e - kC, \tag{5.30}$$

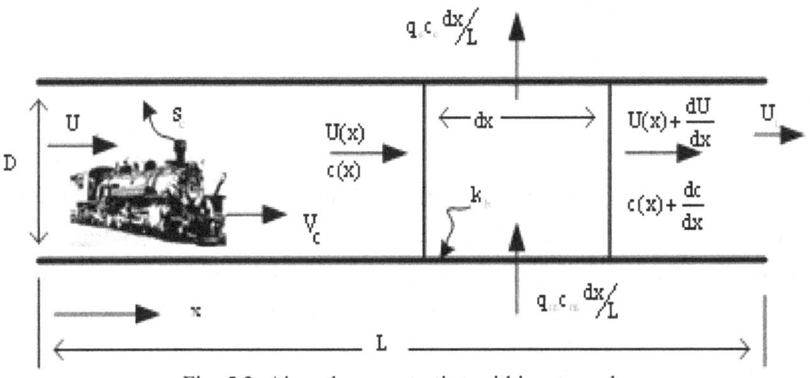

Fig. 5.3 Air and concentration within a tunnel.

where $s = S/LA$ ($\mu g/m^3$-min) and $k = 4k_d/D$ (min^{-1}), D is the tunnel diameter, C_m is the contaminant, S is the source (mg/hr), and k_d (m/s) is the rate at which contaminant is deposited on the tunnel walls. Combining Eqs. 5.28–5.30, the equation for contaminant within the tunnel is

$$U\frac{dC}{dx} = s + q_m C_m - C(k + q_m), \tag{5.31}$$

which can be rewritten using Eq. 5.30 as

$$\frac{dC}{(q_m C_m + s) - (k + q_m)C} = \frac{dx}{U_o + (q_m - q_e)x}. \tag{5.32}$$

If q_m and q_e are constant (unequal and nonzero), Eq. 5.32 can be integrated to

$$C(x) = \frac{s + q_m C_m}{k + q_m} + \left[c_o - \left(\frac{s + q_m C_m}{k + q_m}\right)\right]\left[\frac{U(x)}{U_o}\right]^{-b}, \tag{5.33}$$

where $b = (k + q_m)/(q_m - q_e)$ and $U(x)/U_o$ can be replaced using Eq. 5.29. If q_e and q_m vary with x, Eq. 5.33 must be solved using a numerical approach. If q_e and q_m are zero, Eq. 5.33 cannot be used and Eq. 5.32 must be integrated directly. When q_e and q_m are equal, the system is balanced. The usual case is for $q_m > q_e$.

5.3. Box Model

When the concentration within an enclosure is nonuniformly distributed, it is inaccurate to assume the enclosure can be treated as a well-mixed region. Although one could utilize partially mixed conditions and use mixing factors, the uncertainty in selecting values for m_f and the tendency of the partially mixed model to still predict spatially uniform concentrations would likely result in large inaccuracies. An alternative approach to the analytical tools utilized in the previous section is the *box model*, also sometimes referred to as the multi-cell well-mixed model.

There are basically four types of mixing that can occur within an enclosure. These are (1) displacement, (2) cavity, (3) mixing, and (4) piston, as shown in Fig. 5.4. In displacement, incoming air *displaces* the existing air within the enclosure. In a cavity, air flows into and out of the enclosure, much like flow over a cavity (which in turn creates a recirculation). In mixing, flow enters at the top and exits at the bottom, creating a mixed region within the enclosure. The last model works much like a piston where air is pumped into a container and then exits out the sides – like the piston and cylinder effect in an automobile.

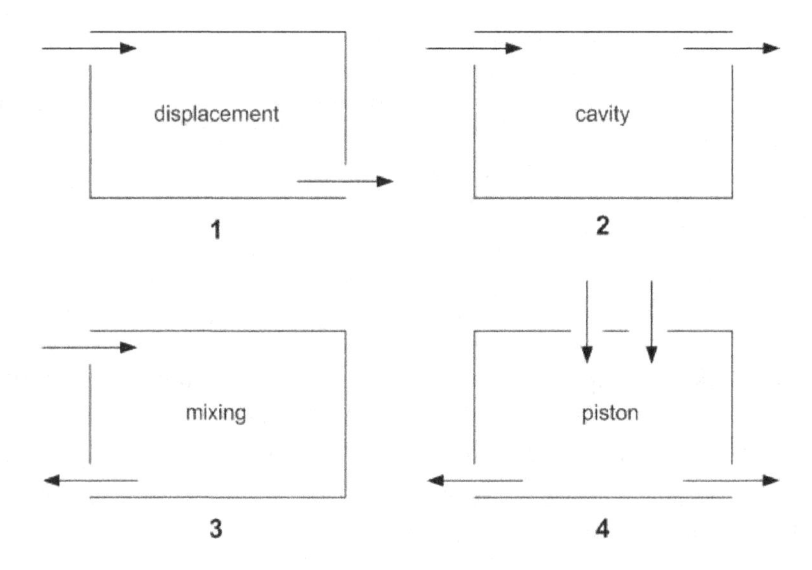

Fig. 5.4 Types of mixing within an enclosure.

Figure 5.5 shows a schematic of a partially mixed enclosure with two sources, two makeup air vents, and two exhaust vents, common to type 3 mixing.

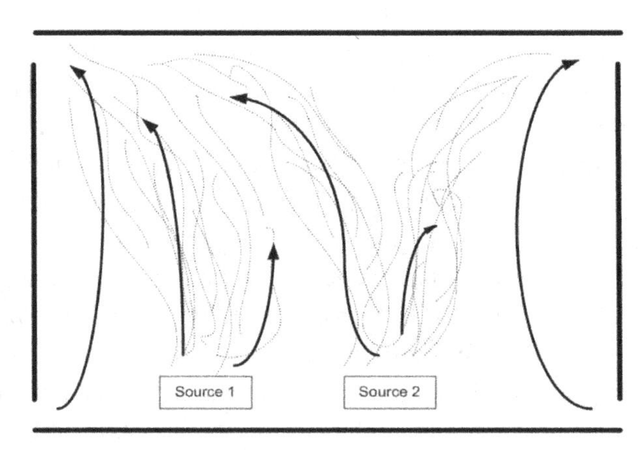

Fig. 5.5 Partially mixed enclosure: Two each of sources, makeup air and exhaust vents.

Utilizing a box model approach, the domain is divided into two cells with contaminant that transfers between each cell. This is shown in Fig. 5.6 for the two-cell model; the user can implement as many cells as desired – in this case, the problem domain is ideal for establishing a two-cell approach.

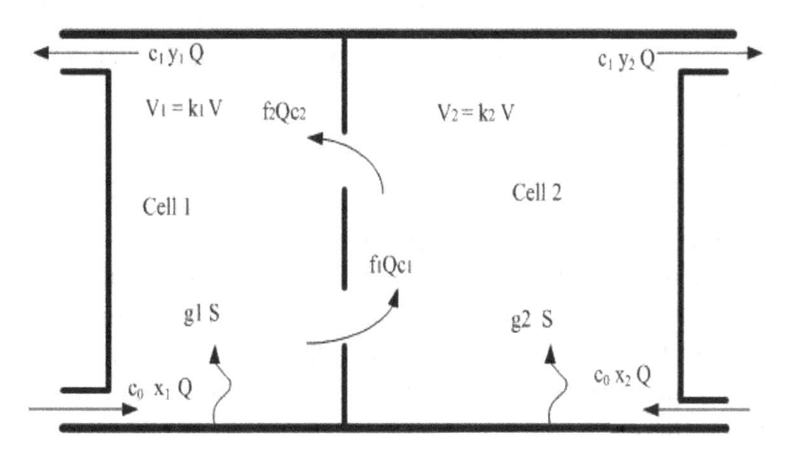

Fig. 5.6 Two-cell box model.

We begin by introducing the volumetric flow rates (Q) and fractions of those rates entering (x) and leaving (y) the cell boundaries.

Entering the enclosure:

$$\begin{aligned} Q_{1,i} &= x_1 Q \\ Q_{2,i} &= x_2 Q \\ Q &= Q_{1,i} + Q_{2,i} = Q(x_1 + x_2) \\ x_1 + x_2 &= 1. \end{aligned} \tag{5.34}$$

Leaving the enclosure:

$$\begin{aligned} Q_{1,o} &= y_1 Q \\ Q_{2,o} &= y_2 Q \\ y_1 + y_2 &= 1. \end{aligned} \tag{5.35}$$

The fractions are obtained from knowing the amounts of makeup air, recirculation, exhaust, and infiltration. The fraction of contaminant in each cell is designated as s_1 and s_2, i.e.,

$$\begin{aligned} S &= S(s_1 + s_2) \\ s_1 + s_2 &= 1, \end{aligned} \tag{5.36}$$

and the volume of cells 1 and 2 expressed as fractions, v_1 and v_2, of the total volume, V, as

$$\begin{aligned} \mathcal{V} &= \mathcal{V}(v_1 + v_2) \\ v_1 + v_2 &= 1. \end{aligned} \tag{5.37}$$

Fractional values for x, y, s, and v are input by the user.

The amount of volumetric flow rate transferred across the internal boundary between cells 1 and 2 is denoted through the use of exchange coefficients, k_1 and k_2, which can vary from less to greater than 1. Performing a conservation of mass balance yields the following expressions for cell 1 and 2,

$$\begin{aligned} \text{Cell 1: } x_1Q + k_2Q - y_1Q - k_1Q &= 0 \\ x_1 - y_1 + k_2 - k_1 &= 0 \\ \text{Cell 2: } x_2Q + k_1Q - y_2Q - k_2Q &= 0 \\ x_2 - y_2 + k_1 - k_2 &= 0. \end{aligned} \tag{5.38}$$

The governing equations for the conservation of mass become

$$\begin{aligned} v_1\mathcal{V}\frac{dC_1}{dt} &= s_1S + x_1QC_{air} + k_2QC_2 - k_1QC_1 - y_1QC_1 \\ v_2\mathcal{V}\frac{dC_2}{dt} &= s_2S + x_2QC_{air} + k_1QC_1 - k_2QC_2 - y_2QC_2, \end{aligned} \tag{5.39}$$

where c_a is the concentration in the air entering cells 1 and 2. The two sets of relations described by Eq. 5.39 can be rewritten to the simpler pair of simultaneous, first-order differential equations by assuming $v = v_1$, $s = s_1$, and $k = k_1$ (and thus $k_2 = k + y - x$)

$$\begin{aligned}\frac{dC_1}{dt} &= A + BC_1 + DC_2 \\ \frac{dC_2}{dt} &= E + FC_2 + GC_1,\end{aligned} \tag{5.40}$$

where the coefficients are defined as

$$\begin{aligned}A &= \left(\frac{N}{v}\right)\left[x_1 c_a + \left(\frac{sS}{Q}\right)\right] \\ B &= -\left(\frac{N}{v}\right)[k + y_1] \\ D &= \left(\frac{N}{v}\right)[x_2 - y_2 + k] \\ E &= \left(\frac{N}{(1-v)}\right)\left[x_2 c_a + \left\{(1-s)\frac{S}{Q}\right\}\right] \\ F &= -\left(\frac{N}{(1-v)}\right)[x_2 + k] \\ G &= \frac{Nk}{(1-v)} \\ N &= \frac{Q}{V}.\end{aligned} \tag{5.41}$$

The general solution to the pair of relations defined by Eqs. 5.40–5.41 is

$$\begin{aligned}C_1(t) &= K_1 \exp(Ntw_1) + K_2 \exp(NTw_2) + C_{1,ss} \\ C_2(t) &= MK_1 \exp(Ntw_1) + LK_2 \exp(NTw_2) + C_{2,ss},\end{aligned} \tag{5.42}$$

where $C_{1,ss}$ and $C_{2,ss}$ are the final (steady-state) cell concentrations given as

$$C_{1,ss} = \frac{AF - ED}{DG - BF}; \quad C_{2,ss} = \frac{EB - AG}{DG - BF}, \tag{5.43}$$

with

$$\begin{aligned}
w_1 &= \left(\frac{1}{2N}\right)\left[(B+F) + \sqrt{(B-F)^2 + 4DG}\right] \\
w_2 &= \left(\frac{1}{2N}\right)\left[(B+F) - \sqrt{(B-F)^2 + 4DG}\right] \\
M &= \frac{Nw_1 - B}{D} \\
L &= \frac{Nw_2 - B}{D} \\
I_1 &= C_1(0) - C_{1,ss} \\
I_2 &= C_2(0) - C_{2,ss} \\
K_1 &= \left[L - \left(\frac{I_2}{I_1}\right)\right]\left[\frac{I_1}{(L-M)}\right] \\
K_2 &= \left[\left(\frac{I_2}{I_1}\right) - M\right]\left[\frac{I_1}{(L-M)}\right],
\end{aligned} \tag{5.44}$$

where $C_1(0)$ and $C_2(0)$ are the initial cell concentrations. It is a simple matter to solve for the equation pair established by Eq. 5.42. The only difficulty is in selecting an appropriate value for the exchange coefficient, k, which is difficult to establish. The best way is to use trial and error or some empirical judgment to determine a range of values for k. Note that as the value of k increases, the exchange between cells increases. When k reaches a value of around 15, the concentration in both cells approaches the equivalent of a single well-mixed cell, i.e., well-mixed conditions can be assumed throughout the entire enclosure.

Although using two cells is crude, it is much better than assuming well-mixed conditions for the problem domain. Of course, one can always add more cells in an effort to refine the problem details and obtain a more accurate solution; however, the complexity of analyzing multiple cells increases proportionally to the square of the number of cells. If one winds up using many cells, it may be best to ultimately go to the use of numerical methods, i.e., CFD. The accuracy of the box model is limited by the inability to establish enough detail to describe the exchange of air among cells, especially if transient solutions are sought.

Example 5.3.1 Box model: Objects are to be cleaned in HCl solutions in one room before final assembly in an adjacent room. The liquid surface area is 75 ft^2 and HCl vapor is emitted at a rate of 0.02 $gm/s\text{-}m^2$. The room with the HCl is 30 ft x 30 ft x 15 ft and has a doorway 10 ft x 15 ft into the adjacent room that is 50 ft x 30 ft x 15 ft. No HCl is generated in the adjacent room.

The plant manager has reservations that placing an air curtain in the doorway will prevent HCl vapor from entering the adjacent room. Each room has its own HVAC system. Each room is well mixed and infiltration and exfiltration are equal to one change per hour.

The ventilation system delivers 600 CFM of outside air to the adjacent room and 600 CFM of contaminated air is removed from the room containing HCl. What are the steady-state concentrations in each room, how fast do the concentrations increase in time, and is the PEL (5 ppm ~ 7 mg/m3) exceeded in each room? Assume that no adsorption occurs and that HCl is initially zero.

Utilizing Eqs. 5.42–5.44, the following values for the various constants, assuming a two-cell model, are calculated as shown in Table 5.1. Using Eq. 5.42, a steady-state concentration value of 175 ppm is obtained with a time constant of 0.5 hr, using a value of k = 15 (which is very conservative but allows one to place an upper bound on the exchange). Using this upper limit, the PEL would be exceeded during the working day.

Figure 5.7 shows concentration values versus time within the two cells using k = 0.2 and k = 1.8. If k = 0, cell 1 is well mixed and cell 2 receives no HCl. Notice that as k increases, the mixing increases. Steady-state concentrations in both rooms approach values that would be predicted assuming well-mixed conditions throughout the rooms. The time it takes for the concentration to become well mixed throughout the rooms is around 2 hours, as indicated in the flattening of the curves for the two cases.

Table 5.1 Values for Various Parameters Used in Box Model Example

Parameter	Value
x_1	0.1875
x_2	0.8125
y_1	0.6875
y_2	0.3125
s	1.0
v	0.375
Q	72,000 CFM
S	502 gm/hr
N	2 hr^{-1}
A	0.0372 gm/hr-m^3
B	-5.33 (k+0.6875) hr^{-1}
D	5.33 (k+0.5) hr^{-1}
E	0
F	-3.2 (k+0.8125) hr^{-1}
G	3.2 (k) hr^{-1}

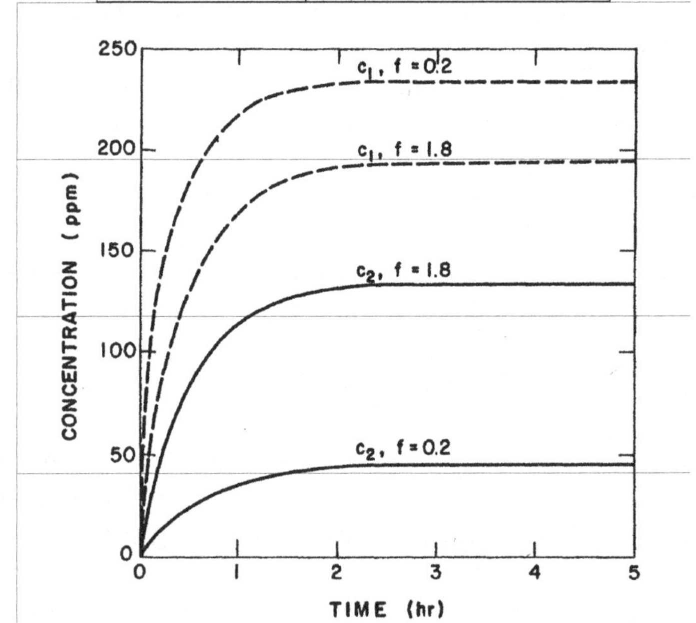

Fig. 5.7 Box model example for mixing of HCL between two rooms (from *Industrial Ventilation*, R. J. Heinsohn, J. Wiley & Sons, New York, 1991, pg. 270).

5.4 Comments

The establishment of a simple model to describe mixing within an enclosure is beneficial for two reasons: (1) the modeling demands that the investigator set up proper boundary conditions for the problem, and (2) the model may permit a simple solution to be obtained for an initially preconceived difficult problem. One should always attempt to obtain a simple solution first – many times one only needs to establish an order of magnitude or "ball park" estimate to determine if a problem needs to be addressed in more detail.

The example problems illustrated in this chapter can be easily solved using one of the many popular mathematical engineering packages commercially available. For example, MAPLE, MATHEMATICA and MATLAB are popular commercial codes that allow the user to input the mathematical expressions and ultimately obtain solutions. An interface between MAPLE and MATLAB (known as the MAPLE Toolbox for MATLAB), allows the user to input the relations and then subsequently produce a MATLAB model, if desired. It is recommended that the reader work through these examples to become familiar with setting up the governing equations and solving differential equations.

Chapter 6

Dynamics of Particles, Gases and Vapors

Contaminants typically appear in the form of either particles or gases and vapors. Particulates range in size from near atomistic levels to those which we can easily see in the air, such as pollen, smoke, or dust. At the smallest scales, we must treat particles as individual entities, requiring the use of molecular hypothesis. As we get to larger sizes and volumes, we can begin to treat the array of particles as a continuum, allowing the use of the more familiar equations of motion commonly used for fluid flow and species transport.

It is important in our study of indoor air pollution that we first understand the fundamental principles associated with particulate motion. We can then proceed to gases and vapors.

6.1 Drag, Shape, and Size of Particles

Analyzing the force on a particle in a flow field reveals the fluid to be exerting a force proportional to the particle's projected area, the square of the relative velocity of the particle to the fluid. This proportionality is known as Newton's resistance equation. In general form Newton's resistance equation is

$$F_d = C_d \frac{\pi}{8} \rho_f d_p^2 |V| V, \tag{6.1}$$

where V is the relative velocity of the particle with diameter d_p having a drag coefficient C_d in a fluid with density ρ_f. The relative velocity is defined as

$$V \equiv (u_x - U_x) \cdot \mathbf{i} + (v_y - V_y) \cdot \mathbf{j} + (w_z - W_z) \cdot \mathbf{k}$$

$$|V| \equiv \left[(u_x - U_x)^2 + (v_y - V_y)^2 + (w_z - W_z)^2 \right]^{0.5},$$

with **i**, **j**, **k** being the unit vectors denoting x, y, and x directions, U_x denoting the fluid velocity in the x-direction, u_x being the particle velocity in the x-direction, etc. This equation is valid for particle motion at subsonic speeds. Particles which have Reynolds number (Re) less than one, where $Re = d_p V \rho_f / \mu \leq 1$, is known as the Stokes regime. The drag force is

$$F_d = 3\pi \mu d_p V. \tag{6.2}$$

when substituted into Eq. 6.1, the coefficient of drag is $C_d = 24/Re$.

If particle size is of the order of the molecular mean free path (usually denoted as λ), the particle does not experience the fluid as a continuum, but as an individual molecule. Particles of this size invalidate the assumption of a no-slip boundary condition for the fluid on the particle's surface used in the Stokes flow analysis. The particle is able to slip through the fluid, reducing the drag experienced by the particle as predicted from a continuum analysis in Stokes' flow regimes. A slip factor (Cunningham slip correction factor) for particle drag corrects the Stokes drag coefficient

$$C_c = 1 + (2\lambda / d_p)(A_1 + A_2 e^{-A_3 d_p / \lambda}). \tag{6.3}$$

The molecular mean free path, λ, is given by (Cooper and Alley, 1994)

$$\lambda = \frac{\mu}{0.499 P \sqrt{8 MW_f / \pi R T}}, \tag{6.4}$$

where μ is the absolute viscosity of the fluid, MW_f is the molecular weight of the fluid, R is the universal gas constant, P is the pressure, and T is the absolute temperature. Any consistent set of units will provide the length of the mean free path. The factors A_1, A_2, A_3 are dimensionless empirical constants for small particle drag (Martin *et al.*, 1983).

The slip factor is used to augment the coefficient of drag in the force equation and the force of drag becomes

$$F_d = \frac{3\pi \mu V d_p}{C_c} . \tag{6.5}$$

Particles of various shapes and sizes are found in indoor environments. Depending on the molecular structure of the mineral or molecules forming the particle, it is possible to predetermine the shapes expected from some compounds, e.g., salt has a cubical shape and fibers are cylindrical in shape.

The Newton's resistance equation and Stokes flow analysis can be adjusted to account for non-spherical particles. By using an equivalent volume for the particle, that is, creating a sphere of equivalent volume that an irregular shaped particle would have if it were spherical Stokes law becomes

$$F_d = 3\pi \mu V d_{pe} , \tag{6.6}$$

where d_{pe} is the equivalent diameter of the particle.

An aerodynamic diameter is an equivalent diameter and is defined as the diameter a spherical water droplet (a spherical particle with unit density) which has the same settling velocity, v_s, as the particle. The mathematical relation for aerodynamic diameter is

$$d_a = \sqrt{\frac{18 \mu v_s}{C_c \rho_{water} g}} . \tag{6.7}$$

Any equivalent set of units can be used to determine the aerodynamic diameter. The settling velocity, a terminal velocity of a particle in calm air, is determined by solving a particle's steady-state rectilinear motion in a gravitational field, i.e., by solving

$$F_g - F_d = m_p \frac{dv_p}{dt} , \tag{6.8}$$

where F_g is the gravitational force exerted on the particle having mass m_p. Then solving this differential equation for the particle's velocity, v_p at steady state yields a terminal settling velocity

$$v_s = \frac{\rho_p d_p^2 C_c}{18 \mu} g . \tag{6.9}$$

6.2 Particle Motion

When the number of particles in the air is low, it is fair to assume that the particles do not influence the velocity field of the air. In other words, the average distance between any two particles is at least 10X the particle diameter. For water droplets, this would correspond to less than 4.2 kg/m^3 in air. Table 6.1 shows upper limits for particle concentration influence on the flow field based on particle diameter and number density.

Table 6.1 Particle Diameter Versus Density for Influencing Flow Field.

Diameter (μm)	particles/m^3
1.0	8×10^{15}
10.0	8×10^{12}
100.0	8×10^{9}

If knowledge of the velocity field of the air (or carrier gas) is known, then particle trajectories can be calculated. For situations when the density of a particle is 1000X greater than the density of air, buoyancy on a particle can be neglected. The motion of a single spherical particle can be expressed using the relation

$$\left(\frac{\pi D_p^3}{6}\right)\rho_p \frac{d\mathbf{v}}{dt} = -\left(\frac{C_d \rho}{2C}\right)\left(\frac{\pi D_p^2}{4}\right)\mathbf{V}|\mathbf{V}| - \left(\frac{\pi D_p^3}{6}\right)\rho_p \mathbf{g}, \qquad (6.10)$$

where **v** is the velocity of the particle, V is the relative velocity (particle–fluid velocity), C is a slip factor (~1 for $D_p \geq 10\mu m$), and **g** is acceleration of gravity. Equation 6.10 is useful when calculating freely falling particles due to gravimetric settling, horizontal motion in quiescent air, and particles traveling through a moving stream.

For a particle settling in quiescent air (U = 0) due to gravitation, motion is only downward. Hence, the vector velocity becomes $\mathbf{v} \equiv -v$ (where v denotes vertical motion). Likewise, the drag coefficient becomes $C_d = 24\mu/\rho D_p v$. Equation 6.11 can be simplified to the following form,

$$\frac{dv}{dt} = g - \frac{v}{\tau C}, \tag{6.11}$$

where $\tau = \rho_p D_p^2/18\mu$ which is known as the relaxation time. If the particle starts from rest, the downward velocity is

$$v(t) = Cg\tau\left[1 - \exp\left(-\frac{t}{\tau C}\right)\right]. \tag{6.12}$$

If $t \gg \tau$, then the settling, or terminal velocity (v_t) of the particle can be calculated using the simple relation

$$v_t = \tau g C, \tag{6.13}$$

assuming that the Reynolds number ($Re = \rho U D_p/\mu$) is low. Figure 6.1 shows particle diameter versus settling velocity for three specific gravities (SG). Note that the settling velocity varies as the square of the particle diameter when $Re \leq 1$.

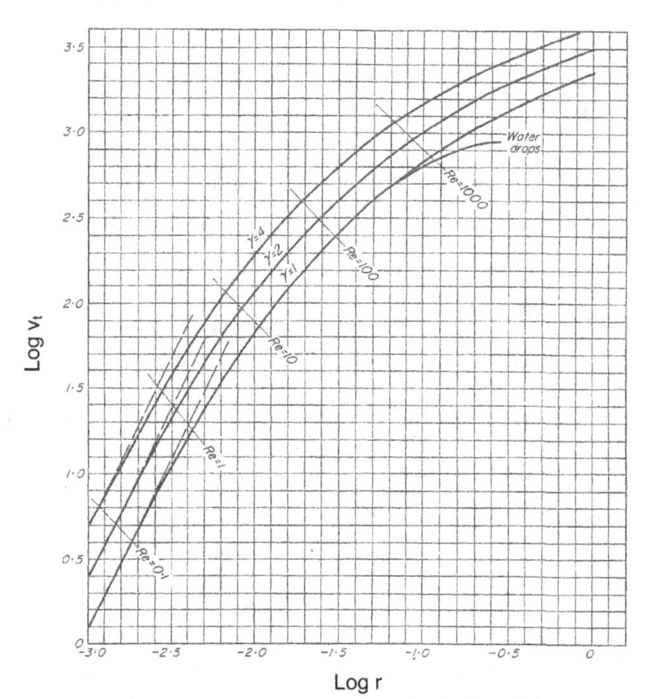

Fig 6.1 Settling velocity of spherical particles for spherical particles and three densities (from *Mechanics of Aerosols*, N. A. Fuchs, 1964, pg. 33).

When the particle is very large and Re > 1000, $C_d \sim 0.4$ and the settling velocity can be found from the relation

$$v_t = \sqrt{\left(\frac{4}{3}\right)\left(\frac{\rho_p D_p g}{0.4\rho}\right)}. \tag{6.14}$$

For a particle moving horizontally in quiescent air, we will assume that the horizontal velocity (u) of a sphere occurs when Re ≤ 1.0. The differential equation for the horizontal motion is

$$\frac{du}{dt} = -\frac{u}{\tau C}, \tag{6.15}$$

which can be integrated to yield

$$u(t) = u(0)\exp\left[-\frac{t}{\tau C}\right], \tag{6.16}$$

and the horizontal displacement (also known as stopping or penetration distance) calculated as

$$\int_0^{x(t)} dx = \int_0^t u dt = \tau C u(0)\left[1 - \exp\left(-\frac{t}{\tau C}\right)\right]. \tag{6.17}$$

The maximum stopping distance is easily found by allowing $\tau >> t$.

For particles traveling in a 2-D moving air stream, Eq. 6.11 must be modified to the form

$$\frac{d\mathbf{v}}{dt} = -\left(\frac{3C_d}{4C}\right)\left(\frac{\rho}{\rho_p D_p}\right)\mathbf{V}|\mathbf{V}| - \mathbf{g}, \tag{6.18}$$

which can be reduced to the following pair of coupled differential equations,

$$\begin{aligned} \frac{du}{dt} &= -\left(\frac{3C_d}{4C}\right)\left(\frac{\rho}{\rho_p D_p}\right)u_r(u_r^2 + v_r^2)^{1/2} \\ \frac{dv}{dt} &= -\left(\frac{3C_d}{4C}\right)\left(\frac{\rho}{\rho_p D_p}\right)v_r(u_r^2 + v_r^2)^{1/2} - g, \end{aligned} \tag{6.19}$$

where u_r and v_r are relative velocity components in the x and y directions, respectively, and

$$C_d = 0.4 + \frac{24}{Re} + \frac{6}{(1+Re^{1/2})}$$
$$Re = \rho D_p \frac{\left[u_r^2 + v_r^2\right]^{1/2}}{\mu}. \qquad (6.20)$$

For the case when the particle's motion is in a flow regime where Re ≤ 1.0, the pair of equations reduce to the much simpler form

$$\frac{du_x}{dt} = -\left(\frac{u_x - U_x}{\tau C}\right)$$
$$\frac{dv_y}{dt} = -\left(\frac{v_y - V_y}{\tau C}\right) - g, \qquad (6.21)$$

which can be integrated, assuming u(0) = v(0) = 0, to

$$u_x(t) = u_x(0)\exp\left[-\frac{t}{\tau C}\right] + U_x(0)\left[1 - \exp\left(-\frac{t}{\tau C}\right)\right]$$
$$v_y(t) = v_y(0)\exp\left[-\frac{t}{\tau C}\right] - (g\tau C)\left[1 - \exp\left(-\frac{t}{\tau C}\right)\right], \qquad (6.22)$$

where U_x and U_y denote components of the air velocity in the horizontal and vertical directions, respectively. If Re values are unknown and the flow regime is well beyond low flow levels, numerical methods (CFD) are required to compute the particle velocities and trajectories.

Particle motion is described by the time dependent convection–diffusion equation. For inviscid analysis or in laminar flow the transport equation can be easily adjusted to account for settling by incorporating a settling velocity into the advection–diffusion equation. Particle trajectories can be conveniently calculated using the Lagrangian form of the transport equation, i.e.,

$$d\mathbf{x} / dt = \mathbf{V}, \text{ or } \mathbf{x}^{n+1} = \mathbf{x}^n + \mathbf{V}\Delta t,$$

which we will return to later.

6.2.1 Deposition of particulate with aerodynamic diameters > 1 μ by settling

The deposition of large particles by diffusion is extremely small as is evidenced by examining the equation for the velocity of deposition through a boundary layer of thickness δ_{part_j} from diffusion alone. In the absence of thermophoretic velocities and turbulent dispersion, deposition velocity is a function of gravitational settling. The equation for species transport with settling becomes

$$\frac{\partial C_j}{\partial t} + u\frac{\partial C_j}{\partial x} + (v - v_{s_j})\frac{\partial C_j}{\partial y} + w\frac{\partial C_j}{\partial z} = \frac{D_j}{\rho}\left[\frac{\partial^2 C_j}{\partial x^2} + \frac{\partial^2 C_j}{\partial y^2} + \frac{\partial^2 C_j}{\partial z^2}\right] + Q_c \quad . \tag{6.23}$$

This equation has an advective velocity term in the y coordinate to account for the direction of gravitational influence (assuming the y-direction). The advective term v-v_s in Eq. 6.23 represents some relaxation of the particulate velocity versus the free stream velocity.

The deposition rate is given by

$$J = v_s \rho_p \, . \tag{6.24}$$

For particles larger than 10 μm and Reynolds number between 2 and 500, the settling velocity is defined as (Cooper and Alley, 1994)

$$v_s = \frac{0.153\, d_p^{1.14} \rho_p^{0.71} g^{0.71}}{\mu \rho^{0.29}} \, . \tag{6.25}$$

Inertial deposition from laminar flow occurs for larger particles which may be carried from the streamline flow onto an obstruction. The distance the particle would be carried from the streamline is dependent on the particle's momentum and size. The trajectory of the particle is initially a function of the fluid's trajectory.

Consider a distance a particle will travel from its inertia. Let that distance be just to a surface – a stopping distance of 'd_s'. The velocity 'v_s' of the particle normal to the surface multiplied by the time 't', the relaxation time, is the stopping distance d_s=v_s t.

The rate of deposition from this stopping distance is determined by the concentration of particles with this relaxation time. As the particle size decreases, the distance traveled from inertia decreases, that is, the relaxation time is decreased. Relaxation time is defined as

$$t = \frac{C_c \rho_p d_p^2}{18\mu}. \tag{6.26}$$

The particulate flux from the stopping distance, 'd_s ', is

$$J_s = v_s C_j |_{d_s}. \tag{6.27}$$

The value of 'C_j', is the value of the concentration at the stopping distance for that particular particulate density and size.

Inertial forces on large particles in turbulent flow are important mechanisms for deposition by impingement. Larger particles are carried into the boundary layer by inertia. The distance particles are carried into the transitional and laminar sublayers depends on the stopping distance which is dependent on the following factors: 1) particle size, 2) particle mass, and 3) degree of turbulence or how energetic the flow.

If molecular diffusion is neglected, the velocity of deposition for particles is given by

$$V_{d+} = \frac{\mu_{turb}}{\nu_{fluid}} \frac{dc_+}{dy_+} = \frac{V_d}{u_*}. \tag{6.28}$$

This equation is true for one-dimensional flow towards a flat plate. This equation provides a good approximation to flow within a cylinder where the radius of a surface is large compared to the scale of the turbulent boundary layer (Davies, 1966).

Consider, a particle travel distance just to a surface as the stopping distance of d_{s+} with a turbulent velocity v_{s+} normal to the surface, then

$$d_{s+} = v_{s+} t_+, \tag{6.29}$$

where $t_+ = t u_*^2 / \nu$ is the nondimensional relaxation time.

The rate of deposition from this distance is determined by the concentration of particles with this relaxation time. As the particle size decreases, the distance traveled from inertial forces decreases, that is, the relaxation time is decreased. Relaxation time is defined as

$$t = \frac{C_c \rho_p d_p^2}{18\mu}. \tag{6.30}$$

The nondimensional particulate flux from the stopping distance d_{s+} is $J_{s+}=v_{s+}\ c_+|_{ds+}$. The value of c_+ is the value of the concentration at the stopping distance for that particular particulate density and size.

6.2.2 Particle motion in electrostatic field

Electrostatic forces can have very significant influence on the motion of aerosols. Most airborne particles are electrically charged, and when in presence of electric potential the resulting forces on the particles cause significant motion. So much so, that this force is utilized by electrostatic precipitators for air cleaning and by aerosol measurement instruments.

Coulomb's law describes the electrostatic force as

$$F_e = \frac{1}{4\pi\varepsilon_o}\frac{q_1 q_2}{r_{12}^2}, \tag{6.31}$$

where 'q_1' is the particles charge, 'q_2' a surfaces charge (or other point source), 'ε_o' the permeability of a vacuum, and 'r_{12}' the distance between the charges.

A field strength E is the electrostatic force produced per unit charge of the particle. This field is then

$$E = \frac{F_e}{q_p}, \tag{6.32}$$

where $q_p = ne$, 'n' being the number or units of electron charge, 'e', 1.6 x 10^{-19} Coulombs.

The work required to move a particle distance 'x' in an electric field per unit charge is

$$W_p = \frac{F_e \Delta x}{q_p}, \tag{6.33}$$

This work is the potential difference in the electric field and is measured in volts, e.g., the voltage between parallel plates in an electrostatic precipitator.

The difference between the drag force and the electrostatic force determines particle acceleration in an electric field

$$F_e - F_d = m_p \frac{dv_p}{dt}. \quad (6.34)$$

6.2.3 Particle motion induced by temperature gradients

A temperature gradient will result in particles moving from the warmer region to the cooler region or surface. This phenomenon is the result of thermophoretic forces on the particles.

6.2.4 Thermophoretic motion for gases and particles with diameter less than the molecular mean free path

When the Knudsen number, Kn = λ/d_p, is greater than 1.0, the thermophoretic velocity is given by

$$V_{Thermo} = -0.55 \frac{\mu}{\rho T} \nabla T, \quad (6.35)$$

where

$$\nabla T = \frac{T_{hot} - T_{cold}}{ds},$$

and T is the ambient or bulk temperature of the fluid (Hinds, 1982).

6.2.5 Thermophoretic transport for particles with diameter greater than the molecular mean free path

When Kn < 1.0, the particle is influencing inertial and thermodynamic states of nearby gas molecules. The thermophoretic velocity is found by equating resistive forces to the thermal force (Hinds, 1982) and is given by

$$V_{Thermo} = -\frac{3}{2} H \frac{\mu}{\rho T} \nabla T C_c, \quad (6.36)$$

where T is the ambient or bulk temperature of the fluid, C_c is the Cunningham slip correction factor, and

$$H=\left[\frac{1}{1+6\lambda/d_p}\right]\left[\frac{\frac{k_f}{k_p}+4.4\lambda/d_p}{1+2\frac{k_f}{k_p}+8.8\lambda/d_p}\right], \tag{6.37}$$

where k_f and k_p are the thermal conduction of the fluid and particle, respectively.

Thermophoretic forces have an insignificant influence on the rate of deposition for particles of one micron physical diameter or larger. For very small particles this velocity would add as a vector function to the settling velocity and the velocity of the air stream.

6.3 Particle Flow in Inlets and Flanges

Contaminants and air are withdrawn by inlets of various shapes and sizes. The effectiveness of an inlet is basically how well it serves to capture contaminant. The locations of dividing streamlines and bounding trajectories of particles can be determined as a first guess using much of the analytical tools previously discussed. The quantitative measure of inlet effectiveness is generally referred to as reach.

The reach defines the boundaries of the region from which the inlet reaches out and captures contaminants. In more definable terms, the reach can be defined as the ratio of the cross-sectional area of the stream tube of contaminants entering the inlet to the cross-sectional area of the stream tube of air entering the inlet. The reach for particles is not always equal to one since some of the particles may not enter the inlet, even though all the air is pulled into the inlet. This is due to particle inertia and deflection. For gases and vapors, the reach is unity.

Figure 6.2 (a–b) shows a set of dividing streamlines and bounding trajectories for a flanged inlet in a uniform flow above the duct. The ideal flow was solved via numerical conformal mapping using the CONFORM software (Ivanov and Trubetskov, 1994).

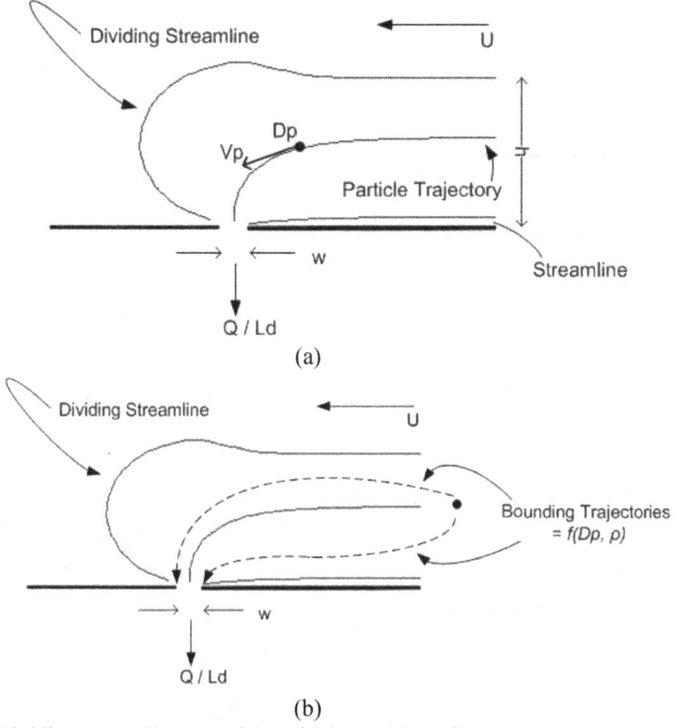

Fig. 6.2 Dividing streamlines, particle velocity and bounding particle trajectories for suction flow into a flanged duct with uniform flow field above.

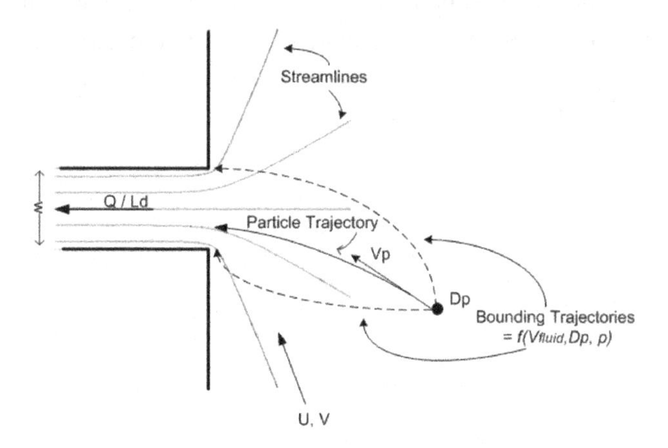

Fig. 6.3 Streamlines, particle velocity and bounding particle trajectories for suction flow into a flanged duct with quiescent flow field at distance.

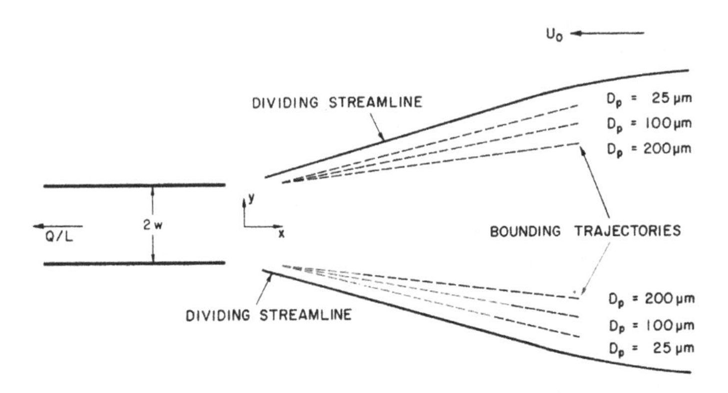

Fig 6.4 Reach of an unflanged inlet for several particle sizes (from *Industrial Ventilation*, R. J. Heinsohn, J. Wiley & Sons, New York, 1991, pg. 537).

To find particle velocity, displacement, and its new location (x, y), Eq. 6.22 must first be solved. Once the velocities are determined, the location of the particle at the end of an interval of time can be found using the simple Lagrangian relations

$$\begin{aligned} x_j &= x_i + \Delta t u(x_i, y_i) \\ y_j &= y_i + \Delta t v(x_i, y_i), \end{aligned} \tag{6.38}$$

where 'I' denotes initial (previous) position and 'j' is the new position. Repeating solution of this pair of equations produces a table of x and y values that can be used to create a trajectory for the particle position. More details on this simple technique is given later in chapter 8 dealing with Lagrangian Particle Transport.

6.4 Comments

The movement of particles and gaseous contaminants are directed principally by the ventilation patterns within a room or building. In the absence of any indoor air movement, the principle mechanism for particle dispersion is diffusion – much like the pattern of waves created when a pebble is thrown into a calm pond. Even a hint of air entering a room has an affect on dictating the movement and direction of contaminant.

A simple experiment one can conduct is to pop popcorn in a microwave and then open the bag – notice how quickly the smell of popcorn spreads within a room, especially if there is discernable air movement within the room.

In this chapter, we have attempted to list some of the more important analytical equations that can be used to determine particle motion in an air stream. These relations can be used to quickly predict general trajectories of particulate contaminant, i.e., an order of magnitude assessment – which may be sufficient in many cases. Greenspan (2005) discusses modeling at the molecular and particle level, and provides simple source codes for predicting motion at the molecular and nanoscale levels. While permitting only a microcosm of activity to be simulated, the modeling presents interesting results that can be combined to produce larger scale (continuum) motion.

Chapter 7

Numerical Modeling – Conventional Techniques

Indoor air quality can have a larger effect on human health than outdoor air quality. The common practice of relating measurements of outdoor pollutants to human exposure can be fundamentally wrong, especially with regards to hazardous material. Direct measurements of indoor air quality are the best way to evaluate the existence and the gravity of contaminants. In some instances, statistical data can be used to estimate flow rates. While such analyses lead to order of magnitude projections, they do not provide sufficient data for ventilation feedback and remediation. In order to obtain accurate assessments and forecasts of the effects on ventilation/air quality, modeling based on solution of the nonlinear equations of fluid motion (Computational Fluid Dynamics, or CFD) must be undertaken.

In order to accurately model the dispersion of contaminants within an indoor environment, it is important to incorporate as much physics as possible. One must be able to model both laminar as well as turbulent fluid motion common to real world situations. As we have seen in Chapter 2, the governing equations are transient, nonlinear PDEs that have no tractable analytical solutions except for the simplest of cases. In order to solve these difficult equations, the equations must be discretized and then numerically solved. There are various numerical techniques that can be used to solve these equations. We will address the more popular and conventional methods now being used, including a brief mention of several of the more widely used commercial CFD codes now in use.

There are three fundamental numerical methods that are commonly used to model flow and species transport within enclosures. The two most popular and prevalent numerical techniques are the finite difference method (FDM) and the finite volume method (FVM). The rapidly rising third technique is the finite element method (FEM), which has a great deal of versatility and applicability. We begin with the simplest of these three methods, the FDM.

7.1 Finite Difference Method

The finite difference method is based on approximating derivatives using truncated Taylor series expansions. Nodal values are found using a mesh to discretize the problem domain. A recursion relation is formed from the approximation that is then solved repeatedly as the solution sweeps over the array of nodes. Higher order terms are truncated in the Taylor series, thus creating an approximation to that derivative that may be accurate to the first order, the second order, or higher, depending on how many terms one wishes to include before truncating the series. In nearly all cases, either first or second-order approximations are typically employed, with the mesh being fine enough to ensure a converged (or decent) solution.

Utilizing discrete distances and increments of time in the Taylor series expansion, a finite difference approximation is made of the original differential equation. For example, looking at the 1-D equation for time dependent isotropic advection–diffusion of a variable, φ,

$$\frac{d\varphi}{dt} + u\frac{\partial \varphi}{\partial x} = k\frac{\partial^2 \varphi}{\partial x^2}, \tag{7.1}$$

we seek the derivatives of each term, found by Taylor series expansion. Representing the derivative of 'φ' with respect to 'x', we can expand the value using either a backward or a forward approximation, i.e.,

$$\varphi_{x-1} = \varphi_x - \frac{\partial \varphi}{\partial x}\Delta x + \frac{1}{2!}\frac{\partial^2 \varphi}{\partial x^2}\Delta x^2 - \ldots, \tag{7.2a}$$

$$\varphi_{x+1} = \varphi_x + \frac{\partial \varphi}{\partial x}\Delta x + \frac{1}{2!}\frac{\partial^2 \varphi}{\partial x^2}\Delta x^2 + \ldots. \tag{7.2b}$$

Rearranging and dropping higher order terms the first order derivative can be represented in a discrete sense using the backward expansion as

$$\frac{\partial \varphi}{\partial x} = \frac{\varphi_x - \varphi_{x-1}}{\Delta x}, \tag{7.3a}$$

with Δx being the discrete difference between x and x-1. Using a forward expansion, the derivative can be expressed as

$$\frac{\partial \varphi}{\partial x} = \frac{\varphi_{x+1} - \varphi_x}{\Delta x}. \tag{7.3b}$$

In this case, Δx is now the discrete difference between x + 1 and x. A truncation analysis (retaining the highest order term) will show that these two relations (Eq. 7.3 a,b) are first order accurate in space.

If we now subtract the backward from the forward Taylor series expression, a second-order accurate discrete derivative can be obtained

$$\frac{\partial \varphi}{\partial x} = \frac{\varphi_{x+1} - \varphi_{x-1}}{2\Delta x}. \tag{7.4}$$

In a similar fashion, if we add the forward and backward expansions together, we can obtain the second derivative term which produces a second-order accurate approximation,

$$\frac{\partial^2 \varphi}{\partial x^2} = \frac{\varphi_{x+1} - 2\varphi_x + \varphi_{x-1}}{\Delta x^2}. \tag{7.5}$$

Notice that Eq. 7.4 requires knowing values at x – 1 and x + 1 to define the derivative; Eq. 7.5 requires an additional value at x. However, these approximations produce values that are an order of magnitude more accurate than the one-sided, first order values.

At this point, we need to introduce the concept of a mesh to establish the nodal locations of the unknown values for φ. Instead of x, let's use i as an indicial marker to denote the x locations and we will use circles to indicate nodes. This can be seen in Fig. 7.1 below. The nodes are now indicated as intervals of i, and represent the generic set of three adjacent nodes commonly found within a computational domain.

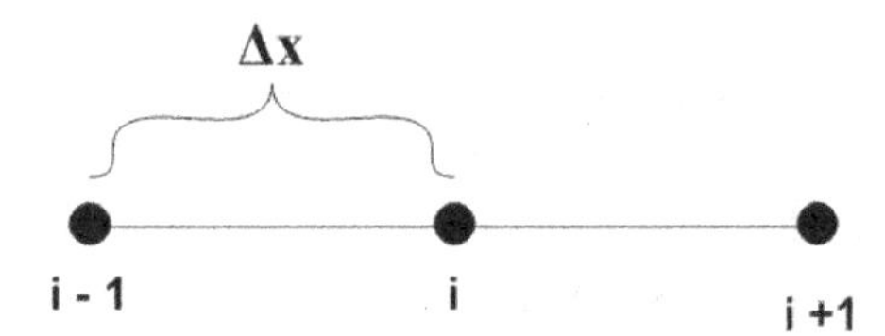

Fig. 7.1 Three node discretization in the x-direction.

Before continuing on, we will address the time-dependent term as a forward in time discretization, and use superscript n+1 to denote the unknown value of φ at the new time, n+1, and superscript n to represent known values at the previous time, n. This procedure is common practice in CFD modeling, and allows us to need to keep track of only previously calculated values at n and unknown values at n+1. A central difference approximation of the time-derivative term would require three levels of values to be kept, which could become excessive in storage – and it turns out we don't need to do this anyway as a Taylor series truncation analysis shows that the discretization produces second order accuracy in time if one uses an implicit time-marching technique. We will discuss this shortly. The time-derivative term becomes

$$\frac{\partial \varphi}{\partial t} = \frac{\varphi_i^{n+1} - \varphi_i^n}{\Delta t}, \tag{7.6}$$

where Δt is the time step.

Substituting Eqs. 7.4, 7.5, and 7.6 into the transient advection–diffusion equation, the discrete representation of the governing equation becomes

$$\frac{\varphi_i^{n+1} - \varphi_i^n}{\Delta t} + u_i \frac{\varphi_{i+1}^k - \varphi_{i-1}^k}{2\Delta x} = k \frac{\varphi_{i+1}^k - 2\varphi_i^k + \varphi_{i-1}^k}{\Delta x^2}. \tag{7.7}$$

This equation is first order accurate in time and second order accurate in space. We now need to briefly address the issue of superscript k and decide on whether to solve the equation explicitly or implicitly.

There are numerous techniques that have been developed over the past 50 years or more dealing with solutions to Eq. 7.7. A great many textbooks exist in the literature that describe the pros and cons of various time-dependent approximations. We shall keep the time-stepping simple and introduce the two basic schemes: *explicit* and *implicit*.

7.1.1 Explicit

If we set k = n in Eq. 7.7, the discretized equation is solved using an explicit marching technique, i.e., there is only one unknown to be solved with everything else being known at time level n. Equation 7.7 becomes

$$\frac{\varphi_i^{n+1}-\varphi_i^n}{\Delta t}=-u_i\frac{\varphi_{i+1}^n-\varphi_{i-1}^n}{2\Delta x}+k\frac{\varphi_{i+1}^n-2\varphi_i^n+\varphi_{i-1}^n}{\Delta x^2}, \tag{7.8a}$$

or

$$\varphi_i^{n+1}=\varphi_i^n+\Delta t\left[-u_i\frac{\varphi_{i+1}^n-\varphi_{i-1}^n}{2\Delta x}+k\frac{\varphi_{i+1}^n-2\varphi_i^n+\varphi_{i-1}^n}{\Delta x^2}\right], \tag{7.8b}$$

which can be simply solved as a repetitive statement in a computer program as one sweeps over the entire set of nodes. In this instance, the solution is first order accurate in time and second order accurate in space. Since this method is explicit, there is a time-step limitation, i.e., Δt cannot exceed a specific limit or the solution will diverge, i.e., become unstable. This time-step limit is based on the larger of the two limiters:

$$\frac{u\Delta t}{\Delta x}\le 1$$

$$\frac{k\Delta t}{2\Delta x^2}\le 1.$$

Usually the dominate limiter is the first term since $u/\Delta x$ is typically much larger than the $k/2\Delta x^2$ term. We will come back to this shortly.

7.1.2 Implicit

If we now set k = n+1 in Eq. 7.7, the discretized equation is solved using an implicit marching technique, i.e., there are three unknowns at i-1, i, and i+1 that must be solved. This now requires the solution of a matrix – in this case, a tridiagonal banded matrix.

Fortunately, the well-known Thomas algorithm is the most efficient solver for tridiagonal matrices and is easy to set up. Again there are numerous textbooks and references that describe this solver. Hence, Eq. 7.7 now becomes

$$\frac{\varphi_i^{n+1}-\varphi_i^n}{\Delta t}=-u_i\frac{\varphi_{i+1}^{n+1}-\varphi_{i-1}^{n+1}}{2\Delta x}+k\frac{\varphi_{i+1}^{n+1}-2\varphi_i^{n+1}+\varphi_{i-1}^{n+1}}{\Delta x^2}, \quad (7.9a)$$

or, in vector form (row times column),

$$\left\lfloor -\frac{u_i\Delta t}{2\Delta x}-\frac{k\Delta t}{\Delta x^2},\quad 1+\frac{2k\Delta t}{\Delta x^2},\quad \frac{u_i\Delta t}{2\Delta x}-\frac{k\Delta t}{\Delta x^2}\right\rfloor \begin{Bmatrix}\varphi_{i-1}\\ \varphi_i\\ \varphi_{i+1}\end{Bmatrix}^{n+1}=\{\varphi_i\}^n, \quad (7.9b)$$

which now must be solved as a set of three unknowns. A tridiagonal matrix is created as one sweeps over the nodes going from $i = 1$ to $i = N$ (total number of nodes). Fortunately, this method does not have a limitation on the time step, i.e., the method is unconditionally stable.

Higher order discretization can be achieved with the use of various components of the Taylor series expansions. Also, notice this equation was developed based on an equal spacing of the discretization and could be modified for nonuniform grid spacing.

7.1.3 Upwinding

Upwinding of the advective term, i.e., using a backward differencing, is sometimes employed since it is a stable discretization even for explicit time stepping. Equation 7.7 now becomes

$$\begin{aligned}\frac{\varphi_i^{n+1}-\varphi_i^n}{\Delta t}&=-\frac{u_i\varphi_i^n-u_i\varphi_{i-1}^n}{\Delta x}+k\frac{\varphi_{i+1}^n-2\varphi_i^n+\varphi_{i-1}^n}{\Delta x^2} \quad \text{for} \quad u_i>0\\ \frac{\varphi_i^{n+1}-\varphi_i^n}{\Delta t}&=-\frac{u_i\varphi_{i+1}^n-u_i\varphi_i^n}{\Delta x}+k\frac{\varphi_{i+1}^n-2\varphi_i^n+\varphi_{i-1}^n}{\Delta x^2} \quad \text{for} \quad u_i<0.\end{aligned} \quad (7.10)$$

The stability constraint is the Courant–Friedrichs–Lewy (CFL) condition, which we saw in the explicit time-marching scheme,

$$C = u\frac{\Delta t}{\Delta x} \leq 1. \tag{7.11}$$

The condition states that a fluid molecule can travel no more than a spatial distance, Δx, in time, Δt. The upwinded term is first order accurate and can produce rather severe numerical diffusion, thus creating a damping of the second order central difference scheme. When C = 1, there is no artificial diffusion; however, this is rarely the case since velocities vary throughout the problem domain.

Another popular technique, called the *donor cell method*, is somewhat second order accurate as a result of central differencing. In this method,

$$\frac{\varphi_i^{n+1} - \varphi_i^n}{\Delta t} = -\frac{u_R \varphi_R^n - u_L \varphi_L^n}{\Delta x} + k\frac{\varphi_{i+1}^n - 2\varphi_i^n + \varphi_{i-1}^n}{\Delta x^2}, \tag{7.12}$$

where

$$u_R = \frac{u_{i+1} + u_i}{2} \quad \text{and} \quad u_L = \frac{u_i + u_{i-1}}{2}$$

$$\varphi_R = \varphi_i \quad \text{for} \quad u_R > 0,\ \varphi_R = \varphi_{i+1} \quad \text{for} \quad u_R < 0 \tag{7.13}$$

$$\varphi_L = \varphi_{i-1} \quad \text{for} \quad u_L > 0,\ \varphi_L = \varphi_i \quad \text{for} \quad u_L < 0.$$

These two upwinding procedures produce stabilizing effects when dealing with step gradients common to the advective term (Roache, 1972; Fletcher, 1991). However, the artificial diffusion imposed by these schemes may not reflect the true characteristic viscosity of the fluid, e.g., the flow of molasses versus the flow of air. Use them carefully.

One must remember that any explicit formulation's time increment is constrained by the CFL) condition and also by diffusion, known as the Fourier number,

$$k\frac{\Delta t}{\Delta x^2} \leq \frac{1}{2}, \tag{7.14}$$

as noted previously.

If φ now becomes a function of (x,y), the transport equation is two-dimensional and additional terms must be included to account for the extra dimension, i.e.,

$$\frac{\partial \varphi}{\partial t} + u\frac{\partial \varphi}{\partial x} + v\frac{\partial \varphi}{\partial y} = k\left(\frac{\partial^2 \varphi}{\partial x^2} + \frac{\partial^2 \varphi}{\partial y^2}\right). \tag{7.15}$$

The discretization must now impose an orthogonality in the x and y directions. This decomposition of the domain into grid points that can be connected by orthogonal lines is referred to as a structured mesh, or grid. For complex domains the representation may suffer if there are curved surfaces or sides oblique to the discretization. A simple 2-D mesh is shown in Fig. 7.2, now with a 2-D molecule denoting three nodes in the x (i values) and y (j values).

The discretized equation produces a double indexed set of variables,

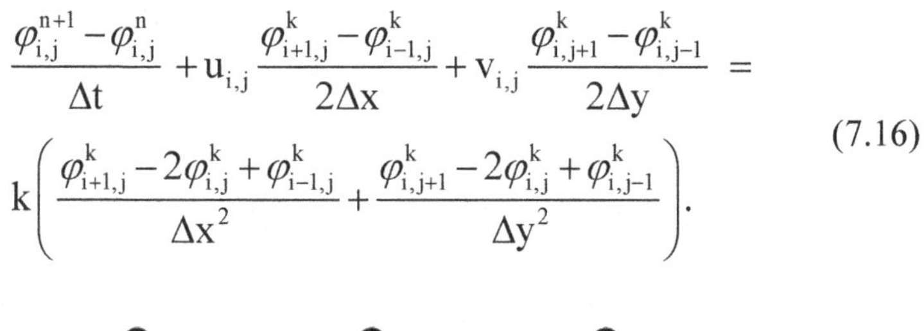

$$\frac{\varphi_{i,j}^{n+1} - \varphi_{i,j}^{n}}{\Delta t} + u_{i,j}\frac{\varphi_{i+1,j}^{k} - \varphi_{i-1,j}^{k}}{2\Delta x} + v_{i,j}\frac{\varphi_{i,j+1}^{k} - \varphi_{i,j-1}^{k}}{2\Delta y} = k\left(\frac{\varphi_{i+1,j}^{k} - 2\varphi_{i,j}^{k} + \varphi_{i-1,j}^{k}}{\Delta x^2} + \frac{\varphi_{i,j+1}^{k} - 2\varphi_{i,j}^{k} + \varphi_{i,j-1}^{k}}{\Delta y^2}\right). \tag{7.16}$$

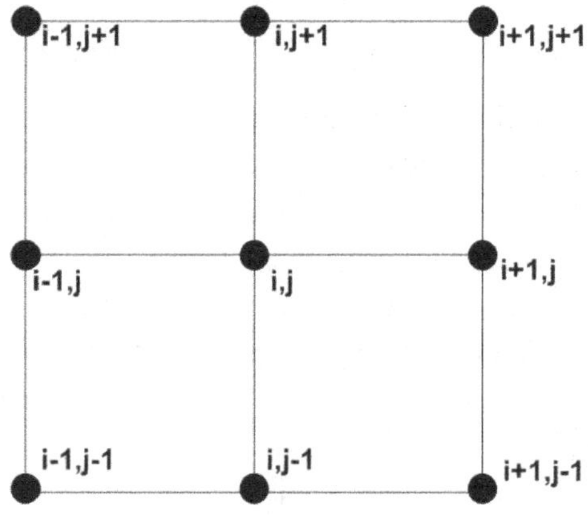

Fig. 7.2 A 2-D structured mesh.

The same procedure is followed in time marching the solution whether explicit or implicit. Since the equation is now 2-D, an implicit solution would require solving a 5-diagonal banded matrix.

There are two popular techniques to do this: the alternating direction implicit (ADI) scheme and the strongly implicit procedure (SIP). The ADI scheme is particularly useful in separating multidimensional systems into time advancement for each dimension and is unconditionally stable. One needs only to solve a tridiagonal matrix using the Thomas algorithm by sweeping in each direction (this works equally well in 3-D). The ADI method can be found in nearly all numerical methods textbooks or on the web.

The SIP solves the 5-banded (or 7-banded in 3-D) matrix, but is more complicated. However, the method can be significantly faster than the ADI). The interested reader is referred to Pepper and Harris (1977) for the SIP technique.

Using a scheme which averages in space the current and future time step is known as a semi-implicit scheme. The most popular method is Crank–Nicolson averaging, and is unconditionally stable. Numerous other time-marching schemes can be found in the set of numerical recipes textbooks published by Cambridge University Press.

Transformations can be constructed for complex domains that fit the boundaries and coordinates to an orthogonal discretization. This Boundary Fitted Coordinate (BFC) transformation (Thompson *et al.*, 1985) can be complicated but allows for the use of FDM to solve problems on complex domains. An orthogonal mesh is created in computational space, allowing simple difference approximations to be utilized. The transformation of an airfoil from physical space to computational space is shown in Fig. 7.3.

The FDM requires orthogonality when establishing a mesh and determining derivatives, i.e., the rows and columns of lines created by the mesh must be perpendicular at the cross points, or nodes. This can occur in either physical space (the problem domain) or computational space (i.e., transformed space). Even the finite element method transforms each individual element from physical space to a unit square in computational space. The difference between the FEM and the FDM is that the FDM (or FVM) requires a global transformation of the physical space and governing equations. The reader is referred to the textbooks by Fletcher (1991), Warsi (1999), Chung (2002), and Anderson (2001).

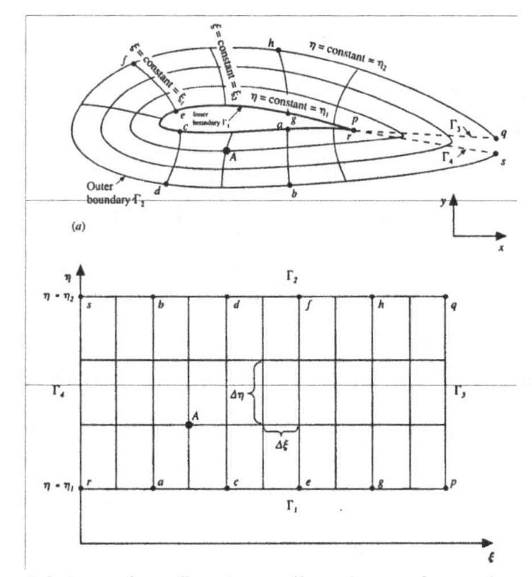

Fig. 7.3 Boundary fitted coordinate) transformation.

Example 7.1.4 FDM simulation of flow in an office complex An office complex is shown in Fig. 7.4 (a–d) below. Determine the flow pattern within the offices using FDM if the flow from the hallway enters the outer office at u = 1 f/s. We will use this office complex as we discuss the other methods.

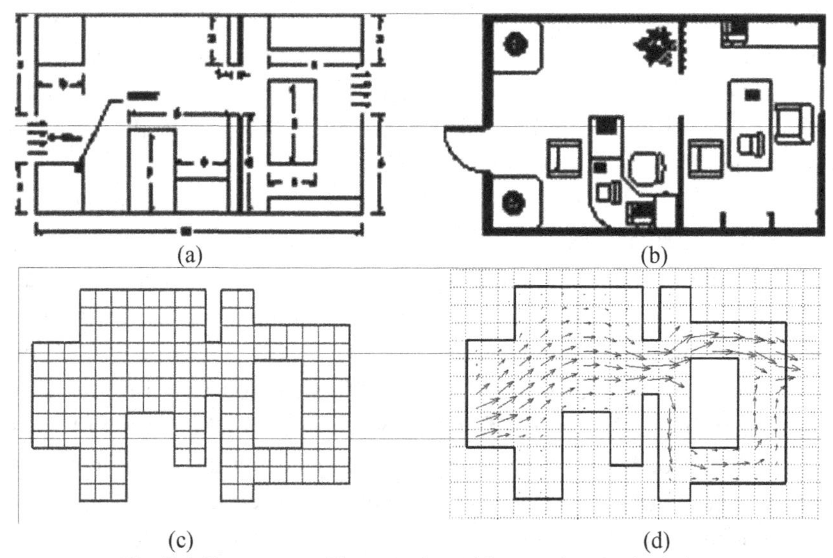

Fig. 7.4 Two-room office complex with open door and window.

In this example problem, the two-dimensional vorticity–streamfunction formulation is used to solve for the interior velocities. The equations for the vorticity–streamfunction formulation stem from the Navier–Stokes (or primitive equations) for fluid motion. The use of the vorticity equation eliminates the pressure term in the equations of motion, thereby reducing the set of equations to two equations with two unknown, i.e., ω and ψ, where u and v velocity components are just the derivatives of ψ. The two equation set is

$$\begin{aligned} &\frac{\partial \omega}{\partial t}+\frac{\partial \psi}{\partial y}\frac{\partial \omega}{\partial x}-\frac{\partial \psi}{\partial x}\frac{\partial \omega}{\partial y}=\frac{1}{\mathrm{Re}}\left(\frac{\partial^2 \omega}{\partial x^2}+\frac{\partial^2 \omega}{\partial y^2}\right) \\ &\omega=-\left(\frac{\partial^2 \psi}{\partial x^2}+\frac{\partial^2 \psi}{\partial y^2}\right), \end{aligned} \tag{7.17}$$

where vorticity and velocities are defined as

$$\omega=\frac{\partial v}{\partial x}-\frac{\partial u}{\partial y}$$

$$u=-\frac{\partial \psi}{\partial y},\ v=\frac{\partial \psi}{\partial x}.$$

The discretized forms of Eq. 7.17 are

$$\begin{aligned} &\frac{\omega_{i,j}^{n+1}-\omega_{i,j}^{n}}{\Delta t}+u_{i,j}\frac{\omega_{i+1,j}^{k}-\omega_{i-1,j}^{k}}{2\Delta x}+v_{i,j}\frac{\omega_{i,j+1}^{k}-\omega_{i,j-1}^{k}}{2\Delta y}= \\ &\frac{1}{\mathrm{Re}}\left(\frac{\omega_{i+1,j}^{k}-2\omega_{i,j}^{k}+\omega_{i-1,j}^{k}}{\Delta x^2}+\frac{\omega_{i,j+1}^{k}-2\omega_{i,j}^{k}+\omega_{i,j-1}^{k}}{\Delta y^2}\right), \end{aligned} \tag{7.18}$$

and

$$-\omega_{i,j}=\left(\frac{\psi_{i+1,j}^{k}-2\psi_{i,j}^{k}+\psi_{i-1,j}^{k}}{\Delta x^2}+\frac{\psi_{i,j+1}^{k}-2\psi_{i,j}^{k}+\psi_{i,j-1}^{k}}{\Delta y^2}\right), \tag{7.19}$$

where

$$u_{i,j}=-\frac{\psi_{i,j+1}-\psi_{i,j-1}}{2\Delta y},\ v_{i,j}=\frac{\psi_{i+1,j}-\psi_{i-1,j}}{2\Delta x}. \tag{7.20}$$

Equation 7.18 can be solved using a Crank–Nicolson implicit ADI) approach. Equation 7.19, which is a steady-state Poisson equation, is best solved using a Successive Over Relaxation (SOR) technique. The SOR is an iterative scheme based on the Gauss–Seidel method.

7.2 Finite Volume Method

The majority of fluid flow simulations are conducted using the finite volume approach (Patankar and Spalding 1972; Patankar, 1980; Anderson *et al.*, 1997; Fletcher, 1991). This is principally because of its ease of use and simplicity in establishing meshes for orthogonal regions (i.e., rectangles). The application of the BFC) technique to model irregular geometries helped in overcoming this handicap (Thompson *et al.*, 1985). However, the computational accuracy of these simple difference schemes is limited to first order (spatially); in addition, such methods require an extensive meshing effort and massive numbers of nodes (especially in three dimensions), and can become quite formidable for non-orthogonal problem domains. Modeling complex 3-D problems using finite volume (or finite difference) methods today may typically require over 10^6 nodes, overwhelming the resources of most current PCs. Such massive problems must be run on large supercomputers, usually configured in a parallel cluster arrangement of many PCs.

The finite volume method (FVM) is really a subset of the Method of Weighted Residuals (MWR) and in this sense is a cousin to the finite element method (FEM) (see Baker and Pepper, 1980). It is an inner product projecting the residual to zero. Since the problem domain is a discrete system, the method seeks to minimize the error or residual, **R**, over the domain. Let the residual equation or relationship be determined by

$$R \equiv L\varphi(x) = 0, \tag{7.21}$$

where $L\varphi(x)$ is the approximation to some differential equation. The approximation of the function $\varphi(x)$ is given by

$$\varphi(x_i) = \sum_{i=1}^{n} N_i \varphi_i , \tag{7.22}$$

which is a polynomial expansion. The term N_i is the test (or shape function), and φ_i is the trial value. For the FVM, this weight is just one or zero depending on if the elemental domain is being evaluated or not, and n is equal to 1.

The method seeks to minimize this residual over a domain. Requiring the residual to be zero on average is accomplished by multiplying the residual equation by an appropriate weighting function, W, and integrating over the entire domain,

$$\int_{\Omega} W L \varphi(x_i)\, d\Omega = 0 \,, \tag{7.23}$$

where Ω denotes the problem domain (x_i = x, y, z).

When applied over a domain, which is discretized into finite volumes, the resulting set of algebraic equations can be solved for the unknowns, that is, the values of φ, i.e.,

$$\int_{\Omega} (R\,,\, W\,) d\Omega = 0 \,, \tag{7.24}$$

where the residual R is the approximate solution and W is some appropriate weight or approximation. The residual is by definition approximate since only an approximation of the solution on the domain is possible with any discrete representation. It is possible, however, to have an exact solution at the nodal points, known as superconvergence.

The difference between the FEM and FVM is in the order of the interpolation polynomial where FEM has at least first order weighting functions and the variables can be represented as functions of higher order polynomials. The finite volume method uses zero order polynomials as both test and weight functions. Setting N_i = W produces the Galerkin method, which is the most popular procedure used in the FEM. The finite volume method is referred to as a subdomain method.

To see how this works, we look at the equation for conservation of mass in integral form

$$L(\varphi) = \nabla \bullet \rho \mathbf{V} = 0 \,, \tag{7.25}$$

where **V** is the velocity vector. Then

$$\int_{\Omega} L(\varphi)\, d\Omega = \int_{\Omega} \nabla \bullet \rho \mathbf{V} d\Omega \,. \tag{7.26}$$

Applying Green's Theorem (this is integration by parts in 1-D), we obtain

$$\int_{\Omega} \nabla \bullet \rho \mathbf{V} d\Omega \equiv \oint_{\Gamma} (\hat{n} \bullet \rho \mathbf{V}) d\Gamma , \qquad (7.27)$$

where Γ denotes the boundary and $\hat{n}$ is the normal vector. This integral equation evaluated over the domain Ω, shown in Fig. 7.5, produces an expression for the conservation of mass given by

$$\rho_r V_r A_r - \rho_l V_l A_l + \rho_t V_t A_t - \rho_b V_b A_b = 0 , \qquad (7.28)$$

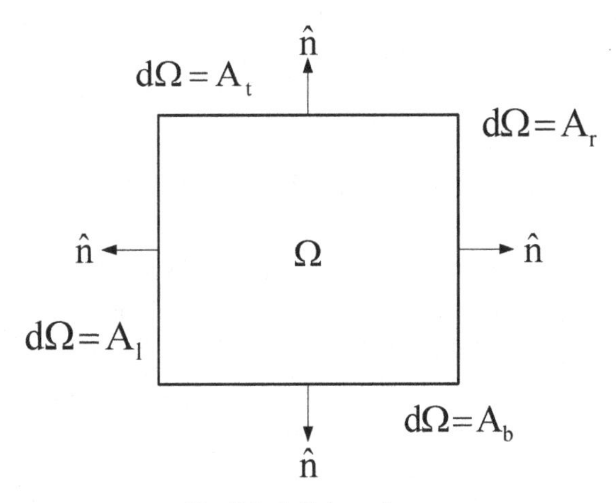

Fig. 7.5 A finite volume.

where subscripts r, l, t, and b denote right, left, top, and bottom faces. The flux of a variable into the volume is evaluated as it crosses one of the boundaries; the flux leaving the volume is calculated at the opposite face. This ensures mass conservation – one of the big advantages of the FVM. If we assume the values of ρ are placed at the center of the cell and the values of velocity (V = u(x), v(y)) at the faces of each cell, an offset grid (or staggered cell) can be created which avoids the difficulties associated with $2\Delta x$ instabilities (Hansen, 1996). This discretization is shown in Fig. 7.6.

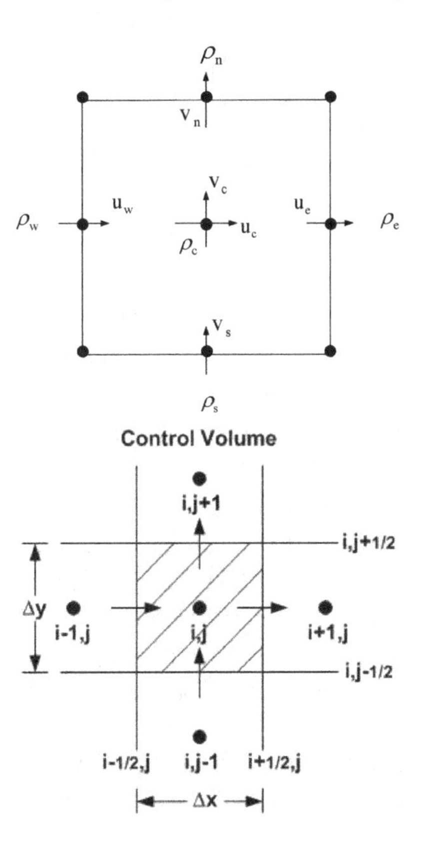

Fig. 7.6 Discretization of a finite volume.

where w, e, s, and n denote west, east, south, and north faces. An upwinding scheme is typically used for the advection terms. Only the center density belongs to the cell and is considered constant throughout the cell. Therefore ρ_w, ρ_e, ρ_s and ρ_n belong to the adjacent cells and are evaluated at the center of those cells. The velocities however, do belong to the points depicted on the faces of the cell. Upwinding in this instance is accomplished as

$$(\rho u)_e = \rho_c \operatorname{Max}(u_c, 0) - \rho_e \operatorname{Max}(-u_e, 0). \tag{7.29}$$

Noticing from Fig. 7.4 that

$$
\begin{aligned}
A_t &= A_b = \Delta y \\
A_e &= A_w = \Delta x,
\end{aligned}
\tag{7.30}
$$

the discretized equation for mass conservation becomes

$$
\begin{aligned}
&\int_\Omega \nabla \cdot \rho \mathbf{V} d\Omega \equiv \oint_\Gamma \left(\hat{n} \cdot \rho \hat{V}\right) d\Gamma \cong \\
&\left[(\rho u)_e - (\rho u)_w\right]\Delta y + \left[(\rho v)_n - (\rho v)_s\right]\Delta x.
\end{aligned}
\tag{7.31}
$$

The momentum equations are developed similarly. If the elements are trapezoidal (2-D) or hexahedral (3-D) the surface areas and normal dot products must be calculated in order to evaluate the correct flux, thereby rendering this system capable of handling a non-orthogonal grid discretization. Such unstructured grids are now commonly found in most of the current commercial CFD packages that use the FVM approach, e.g., ANSYS (FLUENT), STAR–CD, CFX, ANSWER, etc.

The discretization error attributed to both the FDM and the FVM are the same, i.e., they are $O(\Delta x)^2$ in space if one performs a truncation error analysis using Taylor series. This can be seen in the following example.

Example 7.2.1 ODE discretization using FVM: Assume that we wish to discretize the following simple ODE equation using both the FDM and the FVM

$$
\frac{d\varphi}{dx} + \frac{d^2\varphi}{dx^2} + \varphi = 0,
\tag{7.32}
$$

We will use central differencing to discretize the derivatives for the FDM and will employ integration over control volumes for the FVM. The mesh is shown in Fig. 7.7.

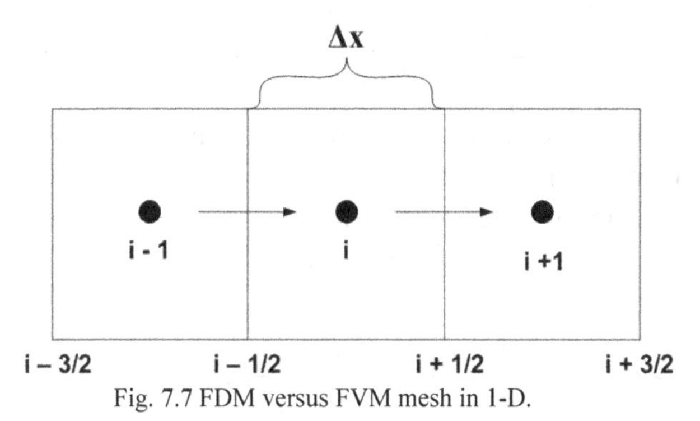

Fig. 7.7 FDM versus FVM mesh in 1-D.

The dark dots denote the node points and the squares surrounding each node are the control volumes with the faces labeled along the bottom interfaces.

7.2.1 FDM

$$\frac{\varphi_{i+1}^{n}-\varphi_{i-1}^{n}}{2\Delta x}+\frac{\varphi_{i+1}^{n}-2\varphi_{i}^{n}+\varphi_{i-1}^{n}}{\Delta x^{2}}+\varphi_{i}=0. \tag{7.33}$$

Equation 7.33 employs simple central differencing for the first and second derivatives. Notice that the single term, φ, is evaluated at node i. This is the usual practice when using FDM.

7.2.2 FVM

We now must integrate over the control volume containing node i. The limits of integration will be from i - 1/2 to i + 1/2. Thus,

$$\int_{i-1/2}^{i+1/2}\frac{d\varphi}{dx}dx+\int_{i-1/2}^{i+1/2}\frac{d^{2}\varphi}{dx^{2}}dx+\int_{i-1/2}^{i+1/2}\varphi dx=0, \tag{7.34}$$

which leads to

$$\varphi|_{i+1/2}-\varphi|_{i-1/2}+\frac{d\varphi}{dx}|_{i+1/2}-\frac{d\varphi}{dx}|_{i-1/2}+\int_{i-1/2}^{i+1/2}\varphi dx=0. \tag{7.35}$$

Using simple averaging, Eq. 7.35 can be rewritten as

$$\frac{\varphi_{i+1}+\varphi_i}{2}-\frac{\varphi_i+\varphi_{i-1}}{2}+\frac{\varphi_{i+1}-\varphi_i}{\Delta x}-\frac{\varphi_i-\varphi_{i-1}}{\Delta x}+\int_{i-1/2}^{i+1/2}\varphi dx=0\,. \quad (7.36)$$

Now the question is what to do about the remaining integral term for φ. Clearly, if we assume φ is constant at node 'i', we would then have $\varphi_i\Delta x$, and the equation reduces exactly to the FDM expression. On the other hand, if we use an integral expression such as Simpson's 1/3 rule, Eq. 7.36 becomes

$$\frac{\varphi_{i+1}-\varphi_{i-1}}{2\Delta x}+\frac{\varphi_{i+1}-2\varphi_i+\varphi_{i-1}}{\Delta x^2}+\frac{1}{3}\left(\varphi_{i-1}+\varphi_i+\varphi_{i+1}\right)=0\,, \quad (7.37)$$

which produces the same discretization in space as the FDM (and order of accuracy). Here we use an integral form for the last term, which retains the same accuracy as the discretization, instead of a constant value at node i.

The big difference between the two methods is the conservation of mass inherent in the FVM, but with the additional hassle of formulating the boundary conditions in the FVM to ensure proper fluxes at the boundaries.

Example 7.2.2 FVM simulation of flow in an office complex (FLUENT): We wish to solve for the flow within the office complex shown in Example 7.1 using the FVM. In this instance, we use the commercial package, ANSYS (FLUENT), which is a popular CFD code widely used in academia and industry. The code is based on the finite volume formulation to solve the primitive form of the equations for fluid flow, heat transfer, and species transport. The code uses the SIMPLE technique developed by Patankar (1980) to resolve the pressure term in the governing equations. FLUENT is relatively easy to set up and can be run on PCs. Figures 7.8 to 7.10 show the mesh, velocity vectors, and pressure contours. The mesh is created using a simple mesh generator to produce a set of orthogonal rows and columns defining the individual volumes. Figure 7.8 shows the interior of the office complex discretized using bilinear volumes.

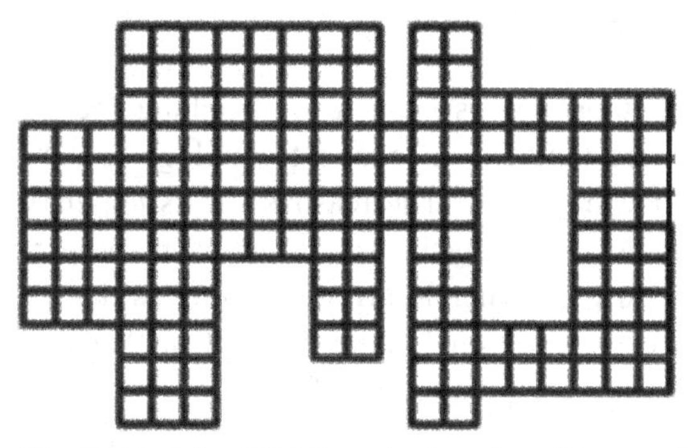

Fig. 7.8 Mesh for FVM discretization of an office complex.

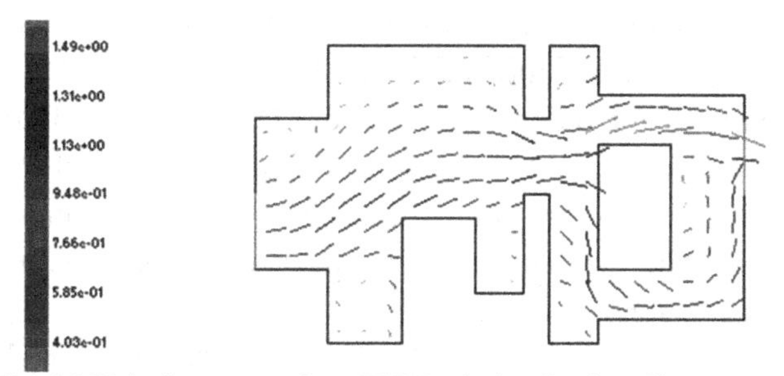

Fig. 7.9 Velocity vectors from FVM solution for the office complex.

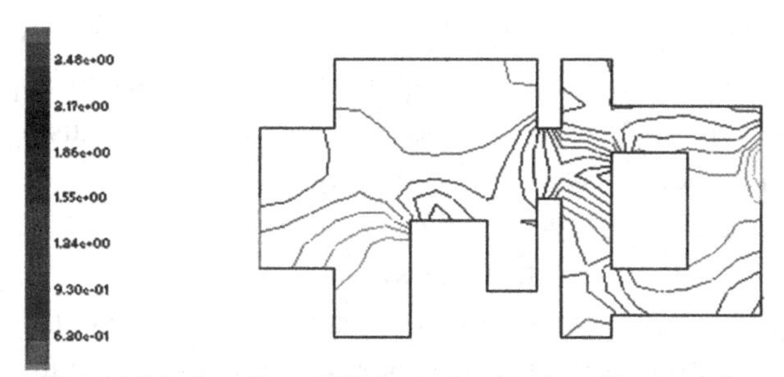

Fig. 7.10 Isobars from FVM solution for the office complex.

There are many popular methods that can be used to discretize a problem domain and solve the resulting equations for fluid flow using the FVM. Some of the early formulation and code listings of the FVM using staggered grids can be found in the literature by Harlow and Welch (1965) and Anderson *et al.* (1997).

7.3 The Finite Element Method

A numerical method that is capable of handling the wide variety of complex problems inherent in today's technology is the finite element method (Zienkiewicz, 1977). The reasons for its popularity include the ability to handle inhomogeneous or variable properties, irregular boundaries, and use of general-purpose algorithms that give high order accuracy. However, traditional finite element methods are not without their faults. The computational effort and storage requirements associated with traditional finite element methods rapidly become excessive when solving fluid flow problems.

The bandwidth generated from the computational mesh and assembly procedure is critical when globally formulating the coefficient matrices. Problems involving a large number of nodes become difficult to solve on even the largest and fastest computers. Pepper (1987) and Pepper and Singer (1990) discuss accurate finite element algorithms that are computationally efficient, and are particularly advantageous in modeling large problems on small computers.

The two most often used ways to formulate the FEM are the Rayleigh–Ritz variational method and the Galerkin Method of Weighted Residuals (MWR), similar to the method used in the FVM. Both approaches use a combination of appropriate functions to approximate the solution. The unknown coefficients are determined using integral statements in such a way as to approximately satisfy the original differential equations. However, there is a major difference between the Rayleigh–Ritz method and the Galerkin method.

The Rayleigh–Ritz method finds the unknown coefficients through an energy minimization process; this process requires a minimum principle. The Galerkin method is based on making the projection of the error in the approximating functions vanish in the finite dimensional space spanned by the functions. This approach allows the Galerkin method to be used in situations when minimum principles do not exist. Such cases occur when convection is the dominant transport mechanism in a fluid system. The Galerkin method is the preferred method of choice for FEM.

In general, the following steps are needed in any finite element approximation to the solution of a differential equation: 1) the equation (or system of equations) and its boundary and initial conditions must be defined to ensure that a well-posed problem is formulated, 2) an element type must be chosen to define the approximation functions to be used in the solution, 3) a mesh must be created that adequately refines regions where large changes in the solution are expected, and that allows the boundary conditions to be properly imposed, 4) the finite element algorithm must be formulated and used to solve the system of algebraic equations, and 5) the error in the approximation must be calculated to determine if the solution is converged or if a more refined solution is needed.

To illustrate finite element methodology, assume a linear operator (L) exists in two dimensions such that

$$L = \frac{\partial^2}{\partial x^2} + \frac{\partial^2}{\partial y^2}, \tag{7.41}$$

One must now reformulate the basic problem in a way appropriate for the application of the FEM. Let an equation exist of the form

$$L\varphi(\mathbf{x}) - f(\mathbf{x}) = 0,$$

where φ and f are functions of (x). Now define the *residual* function as

$$R(\varphi, x) \equiv L\varphi(x) - f(x). \tag{7.42}$$

It then follows that if φ^* is the solution to the differential equation, then $R(\varphi^*,\mathbf{x}) \equiv 0$. However, if φ is only an approximation to the solution, the residual provides a measure of the error in the satisfaction of the equation.

If we now multiply Eq. 7.41 by a weighting function, W, defined over domain (space) Ω, integrate over Ω, and set the integral equal to zero, we obtain the *weighted residual* form

$$\int_{\Omega} W(x) R(\varphi, x)\, d\Omega \equiv \int_{\Omega} W(L\varphi - f)\, d\Omega = 0. \tag{7.43}$$

Hence,

$$-\int_{\Omega} \left[w \left(\frac{\partial}{\partial x}\left(k \frac{\partial \varphi}{\partial x} \right) + \frac{\partial}{\partial y}\left(k \frac{\partial \varphi}{\partial y} \right) \right) + w f \right] d\Omega = 0. \tag{7.44}$$

We now apply Green's theorem to the second integral terms. This operation creates a *weakened* form of the second derivative terms, i.e., the second derivatives are reduced to first derivatives plus terms evaluated at the limits of integration. Although the equation is weakened, it is still valid and applies to the governing equation we wish to solve. Thus, Eq. 7.42 now becomes

$$\int_{\Omega}\left[\frac{\partial}{\partial x}\left(kW\frac{\partial \varphi}{\partial x}\right)+\frac{\partial}{\partial y}\left(kW\frac{\partial \varphi}{\partial y}\right)\right]d\Omega = \int_{\Gamma}\left(kW\frac{\partial \varphi}{\partial x}dy - kW\frac{\partial \varphi}{\partial y}dx\right). \tag{7.45}$$

Using the fact that the components of the unit outward normal to Γ are $n_x = dy/d\Gamma$ and $n_y = - dx/d\Gamma$, the line integrals in Eq. 7.38 become

$$\int_{\Gamma}\left(kW\frac{\partial \varphi}{\partial x}n_x + kW\frac{\partial \varphi}{\partial y}n_y\right)d\Gamma = \int_{\Gamma} kW\frac{\partial \varphi}{\partial n}d\Gamma, \tag{7.46}$$

where n denotes the normal to the surface. Hence, Eq. 7.39 can be rewritten as

$$\int_{\Omega}\left[k\frac{\partial W}{\partial x}\frac{\partial \varphi}{\partial x} + k\frac{\partial W}{\partial y}\frac{\partial \varphi}{\partial y} - Wf\right]d\Omega + \int_{\Gamma} W\left(-k\frac{\partial \varphi}{\partial n}\right)d\Gamma = 0. \tag{7.47}$$

Weak formulations for any second order linear differential operator can be obtained in the manner described above. Nonlinear problems must be treated on a case by case basis, but we can always generate a weak form.

7.3.1 One-dimensional elements

7.3.1.1 *Linear element*

Consider a piecewise polynomial approximation of the domain $0 < x < 1$. Divide the domain into two equal intervals, and seek a solution that is linear over each of the subintervals. A linear function φ between two nodal points x_i and x_{i+1} can be written as

$$\varphi(x) = \left[\frac{x_{i+1} - x}{x_{i+1} - x_i}\right]\varphi_i + \left[\frac{x - x_i}{x_{i+1} - x_i}\right]\varphi_{i+1} \quad . \tag{7.48}$$

Thus,

$$\varphi(x) = \sum N_i(x)\varphi_i = N_1\varphi_1 + N_2\varphi_2 + N_3\varphi_3, \tag{7.49}$$

where

$$N_1(x) = \begin{cases} 1 - 2x & 0 \le x \le \frac{1}{2} \\ 0 & \text{otherwise} \end{cases}, \tag{7.50a}$$

$$N_2(x) = \begin{cases} 2x & 0 \le x \le \frac{1}{2} \\ 2 - 2x & \frac{1}{2} \le x \le 1 \\ 0 & \text{otherwise} \end{cases}, \tag{7.50b}$$

$$N_3(x) = \begin{cases} 2x - 1 & \frac{1}{2} \le x \le 1 \\ 0 & \text{otherwise} \end{cases}. \tag{7.60c}$$

The functions $N_i(x)$, $i = 1, 2, 3$, are shown in Fig. 7.11 and are called *shape* functions, *trial* functions, or *basis* functions.

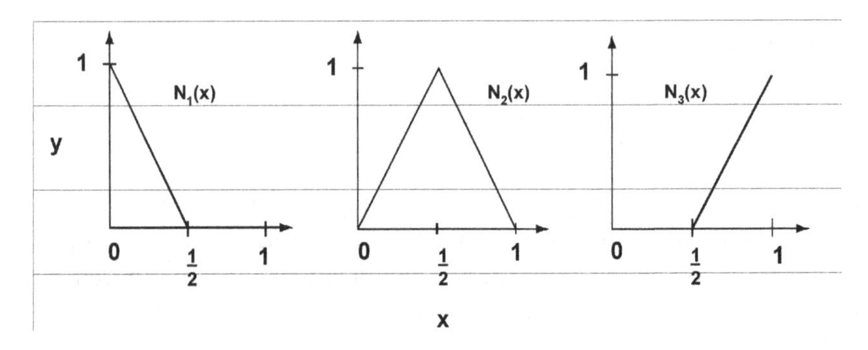

Fig. 7.11 Linear 1-D shape functions for a two-element approximation.

The functions have *local support*, i.e., they vanish outside a maximum of two elements. The implementation of Dirichlet conditions is trivial, e.g., to impose $\varphi(0) = 0$, simply set $\varphi_1 = \varphi_1(x_1) = \varphi(0) = 0$. In a Galerkin formulation, one sets $W_i(x) = N_i$, and the linear shape functions take the generic form over each element shown in Fig. 7.12.

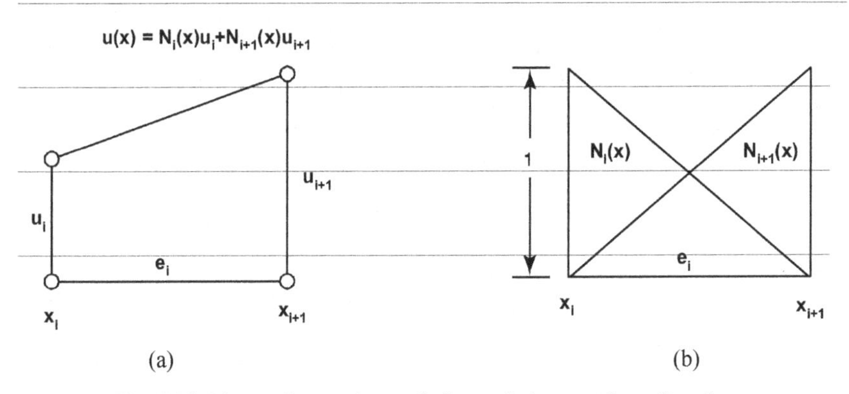

(a) (b)

Fig. 7.12 Linear element interpolation and element shape functions.

7.3.1.2 *Quadratic and higher order elements*

To obtain higher order elements, we must introduce more nodes in the elements. For example, if we desire to use quadratic polynomials over an element, a function $\varphi(x)$ will be approximated as

$$\varphi(x) \cong a + bx + cx^2 \qquad 0 \le x \le h, \tag{7.51}$$

which contains three unknown parameters; h is commonly used in FEM to denote element length (i.e., Δx). To determine the shape functions we place three nodes within the element, one at each end of the interval and one at the midpoint. Note that it is not required that the middle node be placed precisely between the two end point nodes – this allows the user to skew values towards one end of the element or the other. In this case, we set the nodes at $x_1 = 0$, $x_2 = h/2$, and $x_3 = h$, which produces the following shape function relations

$$\begin{aligned} N_1(x) &= 1 - \frac{3x}{h} + \frac{2x^2}{h^2} \\ N_2(x) &= \frac{4x}{h}\left(1 - \frac{x}{h}\right) \\ N_3(x) &= \frac{x}{h}\left(\frac{2x}{h} - 1\right), \end{aligned} \tag{7.52}$$

when $0 \le x \le h$ and zero otherwise.

Figure 7.13 shows the local quadratic shape functions. A finite element approximation based on quadratic elements is more accurate than one based on linear elements.

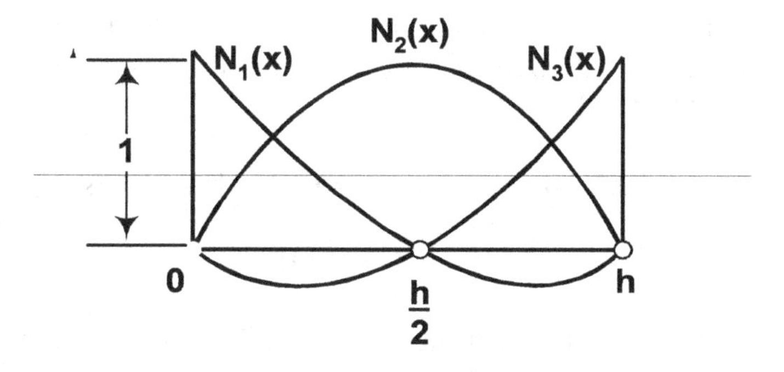

Fig. 7.13 One-dimensional quadratic element and shape functions.

A cubic element consists of two interior nodes located at a distance of h/3 to each end.

There are two ways in which a finite element approximation to any problem can be improved. The first consists in increasing the number of elements used in the mesh, therefore decreasing the size of h and, consequently, the error. This is called the h-method and relies on decreasing the size of the mesh to achieve better accuracy, utilizing always the same element. The second possibility is to keep the number of elements fixed and to increase the degree of the interpolation polynomials in the elements. In this way the number of nodes is increased and so is the order of the element. This is called the p-method. A combination of both can also be used, and this is referred to as the h-p method.

There is a very significant improvement going from linear to quadratic elements. However, the gains going from quadratic to cubic elements are marginal. On the other hand, the calculation cost increases considerably as we increase the order of the elements. To obtain the element stiffness matrices requires significantly more operations for the higher order elements. However, more important is the fact that the bandwidth of the coefficient matrix becomes larger with higher order elements. In many cases a few quadratic elements will yield solutions of much better accuracy than a much larger number of linear elements, and their use is therefore desirable.

Before moving on to two-dimensional elements, let us review the formulation of the 1-D ODE previously used to describe the difference between the FDM and the FVM.

Example 7.3.1 ***ODE discretization using FEM:*** The ODE is

$$\frac{d^2\varphi}{dx^2}+\frac{d\varphi}{dx}+\varphi=0\,. \tag{7.53}$$

We will define two adjacent elements with nodes, i – 1 to i and i to i +1, as shown in Fig. 7.14 (just like Fig. 7.1), and denote them as Δx- and Δx+.

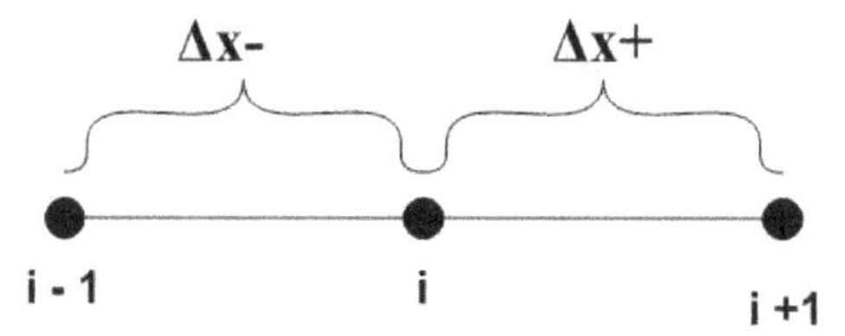

Fig. 7.14 Two linear elements.

Applying the Method of Weighted Residuals, Eq. 7.53 becomes

$$\int_\Omega \frac{d^2\varphi}{dx^2} W dx + \int_\Omega \frac{d\varphi}{dx} W dx + \int_\Omega \varphi dx = 0\,, \tag{7.54a}$$

or

$$-\int_\Omega \frac{d\varphi}{dx}\frac{dW}{dx} dx + W\frac{d\varphi}{dx}_\Gamma + \int_\Omega \frac{d\varphi}{dx} W dx + \int_\Omega \varphi W dx = 0\,, \tag{7.54b}$$

where

$$\varphi = N_{i-1}\varphi_{i-1} + N_i\varphi_i \ \text{ with } N_{i-1} = \frac{x_i - x}{\Delta x},\, N_i = \frac{x - x_{i-1}}{\Delta x}\,.$$

Now set W = N and assemble (perform) the operation over element Δx-, resulting in the expression

$$\begin{aligned} &-\int_\Omega \frac{dN_{i-1}}{dx}\frac{dN_i}{dx} dx \begin{Bmatrix}\varphi_{i-1}\\ \varphi_i\end{Bmatrix} + N_i \frac{d\phi}{dx}_\Gamma + \\ &\int_\Omega \frac{dN_{i-1}}{dx} N_i dx \begin{Bmatrix}\varphi_{i-1}\\ \varphi_i\end{Bmatrix} + \int_\Omega N_{i-1}N_i dx \begin{Bmatrix}\varphi_{i-1}\\ \varphi_i\end{Bmatrix} = 0. \end{aligned} \tag{7.55}$$

We can use a simple exact integration scheme (which works nicely for a simple element)

$$\int N_1^a N_2^b dx = \Delta x \frac{a!b!}{(a+b+1)!} \rightarrow \text{exact integration}\cdot$$

Notice how the integral terms are evaluated, e.g.,

$$\int_\Omega N_{i-1}N_i dx = \Delta x \begin{bmatrix} \frac{2!0!}{(1+2+0)!} & \frac{1!1!}{(1+1+1)!} \\ \frac{1!1!}{(1+1+1)!} & \frac{0!2!}{(1+2+0)!} \end{bmatrix} = \frac{\Delta x}{6}\begin{bmatrix} 2 & 1 \\ 1 & 2 \end{bmatrix}$$

$$\int_\Omega \frac{dN_{i-1}}{dx}\frac{dN_i}{dx}dx = \frac{1}{\Delta x}\begin{Bmatrix} -1 \\ 1 \end{Bmatrix}\begin{bmatrix} -1 & 1 \end{bmatrix} = \frac{1}{\Delta x}\begin{bmatrix} 1 & -1 \\ -1 & 1 \end{bmatrix}.$$

Application of the above integral scheme allows the user to establish a matrix equivalent equation for the element that spans Δx-. A similar procedure is also used to establish the matrix equation for the other elements. The relation for Δx- consists of a series of 2 x 2 matrices, evaluated from the expression

$$-\frac{1}{\Delta x_-^2}\int \begin{Bmatrix} -1 \\ 1 \end{Bmatrix}\lfloor -1 \;\; 1 \rfloor dx \begin{Bmatrix} \varphi_{i-1} \\ \varphi_i \end{Bmatrix} + \frac{1}{\Delta x_-}\int \begin{Bmatrix} N_{i-1} \\ N_1 \end{Bmatrix}\lfloor -1 \;\; 1 \rfloor dx \begin{Bmatrix} \varphi_{i-1} \\ \varphi_i \end{Bmatrix} + \int \begin{Bmatrix} N_{i-1} \\ N_1 \end{Bmatrix}\lfloor N_{i-1} \;\; N_i \rfloor dx \begin{Bmatrix} \varphi_{i-1} \\ \varphi_i \end{Bmatrix} = 0. \quad (7.56)$$

Continuing to perform the matrix multiplications within the integral terms, we obtain the modified form of Eq. 7.47

$$-\frac{1}{\Delta x_-^2}\int \begin{bmatrix} 1 & -1 \\ -1 & 1 \end{bmatrix} dx \begin{Bmatrix} \varphi_{i-1} \\ \varphi_i \end{Bmatrix} + \frac{1}{\Delta x_-}\int \begin{bmatrix} -N_{i-1} & N_{i-1} \\ -N_i & N_i \end{bmatrix} dx \begin{Bmatrix} \varphi_{i-1} \\ \varphi_i \end{Bmatrix} + \int \begin{bmatrix} N_{i-1}^2 & N_{i-1}N_i \\ N_{i-1}N_i & N_i^2 \end{bmatrix} dx \begin{Bmatrix} \varphi_{i-1} \\ \varphi_i \end{Bmatrix} = 0. \quad (7.57)$$

Evaluating the integral terms, we obtain the standard, generic matrix expression for a linear element (even though acting over element Δx-, we will get the same result for element Δx+),

$$-\frac{1}{\Delta x_-}\begin{bmatrix} 1 & -1 \\ -1 & 1 \end{bmatrix}\begin{Bmatrix} \varphi_{i-1} \\ \varphi_i \end{Bmatrix} + \frac{1}{2}\begin{bmatrix} -1 & 1 \\ -1 & 1 \end{bmatrix}\begin{Bmatrix} \varphi_{i-1} \\ \varphi_i \end{Bmatrix} + \frac{\Delta x_-}{6}\begin{bmatrix} 2 & 1 \\ 1 & 2 \end{bmatrix}\begin{Bmatrix} \varphi_{i-1} \\ \varphi_i \end{Bmatrix} = 0 \,. \quad (7.58)$$

If we perform the same procedure over element Δx+, we obtain the following relation

$$-\frac{1}{\Delta x_+}\begin{bmatrix}1 & -1\\ -1 & 1\end{bmatrix}\begin{Bmatrix}\varphi_i\\ \varphi_{i+1}\end{Bmatrix}+\frac{1}{2}\begin{bmatrix}-1 & 1\\ -1 & 1\end{bmatrix}\begin{Bmatrix}\varphi_i\\ \varphi_{i+1}\end{Bmatrix}+\frac{\Delta x_+}{6}\begin{bmatrix}2 & 1\\ 1 & 2\end{bmatrix}\begin{Bmatrix}\varphi_i\\ \varphi_{i+1}\end{Bmatrix}=0. \quad (7.59)$$

We now must *assemble* the results of these two integral expressions into the overall mesh, i.e., we are now the computer doing the summation of results for the two elements

$$\begin{Bmatrix}\varphi_{i-1}\\ \varphi_i\end{Bmatrix} \text{ and } \begin{Bmatrix}\varphi_i\\ \varphi_{i+1}\end{Bmatrix}.$$

Performing this operation, we obtain the following 3 x 3 matrix

$$-\begin{bmatrix}\frac{1}{\Delta x_-} & -\frac{1}{\Delta x} & 0\\ -\frac{1}{\Delta x_-} & \frac{1}{\Delta x_-}+\frac{1}{\Delta x_+} & -\frac{1}{\Delta x_+}\\ 0 & -\frac{1}{\Delta x_+} & \frac{1}{\Delta x_+}\end{bmatrix}\begin{Bmatrix}\varphi_{i-1}\\ \varphi_i\\ \varphi_{i+1}\end{Bmatrix}+$$

$$\frac{1}{2}\begin{bmatrix}-1 & 1 & 0\\ -1 & 0 & 1\\ 0 & -1 & 1\end{bmatrix}\begin{Bmatrix}\varphi_{i-1}\\ \varphi_i\\ \varphi_{i+1}\end{Bmatrix}+ \quad (7.60)$$

$$\frac{1}{6}\begin{bmatrix}2\Delta x_- & \Delta x_- & 0\\ \Delta x_- & 2\Delta x_- + 2\Delta x_+ & \Delta x_+\\ 0 & \Delta x_+ & 2\Delta x_+\end{bmatrix}\begin{Bmatrix}\varphi_{i-1}\\ \varphi_i\\ \varphi_{i+1}\end{Bmatrix}=0.$$

Notice that we can assume any element length for Δx- and Δx+, i.e., the method produces a scheme which automatically handles unstructured grids. Setting Δx- = Δx+ = Δx, produces a structured mesh, and we recover the same discretized expression for the FDM (assuming only φ_i at node i) or the FVM, and with the same discretization error. Extracting only the central terms (those involving the middle node),

$$\frac{1}{6}\left[\varphi_{i-1}\cdot\Delta x_- + \varphi_i\cdot(\Delta x_- + \Delta x_+) + \varphi_{i+1}\cdot\Delta x_+\right]+\frac{1}{2}\left[-\varphi_{i-1}+\varphi_{i+1}\right]$$
$$+\left[\frac{\varphi_{i-1}}{\Delta x_-}-\varphi_i\left(\frac{1}{\Delta x_-}+\frac{1}{\Delta x_+}\right)+\frac{\varphi_{i+1}}{\Delta x_+}\right]=0. \quad (7.61)$$

If Δx- = Δx+,

$$\frac{\Delta x}{6}\left[\varphi_{i-1}+2\varphi_i+\varphi_{i+1}\right]+\frac{\varphi_{i+1}-\varphi_{i-1}}{2}+\frac{\varphi_{i+1}-2\varphi_i+\varphi_{i-1}}{\Delta x}=0. \quad (7.62)$$

The advantages come when the FEM is used to solve the transient equations, subsequently producing a tridiagonal system that is fourth order accurate in space and second order accurate in time, if using a Crank–Nicolson time-marching scheme (Pepper and Baker, 1980).

7.3.2 Two-dimensional elements

7.3.2.1 *Triangular elements*

The simplest two-dimensional figure that defines an area is the triangle. The triangular element is obtained defining a linear interpolation field of the form

$$\varphi(x,y)\cong a+bx+cy, \quad (7.63)$$

and placing the nodes at the corners of the triangle. The shape functions can be written in terms of the nodal coordinates as

$$\begin{aligned} N_1(x,y)&=\frac{1}{2A}\left[x_2y_3-x_3y_2+(y_2-y_3)x+(x_3-x_2)y\right]\\ N_2(x,y)&=\frac{1}{2A}\left[x_3y_1-x_1y_3+(y_3-y_1)x+(x_1-x_3)y\right], \quad (7.64)\\ N_3(x,y)&=\frac{1}{2A}\left[x_1y_2-x_2y_1+(y_1-y_2)x+(x_2-x_1)y\right] \end{aligned}$$

where the area A is given by

$$2A=(x_2y_3-x_3y_2)+(x_3y_1-x_1y_3)+(x_1y_2-x_2y_1), \quad (7.65)$$

and the nodes are numbered counterclockwise as in Fig. 7.15. These elements are discussed by Pepper and Heinrich (1992).

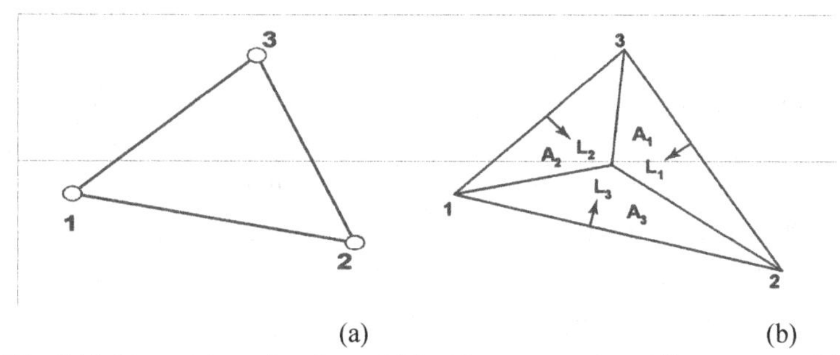

(a) (b)

Fig. 7.15 Linear triangular element (a) and area natural coordinate system (b).

The shape functions can be more easily obtained if we use area coordinates. Joining any point P in the triangle to the vertices of the triangle, the three areas, A_1, A_2, and A_3, can be defined as shown in Fig. 7.15.

A coordinate system that uniquely represents every point in the triangle is given by

$$L_i = \frac{A_i}{A} \qquad i = 1,2,3. \tag{7.66}$$

If the nodes are uniformly distributed along the element sides, the shape functions are easily constructed in this coordinate system, also called the *natural* coordinate system for the triangle.

The shape functions in natural coordinates are independent of the shape of the triangle. This is particularly appealing when we are dealing with highly irregular geometries that may require a large variety of very differently shaped triangles. The ability of the triangle to discretize any kind of geometric figure with relative ease is the main reason for the wide use of triangular elements. From this point of view, triangular elements are always better than quadrilateral elements. Very powerful mesh generators have been developed based on the triangular geometry that can automatically discretize extremely complex regions. Although in the last years much progress has been made in this area using quadrilateral elements, these mesh generators still lack the versatility and degree of automation of those based on triangles.

7.3.2.2 *Quadrilateral elements*

A quadrilateral element is defined by four corner points and therefore is no longer linear, which makes it more complex than a triangular element. However, there is a greater variety of quadrilateral elements, and in general, they can offer many advantages over the use of triangles.

The simplest way to obtain rectangular elements consists in taking the product (also referred to as the tensor product) of one-dimensional elements. In this fashion we generate the family of Lagrangian elements that are bilinear, biquadratic, etc., and contain 2^2, 3^2, 4^2, ... nodes as shown in Fig. 7.16. To obtain the shape functions we only need to know the form of the shape functions, in one dimension, and the two-dimensional function at a node is obtained as the product of the one-dimensional functions that would correspond to that node in the x and y directions, respectively.

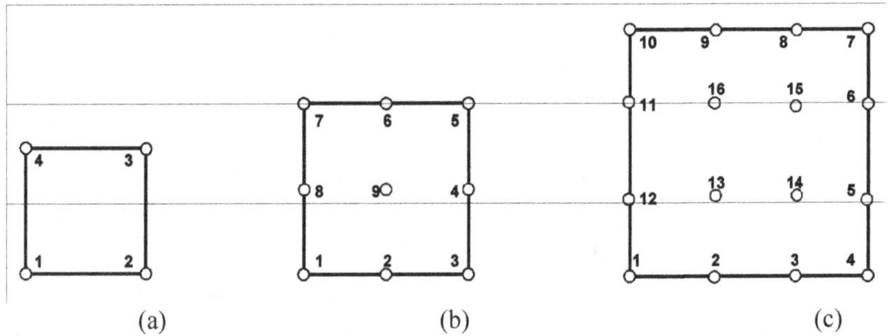

Fig. 7.16 2-D quadrilateral elements, (a) bilinear, (b) biquadratic, and (c) bicubic.

Another important family of rectangular elements is known as the *serendipity* elements. These elements differ from the Lagrangian family in that they do not contain any interior nodes. Examples of the eight-noded quadratic and the 12-noded cubic elements are shown in Fig. 7.17.

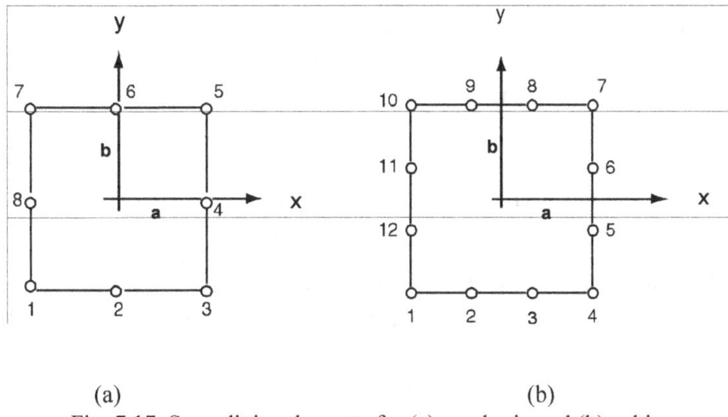

(a) (b)
Fig. 7.17 Serendipity elements for (a) quadratic and (b) cubic.

7.3.2.3 *Isoparametric elements*

Rectangular elements offer advantages over triangular elements. However, rectangular geometry is very restrictive and general quadrilateral elements must be used to deal with more general geometry. To resolve this difficulty, the concept of *isoparametric* transformations was introduced by Irons (1968) to general rectangular elements.

The idea is based on performing a local (element by element) transformation between a general quadrilateral element in the global coordinate system and a "parent" rectangular element defined in a $\xi - \eta$ coordinate system in the square $-1 \leq \xi, \eta \leq 1$ as depicted in Fig. 7.18 for a four-noded element.

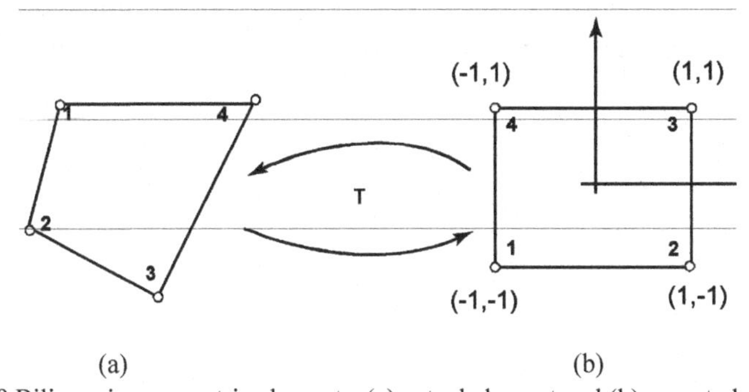

(a) (b)
Fig 7.18 Bilinear isoparametric elements, (a) actual element and (b) parent element.

Transformation of integrals must now be done, i.e.,

$$\iint_{\Omega} F\left(\varphi(x,y),\frac{\partial\varphi}{\partial x},\frac{\partial\varphi}{\partial y}\right)dxdy =$$
$$\int_{-1}^{1}\int_{-1}^{1} F\left(\varphi\big(x(\xi,\eta),y(\xi,\eta)\big),\frac{\partial\varphi}{\partial x},\frac{\partial\varphi}{\partial y}\right)|\det J|\,d\xi d\eta, \tag{7.67}$$

where **J** is the Jacobian of the transformation.

The isoparametric transformation itself is easily obtained using the relation

$$\begin{Bmatrix} x \\ y \end{Bmatrix} = \sum_{i=1}^{N} N_i(\xi,\eta)\begin{Bmatrix} x_i \\ y_i \end{Bmatrix}, \tag{7.68}$$

where N is the number of nodes in the element and Ni(ξ,η) are the shape functions for the corresponding parent element, e.g., $-1 \le \xi,\eta \le 1$. Actually, this transformation is the inverse of what is really needed, since it maps the parent element, not the actual element. It is clearly defined once the coordinates of the nodes (xi, yi) in the global system and the shape functions Ni(ξ,η) in the square parent system of coordinates are known. Similarly, cubic isoparametric elements can be defined; however, in transport applications, quadratic isoparametric elements are used much less than bilinear elements and cubics are hardly ever considered.

For triangular elements the concepts described above are applied using the right triangle shown in Fig. 7.19 as the parent element; in this case, $0 \le \xi \le 1$and $0 \le \eta \le 1$. The shape functions for the linear and quadratic parent elements are not difficult to find. For the linear element these are

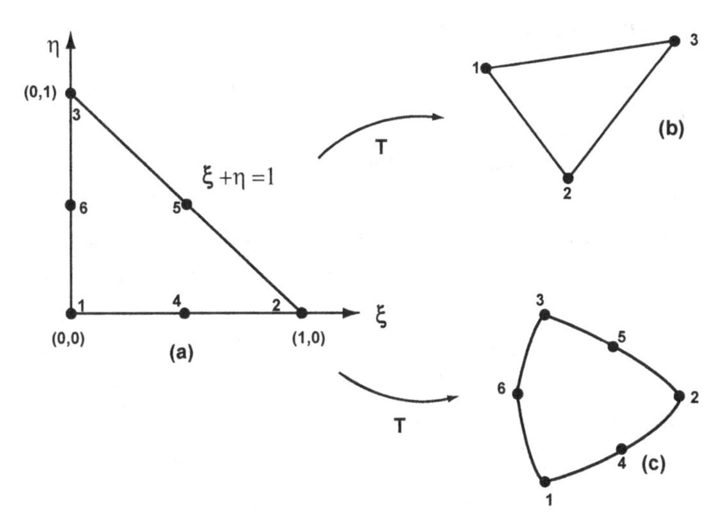

Fig. 7.19 Triangular isoparametric elements for (a) parent element, (b) linear element, and (c) curved quadratic triangle.

For higher order elements these conditions are more difficult to determine and are very sensitive to the location of midside and interior nodes. Serendipity elements do not contain interior nodes, and for this reason the quadratic serendipity element, in particular, has been a very popular element to deal with curved boundaries.

A *subparametric* transformation uses lower order interpolation functions for the geometric transformations than for interpolation. For example, a biquadratic element can be associated with a bilinear transformation if the sides of the elements are always straight lines. In the same note, *superparametric* elements use geometric transformations that are higher order than the interpolation.

When curved boundaries are fitted, curved isoparametric elements can produce enormous improvements in some classes of problems (Zlamal, 1973). However, for problems such as convective flows and heat and mass transfer, such techniques must be applied carefully. Remember that the use of isoparametric elements requires *numerical integration* because the determinant of the Jacobian transforms the integral into a rational function, even though the shape functions are polynomials.

7.3.3 Three-dimensional elements

The transition to three-dimensional problems does not involve any new concepts. Area and line integrals are replaced with volume and surface integrals. The natural extension of the one-dimensional linear and two-dimensional bilinear elements is the eight-noded trilinear or *brick* element shown in Fig. 7.20. This is also a Lagrangian element and the shape functions are easily obtained as products of one-dimensional linear functions.

This can be seen in the following set of relations. One need only multiply the linear, one-dimensional shape functions for node 1 in each coordinate direction to obtain N_1, i.e.,

$$\begin{aligned}
N_1(x,y,z) &= \overline{N}_1(x)\overline{N}_1(y)\overline{N}_1(z)\\
N_2(x,y,z) &= \overline{N}_2(x)\overline{N}_1(y)\overline{N}_1(z)\\
N_3(x,y,z) &= \overline{N}_2(x)\overline{N}_2(y)\overline{N}_1(z)\\
N_4(x,y,z) &= \overline{N}_1(x)\overline{N}_2(y)\overline{N}_1(z)\\
N_5(x,y,z) &= \overline{N}_1(x)\overline{N}_1(y)\overline{N}_2(z)\\
N_6(x,y,z) &= \overline{N}_2(x)\overline{N}_1(y)\overline{N}_2(z)\\
N_7(x,y,z) &= \overline{N}_2(x)\overline{N}_2(y)\overline{N}_2(z)\\
N_8(x,y,z) &= \overline{N}_1(x)\overline{N}_2(y)\overline{N}_2(z)
\end{aligned} \quad . \tag{7.69}$$

The other shape functions for the remaining nodes follow similarily. The resulting hexahedral, or brick, element is created, as seen in Fig. 7.20.

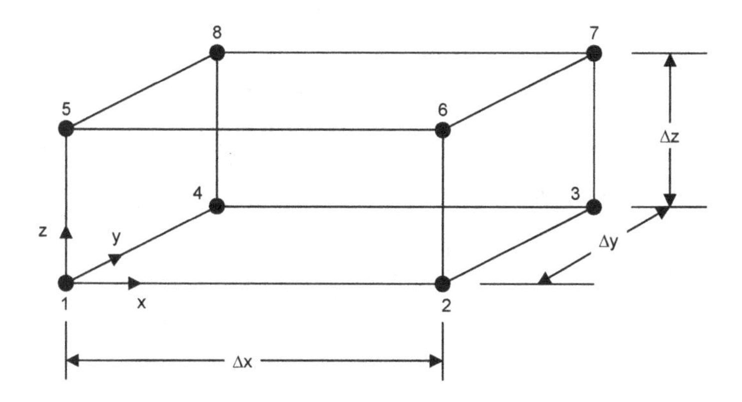

Fig. 7.20 Trilinear "brick" hexahedral element.

Three-dimensional tetrahedral elements are the natural extension of two-dimensional triangles. The simplest of these elements are the linear and quadratic tetrahedrons depicted in Fig. 7.21. A natural coordinate system is defined for tetrahedral elements by means of the four internal volumes determined when any interior point is connected with the four vertices of the tetrahedron – in a manner similar to that used in two dimensions to define the area coordinates. In this case, they are referred to as volume coordinates. These are discussed by Pepper and Heinrich (2002).

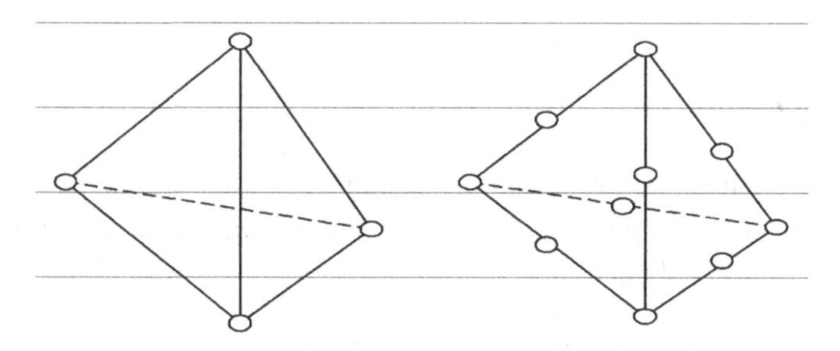

Fig. 7.21 Linear and quadratic tetrahedral elements.

Isoparametric transformations are readily defined for the three-dimensional elements using the following relations

$$\begin{bmatrix} x \\ y \\ z \end{bmatrix} = \sum_{i=1}^{N} N_i(\xi,\eta,\zeta) \begin{bmatrix} x_i \\ y_i \\ z_i \end{bmatrix}. \tag{7.70}$$

The Jacobian matrix is

$$J = \begin{bmatrix} \frac{\partial x}{\partial \xi} & \frac{\partial y}{\partial \xi} & \frac{\partial z}{\partial \xi} \\ \frac{\partial x}{\partial \eta} & \frac{\partial y}{\partial \eta} & \frac{\partial z}{\partial \eta} \\ \frac{\partial x}{\partial \zeta} & \frac{\partial y}{\partial \zeta} & \frac{\partial z}{\partial \zeta} \end{bmatrix}, \tag{7.71}$$

and the derivatives of the shape functions are obtained from

$$\begin{bmatrix} \frac{\partial N_i}{\partial x} \\ \frac{\partial N_i}{\partial y} \\ \frac{\partial N_i}{\partial z} \end{bmatrix} = J^{-1} \begin{bmatrix} \frac{\partial N_i}{\partial \xi} \\ \frac{\partial N_i}{\partial \eta} \\ \frac{\partial N_i}{\partial \zeta} \end{bmatrix}. \tag{7.70}$$

7.3.4 Quadrature

In finite element approximations using isoparametric elements, the integrals are always performed over an element in the parent coordinate system $-1 \le \xi \le 1$; hence

$$I(g) = \int_a^b g(x)dx \to I(f) = \int_{-1}^{1} f(\xi)\,d\xi. \tag{7.72}$$

Approximations of the form

$$I_n(g) \cong \sum_{i=1}^{n} w_i g(x_i), \tag{7.73}$$

are called numerical quadrature or numerical integration formulae. The points x_i are called quadrature points, and the coefficients w_i are called quadrature weights.

A quadrature formula with degree of precision m integrates polynomials of degree ≤ m exactly. The maximum degree of precision that can be achieved with a quadrature rule that uses n integration points is m = 2n – 1, and the quadrature formulae that achieve this accuracy are known as Gaussian quadratures. The integration formulae are referred to as Gauss–Legendre quadratures.

In two and three dimensions, one must evaluate double and triple integrals over areas and volumes, respectively. The easiest way is to evaluate the variables one by one in succession, e.g., fix the independent variable η and define

$$F(\eta) \equiv \sum_{i=1}^{n} w_i f(\xi_i, \eta) \cong \int_{-1} f(\xi, \eta)\, d\xi, \tag{7.75}$$

then

$$\int_{-1} F(\eta)\, d\eta \cong \sum_{j=1}^{m} w_j F(\eta_j) = \sum_{j=1}^{m} w_j \left(\sum_{i=1}^{n} w_i f(\xi_i, \eta_j) \right). \tag{7.76}$$

Hence,

$$I_{nm} = \sum_{i=1}^{n} \sum_{j=1}^{m} w_i w_j f(\xi_i, \eta_j), \tag{7.77}$$

where m and n are not necessarily equal. A quadrature formula with degree of precision 3 is obtained if one uses a Gauss quadrature with n = 2 in each direction, that is,

$$\int_{-1} \int_{-1} f(\xi, \eta)\, d\xi\, d\eta \cong I_{22} = \sum_{i=1}^{2} \sum_{j=1}^{2} w_i w_j f(\xi_i, \eta_j), \tag{7.78}$$

where ξ_i, η_j are sampling points and w_i are the weights (which can be found in nearly every FEM textbook). This simple formula allows ones to accurately approximate the integral relations that are produced from the Method of Weighted Residuals technique and resulting weak statement.

7.3.5 Time dependence

The most commonly used time integration algorithm in the FEM is the θ method, which consists in approximating the time derivative by the backward difference

$$\dot{\varphi} \cong \frac{1}{\Delta t}\left(\varphi^{n+1} - \varphi^{n}\right), \tag{7.79}$$

where $\varphi^n \equiv \varphi(x,\varphi_n)$ denotes the value of a variable at time $t = t_n$, Δt is the time-step increment, and $t_{n+1} = t_n + \Delta t$. The variable φ is then defined by

$$\varphi = \theta\varphi^{n+1} + (1-\theta)\varphi^{n}, \tag{7.80}$$

where the relaxation parameter θ is normally specified to be a value between 0 and 1 and is used to control the accuracy and stability of the algorithm. This method falls in the general category of *one-step* methods, in which the solution at each step is advanced to time t_{n+1} from known values at time step t_n. A quick review of the FDM and FVM schemes show that a similar procedure can be used to obtain the transient discetizations. In fact, the FEM borrows this concept from the FDM approach for marching solutions in time.

Substituting Eqs. 7.79 and 7.80, the one-dimensional transient diffusion equation becomes

$$\frac{\partial\varphi}{\partial t} = k\frac{\partial^2\varphi}{\partial x^2} + Q, \tag{7.81}$$

or

$$\begin{aligned}\left(\frac{1}{\Delta t}\mathbf{C} + \theta\mathbf{K}\right)\varphi^{n+1} &= \\ \left(\frac{1}{\Delta t}\mathbf{C} - (1-\theta)\mathbf{K}\right)\varphi^{n} &+ \theta\mathbf{Q}^{n+1} + (1-\theta)\mathbf{Q}^{n},\end{aligned} \tag{7.82}$$

where $\mathbf{Q}$ has been assumed to be a function of time and approximated over the interval $t_n \le t \le t_{n+1}$ using Eq. 7.80, and the mass matrix, $\mathbf{C}$, and stiffness matrix, $\mathbf{K}$, are evaluated from the integral relations

$$\begin{aligned}
&\frac{\partial\varphi}{\partial t} \equiv \mathbf{C} = \int N_i N_j dx \\
&\frac{\partial^2}{\partial x^2} \equiv \mathbf{K} = k\int \frac{\partial N_i}{\partial x}\frac{\partial N_j}{\partial x}dx
\end{aligned} \tag{7.83}$$

If 1-D, linear elements are used in the space discretization and Q is independent of x, the resulting element equations are

$$\begin{aligned}
&\left\{\frac{\rho c_v h}{6\Delta t}\begin{bmatrix}2 & 1\\ 1 & 2\end{bmatrix}+\frac{\theta k}{h}\begin{bmatrix}1 & -1\\ -1 & 1\end{bmatrix}\right\}\begin{Bmatrix}\varphi_1^{n+1}\\ \varphi_2^{n+1}\end{Bmatrix}= \\
&\left\{\frac{\rho c_v h}{6\Delta t}\begin{bmatrix}2 & 1\\ 1 & 2\end{bmatrix}-\frac{(1-\theta)k}{h}\begin{bmatrix}1 & -1\\ -1 & 1\end{bmatrix}\right\}\begin{Bmatrix}\varphi_1^{n}\\ \varphi_2^{n}\end{Bmatrix}, \\
&+\frac{\theta h Q^{n+1}}{2}\begin{Bmatrix}1\\ 1\end{Bmatrix}+\frac{(1-\theta)hQ^n}{2}\begin{Bmatrix}1\\ 1\end{Bmatrix}
\end{aligned} \tag{7.84}$$

where h denotes the element size. The values $\theta = 0$, 0.5, and 1.0 are most commonly used and, except for the presence of the mass matrix, correspond to the Euler, Crank–Nicolson, and backward implicit methods, respectively. However, the appearance of the mass matrix modifies the algorithms. Therefore they are referred to as Euler–Galerkin when $\theta = 0$, Crank–Nicolson–Galerkin when $\theta = 0.5$, and backward implicit Galerkin when $\theta = 1.0$.

A truncation error analysis shows that the methods converge as first-order methods $O(\Delta t)$ when $\theta = 0.0$ and 1.0; the Crank–Nicolson–Galerkin method is second order $O(\Delta t^2)$ in time, and for other values of θ between 0 and 1, convergence takes place at intermediate rates between first and second order.

7.3.6 Petrov–Galerkin method

In order to improve accuracy in time, one can construct weighting functions that are parabolic in time, e.g., the time variation $\varphi(t)$ is

$$T(t) = \frac{4t}{\Delta t}\left(1-\frac{t}{\Delta t}\right) \tag{7.85}$$

The weighting functions become $N_i(x)\varphi(t)$.

In the one-dimensional case, the truncation error can be interpreted in a difference equation as an added diffusion, and obtain improved algorithms by introducing a balancing diffusion dependent on a parameter α. In this case, a balancing dispersion term of the form $\beta d(\partial^3\phi/\partial x^2\partial t)$ can be added, where the coefficient d must be proportional to $uh\Delta t$ for dimensional consistency. Looking at the modified advection–diffusion equation, one obtains

$$\frac{\partial\varphi}{\partial t}+u\frac{\partial\varphi}{\partial x}-\left(D+\frac{\alpha uh}{2}\right)\frac{\partial^2\varphi}{\partial x^2}+\beta d\frac{\partial^3\varphi}{\partial x^2\partial t}=0. \qquad (7.86)$$

Now apply the Galerkin method and operate on the weak form. The Petrov–Galerkin weights are

$$w_i(x,t)=M_i(x,t)+\frac{\alpha h}{2}\frac{\partial M_i(x,t)}{\partial x}+\frac{\beta h\Delta t}{4}\frac{\partial^2 M_i(x,t)}{\partial x\partial t}. \qquad (7.87)$$

The functions $M_i(x,t)$ in $0 \le x \le h$, $0 \le t \le \Delta t$ are

$$M_1(x,t)=M_4(x,t)=4(1-\frac{x}{h})\frac{t}{\Delta t}(1-\frac{t}{\Delta t})$$

$$M_2(x,t)=M_3(x,t)=4\frac{x}{h}\frac{t}{\Delta t}(1-\frac{t}{\Delta t}).$$

After some algebra (see Heinrich and Pepper, 1999), α and β are

$$\alpha=\coth\frac{\gamma}{2}-\frac{2}{\gamma} \qquad \beta=\frac{c}{3}-\frac{2\alpha}{\gamma c}. \qquad (7.88)$$

If $\beta = 0$, the algorithm reduces to applying the Petrov–Galerkin weights for a steady-state equation with a second-order time-stepping scheme, and in this case it is only second order accurate in space. In the limiting case when $\gamma \rightarrow 0$, the expression for β is undefined. Physically u must go to zero. In this case the algorithm reduces to the Crank–Nicolson–Galerkin scheme because $\gamma \rightarrow 0$ if $u \rightarrow 0$.

The application of Petrov–Galerkin weighting has been found to be very attractive, and is commonly used in many FEM methods where advection terms are dominant. The method is especially attractive when combined with adaptive mesh techniques for solving compressible flows.

7.3.7 Mesh generation

There are basically two types of meshes: structured and unstructured. A structured mesh consists of horizontal and vertical lines that cross orthogonally at intersections called nodes. This constraint is best achieved by discretizing a physical domain that is defined by square or rectangular boundaries – the physical domain becomes the computational domain as well. Many of the early numerical simulations were conducted on problems that were first reduced to rectangular physical systems of interest. Curved boundaries are simplistically represented by staircase–like steps in the mesh. An example of a structured mesh is shown in Fig. 7.22.

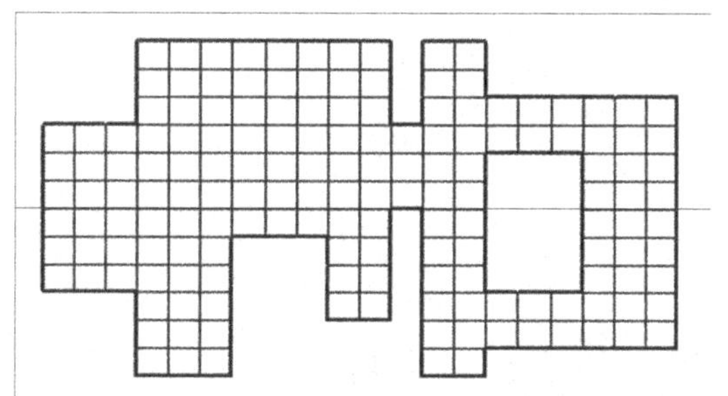

Fig. 7.22 Office complex discretized using FDM.

The computational domain is divided into small domains called grid or mesh cells, or elements, over which the governing equations are discretised. The mesh represents the geometrical shape of the domain, and must be fine enough to permit adequate resolution of the flow. Creating a suitable mesh depends upon the expected behavior of the flow and transport. Unless some form of adaptation (usually refinement) is employed, the process is typically trial and error.

Creating an acceptable mesh may require a number of attempts with further improvements as the calculations proceed in time. It is important to create a mesh-independent solution, i.e., a solution which does not significantly vary as the mesh is refined. This usually requires several solutions using finer meshes until the solution is essentially invariant.

Grids typically fall within three types:

(1) Structured Cartesian grids, where the grid lines are continuous across the domain and the grid cells are quadrilaterals or hexahedrals;

(2) Structured Curvilinear or Body-Fitted grids, where the grid lines follow the computational domain boundary. This can be seen in Fig. 7.3;

(3) Unstructured mesh, usually constructed from tetrahedral or more complex-shaped cells; there are no clearly defined grid lines which are discontinuous across the domain.

An example of an unstructured mesh is shown in Fig. 7.23.

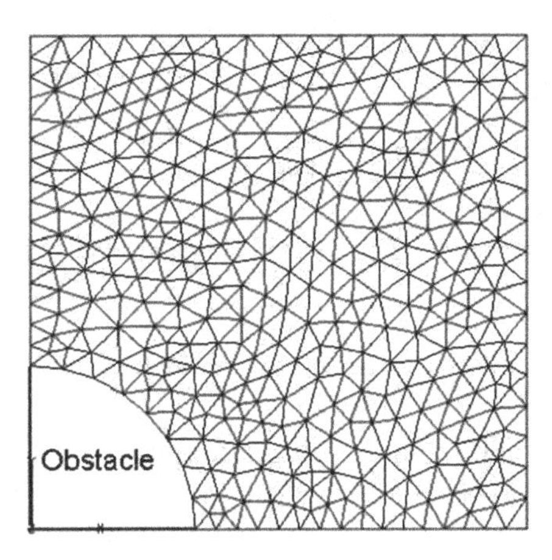

Fig. 7.23 Example of an unstructured two-dimensional grid.

In unstructured meshes, a 2-D physical domain is discretized by a set of seemingly randomly placed nodes that are connected to other nodes via triangular or quadrilaterally shaped subdomains, or elements. The most common types of elements are linear three-noded triangles or linear four-noded quadrilaterals, as discussed earlier. Other popular types of elements include quadratic and cubic triangles and quadrilaterals. In three dimensions these become tetrahedrals and hexahedrals, respectively. The generation of unstructured meshes requires more thought and effort than structured meshes. In general, one puts more nodes (or elements) near surfaces and in regions where activity (or steep gradients) is likely to occur. Many times, the user must guess as to where the most nodes should be placed, ultimately necessitating the generation of a second mesh (and comparing solutions for accuracy). It is up to the user to specify the mesh density (number of nodes and elements), which is best achieved through experience.

Unstructured meshes (especially triangles) are better suited for complex geometries as they can be adapted to any shape. However, unstructured meshes also have several disadvantages. The associated discretised equations are more complicated than is the case with structured grids, when dealing with FDM and FVM methods. As a result, the system of equations can be more difficult to solve and the solution obtained less accurate. Creating a geometry and a grid for a complex space is the most time-consuming task for the CFD practitioner.

There is no reason one should end up using severely distorted elements to discretize a domain. In fact, the interior of most domains can be meshed using non-distorted elements; as one approaches the boundaries, several slightly distorted elements can be constructed. Curved sides should only be employed on curved boundaries, and the curvature should be rather mild ($\leq 30^\circ$ arc). When this is not possible, more elements should be utilized.

If the physical domain has all boundary sides straight, with no internal curved surface (e.g., hole), any type of element will match the boundary exactly. Likewise, boundaries defined by higher degree polynomials can also be matched exactly with corresponding higher order elements. However, non-polynomial curvature cannot be matched exactly by polynomial elements; hence the domain boundary becomes altered to the outer edge of the defining element. Suppose one wishes to use linear elements to prescribe a boundary. In order to reduce the error associated with the area omissions, more linear elements are required. This leads to the decision by the user whether to increase the number of lower order elements or use a higher order element. The quadratic yields about 1% geometric error while the linear element produces about 29% error for a 90° arc (Burnett, 1987). In practice, preprocessing typically requires several refinements of the mesh, including boundary matching, before a suitable solution is achieved.

Hybrid meshes are a compromise, combining structured and unstructured cells. Figure 7.24 shows structured cells used near a wall boundary, coupled with unstructured grids. This allows one to more accurately capture the transfer of heat, mass, or momentum at the walls.

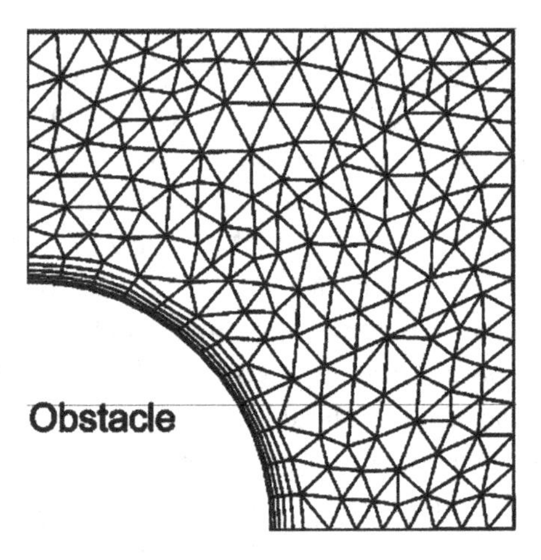

Fig. 7.24 An example of a hybrid two-dimensional mesh.

An alternative is to inflate unstructured cells, i.e. to create thin cells with surfaces parallel to a geometry. This is seen in Fig. 7.25.

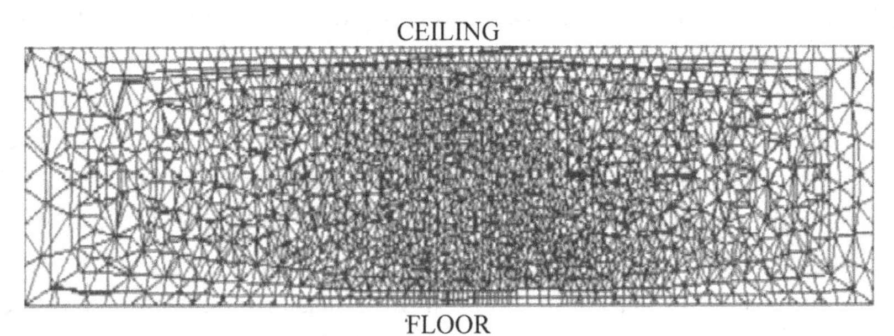

Fig. 7.25 Example of a compressed center and inflated wall boundaries.

Many variations are possible, but the above approaches are the most commonly used and are the ones embodied in most commercial CFD codes.

The size of the cells is important and needs to be chosen with care since it can significantly influence the solution. A cell should ideally be smaller than the length scale of the key flow feature. However, in a complex geometry, a wide range of length scales exist at different locations. Choosing a grid of the size of the smallest length scale, then applying this uniformly over the overall geometry leads to a large number of grid cells and results in excessive computing times or storage limitations.

To overcome this problem, one can cluster cells at specific locations as determined by the user. Figure 7.26 shows a grid that is compressed where the flow changes rapidly, i.e., where large gradients occur. Figure 7.25 showed compression of an unstructured mesh where a source was located.

When creating a mesh, the CFD practitioner should have an intuitive idea of the flow behavior prior to the simulation. Obviously, this can prove to be difficult in many situations, requiring several trial and error estimates before a decent mesh is acquired.

The grid size also has to be consistent with the modeling approach chosen. For example, predefined functions are often employed in combination with a k-ε turbulence model to calculate the velocity profiles near walls. These functions model the very small scale mechanisms of heat and momentum transfers occurring in a very thin region close to solid boundaries. They make it possible, and even necessary, to set the size of the cells next to walls larger than this region.

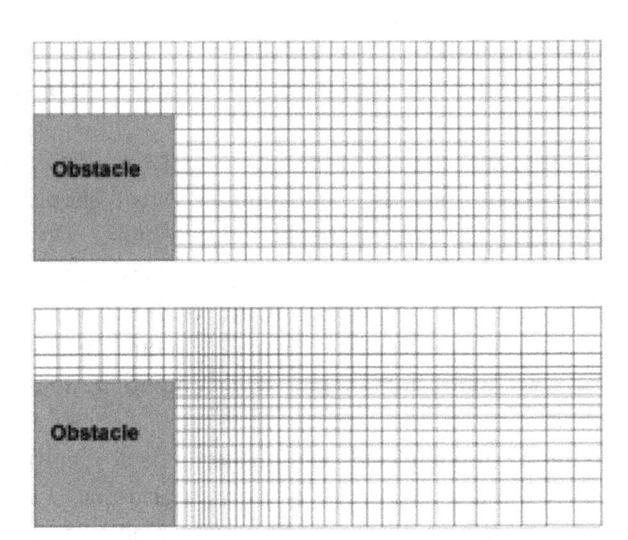

Fig. 7.26 Comparison of uniform (top) and compressed (bottom) structured two-dimensional grids.

7.3.8 Bandwidth

In 2-D and 3-D meshes, the node number pattern dictates the bandwidth of the assembled global matrix. Unless the user is employing an explicit marching scheme, the naturally implicit nature of the finite element method creates a banded, sparse matrix that may or may not be symmetric. Hence it behooves the user to minimize the bandwidth of the matrix to reduce storage and computer time. Finding the optimal minimum pattern can be difficult; however, any effort to achieve a near-optimal pattern is worth trying.

There are many algorithms available that automatically renumber a mesh to minimize its bandwidth; for frontal solvers, the wave front is minimized by renumbering the elements; the procedures are similar for nodal or elemental renumbering. The user must create the starting nodes, i.e., an initial mesh, which then gets reordered. These minimization routines are commonly used in many commercial finite element codes.

Renumbering the nodes (or elements) of a mesh allows one to minimize the storage size required by the matrix solver and reduce the number of operations required by the final system, which ultimately reduces the CPU time. There are many methods that perform this renumbering operation, most of them automatic. A detailed discussion on the advantages and disadvantages of the various methods is given by Marro (1980) and George (1991). A more recent update on the application of renumbering schemes is given by Carey (1997).

7.3.9 Adaptation

Mesh adaptation is becoming widely used and has begun to appear in many commercial finite element codes (although primarily in structural codes). Several of the commercial fluid flow codes now support adaptation – ANSYS, STAR–CD, COMSOL, ANSWER, and GWADAPT, which is a code for porous media flow (Pepper and Stephenson, 1995). The application of adaptation is employed in the office complex example using COMSOL that is discussed in Appendix B. Such codes automatically refine the mesh in regions where increased accuracy is required.

It has been amply demonstrated that adaptation leads to computations of better solutions, producing optimal meshes in forms of size (number of nodes), nodal positions, and element properties (e.g., shape, orthogonality). Refer to the texts by Babuska *et al.* (1983) and Zienkiewicz and Taylor (1989) for detailed discussions on the mathematical aspects of adaptation.

There are basically three types of adaptive techniques in use today: r-refinement, h-refinement, and p-refinement. In r-refinement a fixed mesh is first established; the elements within the mesh are then moved, shrunk, or expanded to accommodate regions where the solution is rapidly changing (or relatively stagnant). This technique has been shown to be effective in some cases; however, the elements can become severely distorted and eventually lead to divergent or less accurate solutions. By far the most popular methods are h- and p-refinement. In h-refinement, elements are subdivided into smaller elements; this technique creates additional nodes and elements, which must be carefully monitored through some form of bookkeeping. In p-refinement the degree of the polynomial is increased to improve the accuracy of the solution, i.e., an element that may have been originally linear is ultimately refined to a cubic, quartic, quintic, or higher order element. A smaller h and a higher p generally yield greater accuracy but slower convergence if too fine a refinement is established. Methods that adapt both h and p together are called h-p refinements. Papers by Pepper and Wang (2006), Shapiro and Murman (1988), Ramakrishnam *et al.* (1990), Oden *et al.* (1986), Pelletier and Hetu (1992), Zienkiewicz *et al.* (1981), and Pepper and Stephenson (1995) are recommended. Other forms of adaptation in the literature include local disenrichment (a form of h-adaptation), which removes one of several points, nested meshes, and multigrid techniques (see Hackbusch and Trottenberg, 1982).

When using h-adaptation, there are basically two choices to be made, *mesh regeneration* or *element subdivision*. Mesh regeneration, or remeshing, requires completely regenerating the entire mesh, either in regions where there is high error or over the complete domain, as shown in Fig. 7.27. The principal advantage of remeshing is that areas can be coarsened where the error is below an allowable amount, thus creating an optimal mesh in which every element has essentially the same level of error. However, the main disadvantage of remeshing is that a high degree of spatial flexibility is necessary when using error estimation procedures. When using element subdivision, every element that exceeds the allowable error threshold is subdivided into smaller elements. This method is particularly effective for four-node quadrilaterals and eight-node hexahedrals. However, the method produces virtual nodes (i.e., constrained midside nodes) that must be handled with care; likewise, only one level of adaptation can be performed at a time.

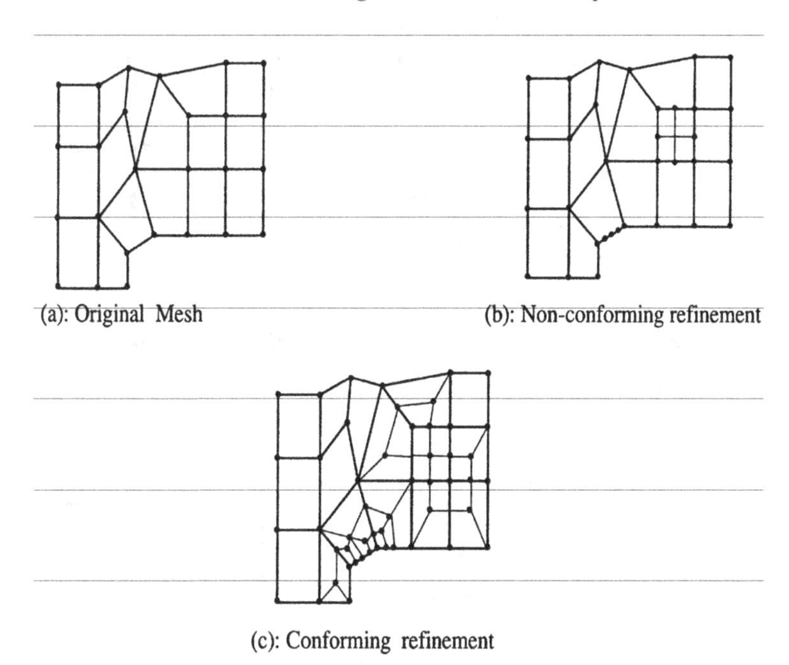

Fig. 7.27 Mesh refinement

Local mesh (h-) refinement and convergence history are shown in Fig. 7.28 (a,b). When using quadrilateral elements, virtual nodes are created which require special treatment to ensure compatibility. The error is asymptotic.

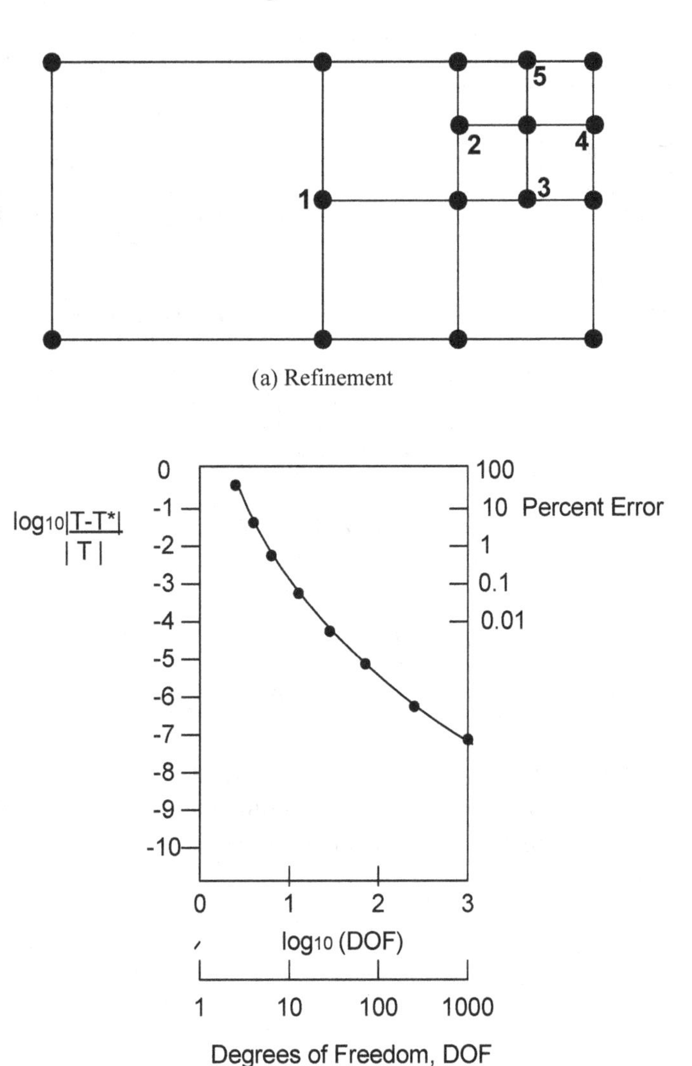

Fig. 7.28 Local h-adaptation (a) refinement) and (b) convergence.

An example of remeshing for three-node triangles and four-node quadrilaterals is shown in Fig. 7.29 (a–d) for a simple heat transfer problem with convective boundary conditions. This benchmark problem is illustrated in more detail in the text by Huang and Usmani (1994), which also includes a set of computer programs for generating adaptive meshes based on the remeshing principle. Figure 7.29 (a) shows the problem domain with three different boundary conditions. For a uniform mesh of 30 triangular elements (Fig. 7.29 b), a temperature error of –24.6% was obtained. When the adaptive procedure is applied, a first level of adaptation (Fig. 7.29 c) produced a norm error of 14.3% (average temperature error of 2% overall) in the triangular mesh; a second level of adaption produced an error norm of 9.8%. In the quadrilateral case, the initial mesh consisting of eight elements (Fig. 7.29 b) produced a norm error of 21.6%; two mesh refinements, shown in Fig. 7.29 (d), yielded 16% and 10%, respectively. If one uses quadratic elements, the error reduces considerably but at a higher computational cost.

7.3.9.1 *Element subdivision*

The starting point for element subdivision is a mesh coarse enough to allow rapid convergence, yet fine enough to allow the flow details to appear. An initial solution is then computed on the crude mesh; it is not necessary to allow this solution to converge completely. The initial solution should not evolve too far before adaptation, or expensive computational time will be used needlessly since the flow features will shift location during the adaptation procedure.

Refinement indicators are computed based on the solution on the initial mesh, and elements that need to be refined are identified. All elements in the mesh that have indicators above a preset refinement threshold value are enriched, whereas those elements that have values below the unrefinement threshold value are coarsened. Refinement proceeds from the coarsest level to the finest level.

After all the mesh changes have been made, the grid geometry is recalculated, the solution is interpolated onto the new grid, and the calculation procedure begun again. For steady-state problems, the entire procedure is repeated until a "converged" mesh is obtained. A converged mesh is a mesh which no longer changes as the solution progresses.

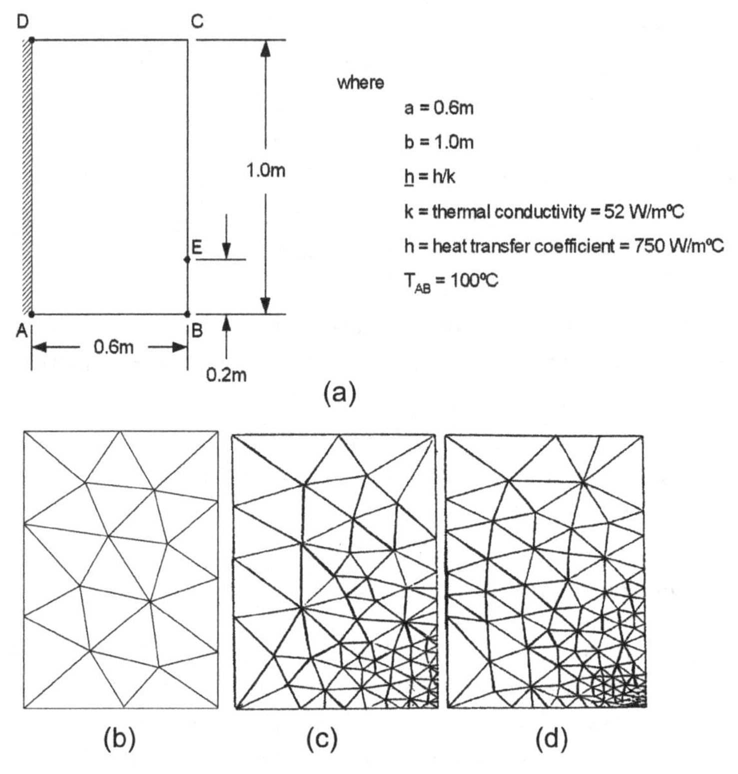

Fig. 7.29 Remeshing example (from Huang and Usmani, 1994).

The calculation procedure continues on the converged mesh until each of the dependent variables converge to a criterion of 10^{-4}. In transient problems, the mesh is adapted as needed to properly capture high gradient features as they evolve in time.

In order to decide which elements to refine or unrefine, an adaptation parameter (A_e) must be defined. There is a great deal of literature indicating possible choices for an adaptation parameter. The two most popular refinement criteria are refinement to minimize error and refinement based on gradients. Both criteria are based upon a key variable which is representative of the solution behavior. Refinement criteria based upon the minimization of maximum errors are generally more complex, and are only as accurate as the method of estimating the error.

The adaptive refinement procedure automatically refines all elements that satisfy a criterion $A_e > R$ and unrefines all elements that satisfy $A_e < U$, where R and U are the refine and unrefine threshold values, respectively. The values of R and U are determined experimentally, based on problem conditions as described by Carrington (2000) and Carrington and Pepper (2002).

The use of quadrilaterals in two dimensions results in midside nodes at the interfaces between the coarse and fine regions of the mesh. In three-dimensions, a face-centered node appears which creates four quadrilateral elements on the face – resulting in eight new hexahedral elements from the original coarse element. These midside nodes are called virtual nodes and require special treatment to obtain a stable, conservative scheme. Figure 7.30 shows a typical interface between a locally fine region and a coarser region. The special treatment used is to set the fluxes and the variable value at node 2 equal to the average of the fluxes and value at nodes 1 and 3 after each iteration.

At the present time, there are many companies selling FEM and CFD related software. Most of the better models include the capability to handle mesh refinement and automatic adaptation. The number of companies will decrease as the smaller companies become absorbed by the larger, more successful firms. The choices are numerous, and sometimes confusing for the buyer interested in obtaining a CFD code.

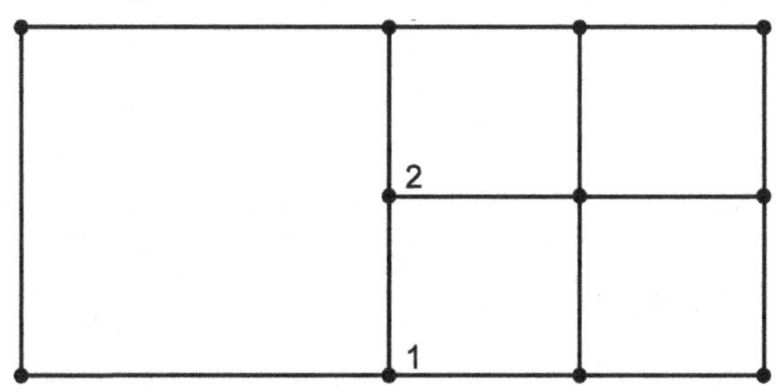

Fig. 7.23 Two-dimensional interface (virtual) node.

We have found the COMSOL code to be quite effective as a general FEM model, reasonably priced, and easy to use. Many of the examples illustrated in this text include COMSOL models that can be accessed on the website.

Example 7.3.9.1. FEM simulation of flow in an office complex: We return once more to the office complex problem, as previously described in Example 7.1.1. This time we will use two FEM models to calculate the flow velocities. In the first model, we will use COMSOL, which is a well-known FEM model that runs on PCs. This is a particularly easy program to use that has a great deal of versatility (Pepper and Wang, 2006), and utilizes h-adaptation, if desired. The second FEM model employs h-p adaptation, and yields exponential convergence (Pepper and Wang, 2007).

Figure 7.31 (a–c) shows the element mesh (using three-node linear triangles), a velocity magnitude contour, and the velocity vectors produced by COMSOL 3.4. The mesh consisted of 274 nodes and 456 elements.

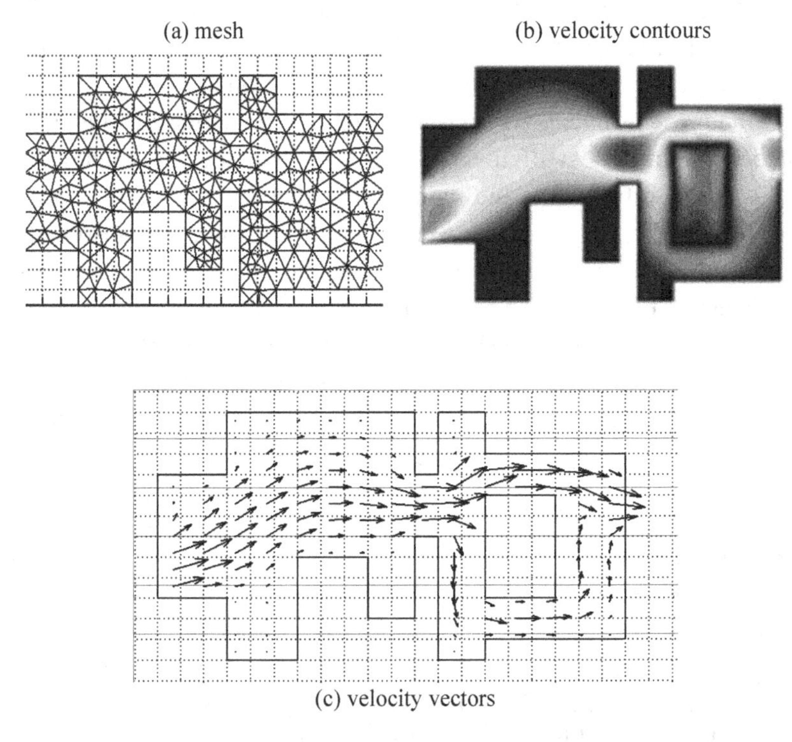

Fig. 7.31 Application of FEM using COMSOL for flow within an office complex.

The second FEM model employs h- and p-adaptation. Together, this combined adaptive technique yields exceptionally accurate results. Although the programming and bookkeeping may be more troublesome, the results are quite good. Figure 7.32 (a–d) shows the initial mesh, which is the same as the original FDM mesh, the final adapted mesh, the velocity vectors, and the flow streamlines. The initial mesh consisted of 213 nodes and 161 elements, and the final mesh contained 736 nodes and 461 elements. The results are clearly comparable with those produced by COMSOL, as well as the FDM and FVM results. This is expected due to the simplicity of the problem geometry. Where the h-p method really shines is when the problem is very complex and there is some question as to where the mesh and calculations should be refined, but accuracy is particularly important.

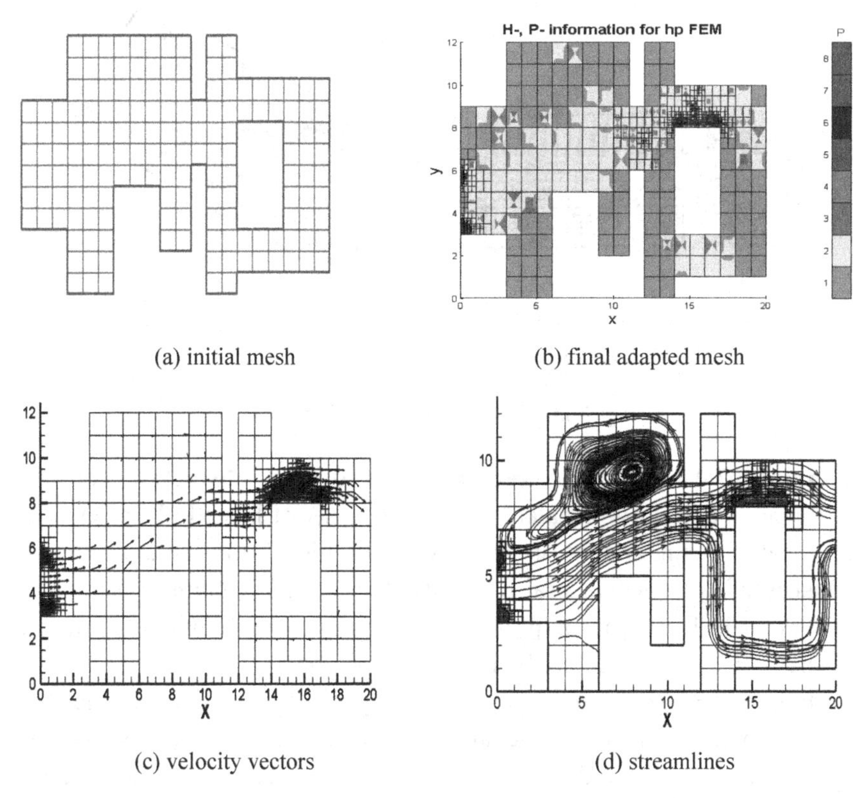

(a) initial mesh

(b) final adapted mesh

(c) velocity vectors

(d) streamlines

Fig. 7.32 FEM h-p adaptation simulation for office complex.

7.4 Further CFD Examples

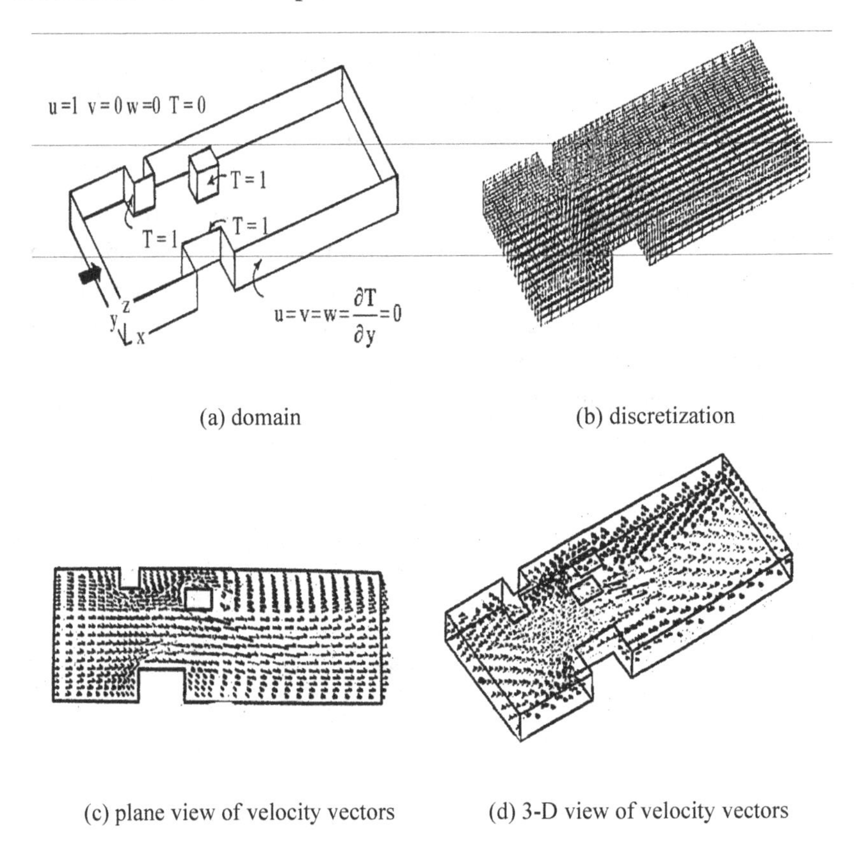

(a) domain (b) discretization

(c) plane view of velocity vectors (d) 3-D view of velocity vectors

Fig. 7.33 Airflow around heated obstacles.

Example 7.4.1 Flow around a set of heated obstacles: For this problem, 3-D airflow is calculated around a set of heated obstacles. The physical domain and mesh are shown in Fig. 7.33 (a,b). The mesh consists of 2868 hexahedral elements; the Reynolds number is $Re=10^3$ and $Pr=1.0$. This type of problem commonly occurs in HVAC where obstructions are encountered within the flow domain.

Figure 7.33 (c,d) gives normal and perspective views of the 3-D velocity vectors within the channel. Recirculation of the flow occurs behind the blocks, and small secondary cells develop in the corners. Thermal plumes emanate from the heated blocks; plume impingement from the left forward block occurs on the small mid-stream block. It is well known that when flow separates at the corners of blocks, horseshoe-like vortices are generated (Hunt *et al.*, 1978).

Example 7.4.2 Air flow over a heated oven within a commercial kitchen: Air enters the kitchen from two ceiling vents (and entrainment from the right open boundary), passes over the heated surface of the oven, and exits through the upper left corner of the exhaust hood (Fig. 7.34 through Fig.7.36). The heated surface acts to enhance the air motion, eventually accelerating the room air out of the domain, and illustrates the ability to model mixed convection (where the flow transitions from motion due strictly to natural convection to strongly forced convection).

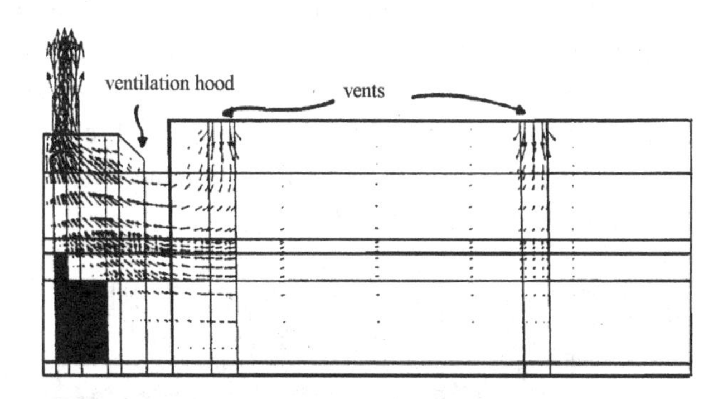

Fig. 7.34 Velocity vectors in side view of kitchen.

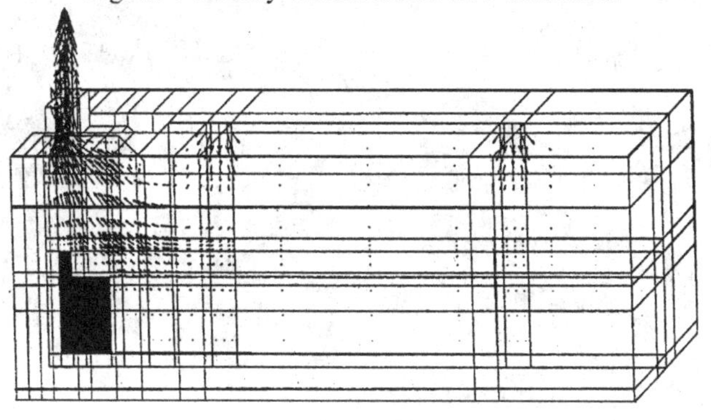

Fig. 7.35 Velocity vectors in 3-D view of kitchen.

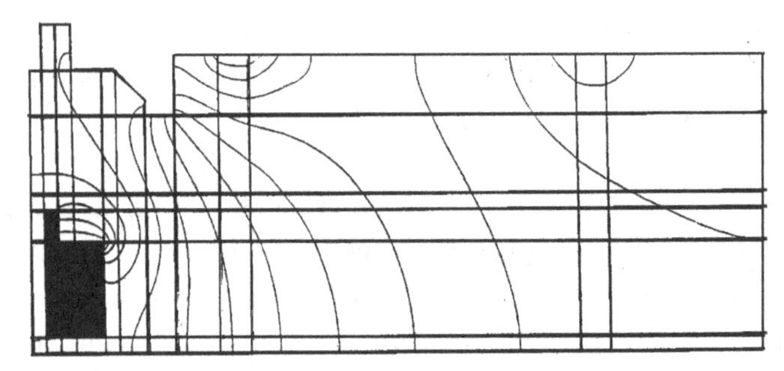

Fig. 7.36 Isotherms over a heated oven in a commercial kitchen.

Example 7.4.3 3-D flow over barriers: This example problem is modeled as a three-dimensional isothermal flow over a set of barriers. Three-dimensional hexahedrals are used to model the domain (Fig. 7.37 and Fig. 7.38). The Petrov–Galerkin technique is used to eliminate numerical oscillations since there is a strong advection component to the problem. In this instance, the finite element method is used to establish the problem domain (Carrington and Pepper, 1998). A three-dimensional simulation of the airflow within the room is first calculated; mid-level velocity streak lines are shown in Fig. 7.38.

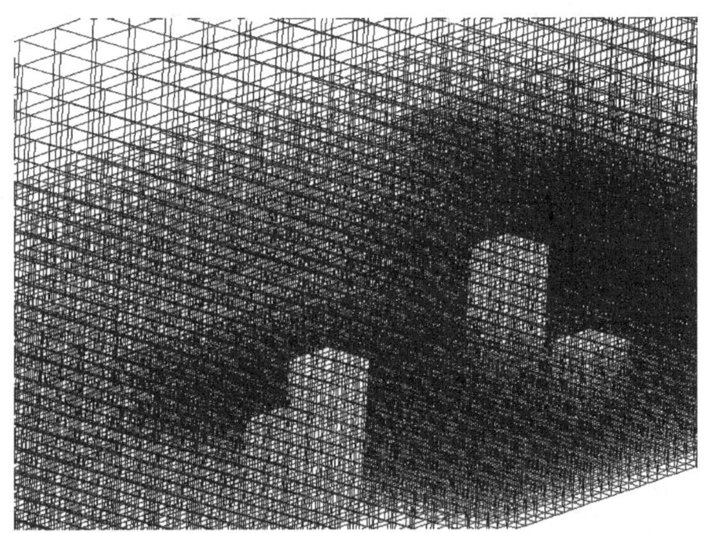

Fig. 7.37 An h-adapted finite element mesh for flow over obstructions.

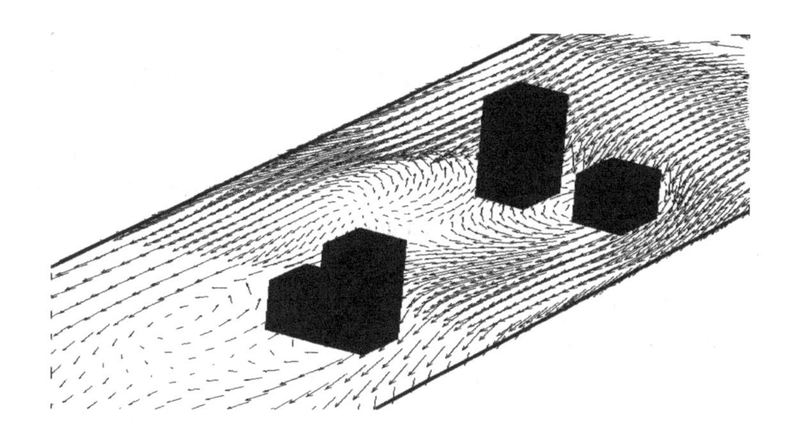

Fig. 7.38 Velocity vectors in horizontal plane of 3-D flow over a set of barriers.

7.5 Model Verification and Validation

Verification and validation are important for any numerical model, especially for commercial CFD codes. According to Roache (1998), the definitions of these two terms are:

Validation: Are the right equations being solved?
Verification: Are the equations being solved correctly?

There are numerous uncertainties that can be attributed to CFD simulations. Obviously, improper use of boundary conditions, using the wrong numerical scheme, or inputting wrong values for various parameters are potential sources of error. While these are fairly obvious and may be easy to find, sometimes more subtle problems occur which can be very difficult to find. There may be bugs in the code itself, or the mesh is just too coarse to yield accurate solutions. You must be sure that you are both solving the correct set of equations and that you have properly discretized the equations using a reliable solver technique. The former error is known as validation and the latter is code validation.

A CFD code should be tested and verified. One must ensure there are no coding errors in discretizing the equations and that the code has been demonstrated to be appropriate for your particular class of problems. In other words, the code is well behaved and produces a reliable solution when compared with experimental data.

Commercial codes are typically verified by the developers within the company. The more widespread use of a commercial code, the better the chances of mistakes being identified and corrected. Needless to say, bugs still tend to exist, even in the oldest and most reliable of software. The process of verification helps to minimize these bugs and reduce their impact to unimportant processes. Examples of well-established commercial CFD codes include: ANSYS (FLUENT), CFX, PHOENICS, STAR–CD, COMSOL, and ANSWER. These codes have been subjected to extensive validation and verification tests over many different applications.

When running a numerical model, the user should check that the results are reasonable. Ideally, it is best to compare predictions with good quality measurements. However, experimental data is generally sparse and may not be available, or even impractical to obtain. An alternative to comparing with experimental data is to check results against well-established correlations for estimating velocities, temperatures, or concentrations. Remember that even a well-documented and validated computer code will likely not match with real world data – there is still a great deal we do not know regarding turbulence and related effects, no matter how good the theory or set of equations. There will always be errors. Many times a good statistical assessment (uncertainty analysis) of the results will help to put the simulations into proper perspective.

Commercial CFD codes are very complex, and can easily exceed over 100,000 lines of code. In fact, PATRAN, which is an old mesh generation package, contained over one million lines of code. Statistically, mistakes are bound to exist, difficult to find, and may be outside the control of the code user. It behooves the use to proceed with care and examine all results in an effort to help pinpoint potential problems.

The most common mistakes are those made by the user, i.e., a wrong value is entered or a wrong key is pressed. Another common source of error, especially when using a commercial CFD code, is that the user does not understand the equations and parameters, or the choice of appropriate submodels, e.g., selection of a turbulence closure scheme. Users do not generally have access to the source code or discretized equations – which would likely be overwhelming and difficult to follow anyway. Code manuals help to identify the equations being solved and the accompanying submodels. However, the user may be confused as to the proper parameters needed for the simulation, i.e., failure to set a parameter will likely result in a default value that may be inappropriate for the problem.

In order to use CFD models reliably, the user should have a good understanding of fluid dynamics, HVAC principles and applications, as well as knowledge of CFD techniques relevant to the application. Experience in using such codes and applying them to air pollution simulation is valuable.

A great deal of work has been done on verifying and validating results obtained with numerical models. Such efforts include comparing results between numerical and analytical models, sensitivity analyses, and seeing how well the numerical model predicts actual results obtained from experimental data. This latter comparison can be fairly tricky if some of the parameters, e.g., exchange coefficients, are not known in the actual experiment. A detailed discussion on model verification and validation can be found in the text by Roache (1998).

Efforts involved in validating and verifying numerical results with experimental data are not trivial – evaluations and comparisons must be carefully considered. Model validation is generally achieved through either field measurements or wind tunnel experiments (e.g. Cermak, 1976).

Assessment techniques include measures of difference, Pearson, Spearman, and Kendall correlations, skewness and kurtosis, tests for normality, and scatter diagrams (Pepper, 1981). Such analyses help to provide insight into the physics of indoor air quality, and enable relations to be constructed to more reliably predict exposures. The incorporation of statistical processes into numerical simulation and assessment is known as stochastic modeling (see Halder and Mahadevan, 2000).

7.6 Comments

The three most popular numerical methods for solving PDEs are the finite difference method, the finite volume method, and the finite element method. There are many "dusty deck" FORTRAN codes still around today based on the FDM and FVM, along with some FEM codes. The development of higher level programming languages including such packages as MAPLE, MATLAB, MATHCAD, MATHEMATICA, C, C++, and JAVA have led to a resurgence as well as an upgrade in algorithms based on these older codes. An understanding of these three fundamental methods should provide a solid basis for development of more advanced numerical models to solve indoor air pollution transport.

Up to several years ago, there were many commercial software companies selling CFD codes. Today, an effort is underway to absorb the smaller companies into the larger companies. For example, ANSYS (an FEM company) absorbed FLUENT (an FVM company), who had previously absorbed FIDAP (an FEM company). The reader need only to do a simple web search to find a plethora of information regarding CFD, and even free software.

CFD is a powerful technique that can provide solutions to the time-dependent, three-dimensional equations for fluid flow and species transport in complex geometries. Under such conditions, simpler analytical and empirical models are just not adequate. CFD has been used extensively to model flows and species transport, e.g., smoke attributed to fires, in modern complex buildings (Hiorns and Sinai, 1999; Mills, 2001; Sinclair, 2001) as well as to examine different generic ventilation configurations (Hadjisophocleous *et al.*, 1999; Klote, 1999).

It is important to remember that solutions obtained by CFD are not exact, i.e., they represent a trend or process occurring during an event. Many assumptions and approximations are usually made during the whole process and some of them can have a significant impact on the results (Gobeau and Zhou, 2003). If the decisions made by the numerical modeler are not based on sound judgment, the results can lead to overestimates of available time for evacuation.

During an emergency evacuation, the reliability of results should not be based solely on the expertise and recommendations of a CFD modeler. Even a CFD expert in fluid flow and species transport may have to make drastic assumptions due to budget and/or time constraints – such as reducing the number of equations to be solved or using a coarse grid. Modeling large complex spaces generally demands large computing resources. During the Kr-85 release from Three Mile Island, a simple analytical model was used to predict the trajectories from the leaking reactor over the first two days (Pepper, 1981). A 3-D CFD related code was eventually used for the following weeks, but required several days to prepare the mesh and set up boundary conditions (Koster and Dickerson, 1990). The point here is that although assumptions might be based on sound judgment – for instance large grid cells away from the source and small cells near the source, they can potentially lead to non-conservative and unrealistic results.

Chapter 8

Numerical Modeling – Advanced Techniques

The most popular numerical methods are the FDM, FVM, and FEM, as we saw in Chapter 7. FDM methods predate the development of computers and the FVM has been around for over 50 years. The FEM, developed in the mid-1950s, became a major tool in the 1970s, and then surged in the late 1980s with the addition of adaptation. During these early years, several other methods began to appear in the literature, but remained fairly invisible since they could only be run on large supercomputers at the time.

The use of Lagrangian particles for modeling dispersion was found to be effective, but computing storage limitations hindered their widespread application (Lange, 1973; Sklarew *et al.*, 1971). Today, PCs are easily capable of running these type codes. During the early 1980s, the Boundary Element Method (BEM) began to appear and began to be noticed as the method reduces the dimension of the problem by one (Brebbia and Dominguez, 1989). Advances in the applications of BEM now make them very attractive, especially for structural analysis. Since the 1990s , the Meshless Method (MM) has been gaining attention and holds promise as the need for a mesh is essentially eliminated (Atluri and Shen, 2002; Pepper, 2005). More recently, efforts have been devoted to models that run at the molecular level, i.e., modeling enough molecules to eventually represent continuum processes (Greenspan, 2005). All of these methods can be used as standalone schemes, or used in conjunction with the three conventional methods as previously discussed.

The methods we will address in this chapter are the Boundary Element Method (BEM), the Lagrangian Particle Transport technique (LPT), the Particle-in-Cell method (PIC), and the Meshless method

(MM). A brief discussion of modeling at the molecular level will be given at the end. There are advantages and disadvantages associated with any numerical scheme. The choice of which scheme to use is typically dictated by the type of problem, and the familiarity with the method by the user. While there are many more methods not discussed, these are currently among the most popular at the present time.

8.1 Boundary Element Method

The boundary element method is a unique numerical scheme which permits rapid and accurate solution of a specific class of equations (Brebbia and Dominguez, 1989). Employing Green's identity, the boundary element method requires only the discretization of the boundary domain – no internal mesh is required as in the finite element method. The BEM reduces the dimensionality of a problem by one, i.e., a two-dimensional problem reduces to a line integral; a three-dimensional problem reduces to a two-dimensional surface formulation. Hence, input data processing consists only of the problem boundary geometry and boundary conditions. Figure 8.1 shows an irregular domain discretized with triangular finite elements and using BEM, with three internal points (if the user wishes to also calculate values at several internal points). The BEM is used in many applications where Laplace or Poisson equations are solved; more recent advances now enable the method to be used for transport equations (Atluri and Shen, 2002).

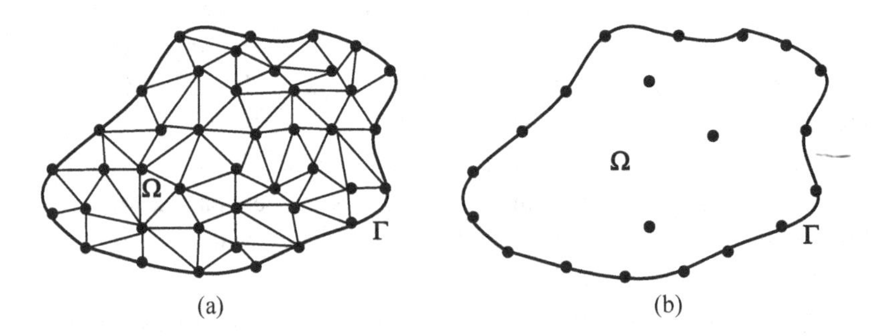

Fig. 8.1 Arbitrary 2-D domain discretized using a) FEM and b) BEM.

To illustrate the method, we begin with the simple operator, L, defined as

$$L\phi = \frac{d^2\varphi}{dx^2} = f(\varphi, x),\ L \equiv \frac{d^2}{dx^2}. \tag{8.1}$$

As we did in the FEM, we will now integrate Eq. 8.1 to create a weak statement. In fact, we will do this twice to create the adjoint of L, which we will denote as L*. The BEM utilizes the adjoint operating on the weight, W, as opposed to the basis function, N, which we saw in the FEM. Performing the integration, we obtain

1st integration – weak statement

$$I = W\frac{d\varphi}{dx}_{\Gamma} - \int_{\Omega}\frac{d\varphi}{dx}\frac{dW}{dx}dx. \tag{8.2}$$

2nd integration

$$I = W\frac{d\varphi}{dx}_{\Gamma} - \varphi\frac{dW}{dx}_{\Gamma} + \int_{\Omega}\varphi\frac{d^2W}{dx^2}dx. \tag{8.3}$$

The adjoint is L* = L (in this case, it is self adjoint), i.e.,

$$L^* = L = \frac{d^2}{dx^2}. \tag{8.4}$$

Example 8.1.1 Determining the adjoint operator: Find the adjoint of the following operator,

$$\mathrm{L} = \frac{\mathrm{d}^2}{\mathrm{dx}^2} - \mathrm{u}\frac{\mathrm{d}}{\mathrm{dx}}. \tag{8.5}$$

1st integration – weak statement

$$\mathrm{I} = \mathrm{W}\frac{\mathrm{d}\varphi}{\mathrm{dx}}|_{\Gamma} - \int_{\Omega}\mathrm{uW}\frac{\mathrm{d}\varphi}{\mathrm{dx}}\mathrm{dx} - \int_{\Omega}\frac{\mathrm{d}\varphi}{\mathrm{dx}}\frac{\mathrm{dW}}{\mathrm{dx}}\mathrm{dx}, \tag{8.6}$$

2nd integration

$$I = W\frac{d\varphi}{dx}_{\Gamma} - \left[\left(\frac{dW}{dx} + uW\right)\varphi\right]_{\Gamma} + \int_{\Omega}\varphi\left(\frac{d^2W}{dx^2} + u\frac{dW}{dx}\right)dx. \tag{8.7}$$

The adjoint is

$$L^* = \frac{d^2}{dx^2} + u\frac{d}{dx}, \tag{8.8}$$

which is not self-adjoint. Notice that two boundary conditions automatically appear in Eq. 8.7, as denoted by the Γ subscript.

Example 8.1.2 BEM solution of a second-order ODE Solve the equation below for the derivative of φ with respect to x at the end points of a domain defined between $a \leq x \leq b$. The equation is

$$\frac{d^2\varphi}{dx^2} - f(\varphi, x) = 0, \tag{8.9}$$

with boundary conditions φ = 0 at x = 0, 1 where a = 0, b = 1. We wish to determine $d\varphi/dx|_a$ and $\varphi|_a$ and $d\varphi/dx|_b$ and $\varphi|_b$. We integrate twice to obtain

$$\int_\Omega W\left[L\varphi - f(\varphi, x)\right]dx = 0, \tag{8.10}$$

or

$$W\frac{d\varphi}{dx}\Big|_a^b - \frac{dW}{dx}\varphi\Big|_{a\Gamma}^b + \int_a^b \varphi\frac{d^2W}{dx^2}dx = \int_a^b Wf(\varphi, x)dx. \tag{8.11}$$

We will now solve Eq. 8.11 by obtaining a homogenous solution and a particular, or fundamental, solution of the equation.

Homogeneous solution:

We now choose W so that L*W = 0. Integrating the second derivative term in Eq. 8.11,

$$W\frac{d\varphi}{dx}\Big|_a^b - \frac{dW}{dx}\varphi\Big|_{a\Gamma}^b = \int_a^b Wf(\varphi, x)dx, \tag{8.12}$$

Fundamental solution:

For the fundamental solution, we set L*W equal to a Kronecker delta function evaluated between the limits of x (defined as ξ), i.e.,

$$L^*W = -\delta(x - \xi), \tag{8.13}$$

Homogeneous solution

For homogeneous solution, we see that

$$\frac{d^2W}{dx^2} = 0, \tag{8.14}$$

We now need two independent solutions. Let $W_{1_1} = x$ and $W_2 = 1$.

a. for $W_1 = x$

$$W\frac{d\varphi}{dx}\Big|_a^b - \frac{dW}{dx}\varphi\Big|_{a\Gamma}^b = \int_a^b Wf(\varphi, x)dx, \tag{8.15}$$

or

$$b\frac{d\varphi}{dx}\Big|_b - a\frac{d\varphi}{dx}\Big|_a - \varphi\Big|_b + \varphi\Big|_a = \int_a^b xf(\varphi, x)dx. \tag{8.16}$$

b. for $W_2 = 1$

$$\frac{d\varphi}{dx}\Big|_b - \frac{d\varphi}{dx}\Big|_a = \int_a^b xf(\varphi, x)dx. \tag{8.17}$$

Imposing the boundary conditions, we wish to find

$$\begin{aligned} \varphi\big|_a = 0, \quad \text{find } \frac{d\varphi}{dx}\Big|_a = ? \\ \varphi\big|_b = 0, \quad \text{find } \frac{d\varphi}{dx}\Big|_b = ?. \end{aligned} \tag{8.18}$$

a. for $W_1 = x$

$$\begin{aligned} b\frac{d\varphi}{dx}\Big|_b - a\frac{d\varphi}{dx}\Big|_a - \varphi\Big|_b + \varphi\Big|_a = \int_a^b xf(\varphi, x)dx \\ (1)\frac{d\varphi}{dx}\Big|_b - (0)\frac{d\varphi}{dx}\Big|_a - 0 + 0 = \int_a^b x \bullet x dx, \end{aligned} \tag{8.19}$$

$$\therefore \frac{d\varphi}{dx}\Big|_b = \frac{1}{3}.$$

b. for W2 = 1:

$$\frac{d\varphi}{dx}|_b - \frac{d\varphi}{dx}|_a = \int_a^b x dx = \frac{1}{2}. \quad (8.20)$$

$$\therefore \frac{d\varphi}{dx}|_a = -\frac{1}{6}.$$

You have now solved the problem using a BEM approach.

We can readily solve a two-dimensional problem using the same approach. For example, assume that we wish to find the adjoint for the simple Laplace equation for temperature, i.e.,

$$\nabla^2 T = 0. \quad (8.21)$$

We multiply Eq. 8.21 by a weight, W(x,y), and integrate,

$$I = \int_\Omega W(x,y)\left[\frac{\partial^2 T}{\partial x^2} + \frac{\partial^2 T}{\partial y^2}\right] dxdy. \quad (8.22)$$

Integrating twice, we obtain

$$\begin{aligned} I = &\int_{y\,\min}^{y\,\max}\left[W\frac{\partial T}{\partial x} - \frac{\partial W}{\partial x}T\right]_{x_1}^{x_2} dy + \int_\Omega T\frac{d^2W}{dx^2}dx + \\ &\int_{x\,\min}^{x\,\max}\left[W\frac{\partial T}{\partial y} - \frac{\partial W}{\partial y}T\right]_{y_1}^{y_2} dx + \int_\Omega T\frac{d^2W}{dy^2}dy. \end{aligned} \quad (8.23)$$

Since $dy = n_x d\Gamma$ and $dx = n_y d\Gamma$ ($\hat{n}$ is usually defined as the vector normal to a surface), Eq. 8.23 can be written as

$$I = \int_\Gamma\left[W\frac{\partial T}{\partial \hat{n}} - T\frac{\partial W}{\partial \hat{n}}\right]d\Gamma + \int_\Omega T\left[\frac{\partial^2 W}{\partial x^2} + \frac{\partial^2 W}{\partial y^2}\right]d\Omega, \quad (8.24)$$

or

$$\int_\Gamma\left[W\frac{\partial T}{\partial \hat{n}} - T\frac{\partial W}{\partial \hat{n}}\right]d\Gamma = \int_\Omega T\left[W\nabla^2 T - T\nabla^2 W\right]d\Omega. \quad (8.25)$$

Let's return for a moment to the fundamental equation for the 2-D Laplace equation, i.e., we wish to solve an equation of the form

$$\frac{\partial^2 \varphi}{\partial x^2} + \frac{\partial^2 \varphi}{\partial y^2} = -\delta(x - \xi_1)\delta(y - \xi_2), \tag{8.26}$$

which represents a concentration located at a point whose coordinates are ξ_1 and ξ_2. The concentration diffuses outwards radially and symmetrically in all directions, assuming a universally unit diffusion coefficient, as shown in Fig. 8.2.

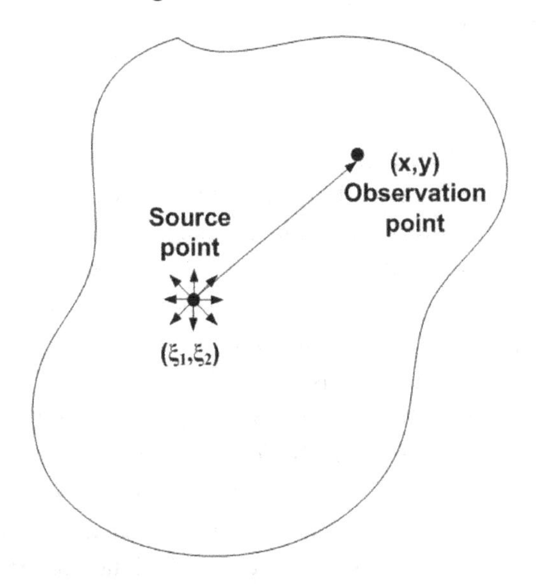

Fig. 8.2 Dispersion within a 2-D domain with source at ξ_1, ξ_2.

It is convenient to work with polar coordinates. Equation 8.26 can be stated in polar coordinates as

$$\frac{1}{r}\frac{\partial}{\partial r}\left(r\frac{\partial \varphi}{\partial r}\right) = -\delta(r), \tag{8.27}$$

which is the axisymmetric form of the Laplacian operator. The solution to this equation is simply

$$\varphi(x, y, \xi_1, \xi_2) = -\frac{1}{2\pi}\ln(r), \tag{8.28}$$

where r is just the distance from the observation point (x,y) in Fig. 8.2 to the source point (ξ_1,ξ_2), defined as

$$r = \sqrt{(x - \xi_1)^2 + (y - \xi_2)^2}\,. \tag{8.29}$$

If the differential equation is 3-D, the axisymmetric Laplacian in spherical coordinates becomes

$$\frac{1}{r^2}\frac{\partial}{\partial r}(r^2 \frac{\partial \varphi}{\partial r}) = -\delta(r)\,, \tag{8.30}$$

and the fundamental solution is

$$\varphi(x, y, z, \xi_1, \xi_2, \xi_3) = \frac{1}{4\pi r}\,. \tag{8.31}$$

The selection of elements along the boundary of the problem domain is left to the user. In 2-D domains, the choices are typically either constant elements or linear elements. In a constant element, the value of φ and $\partial\varphi/\partial n$ are assumed to be constants over each element and the values determined at the mid-points of each element, which are referred to as nodes. In a linear element, the variable and its gradient are assumed to be a linear function of the distance measured along the element. The nodes are now located at the end points of the element (e.g., two end points in a 1-D linear element). One can even use higher order elements if desired. The advantage of using constant elements is reduced storage requirements, but at the expense of reduced accuracy compared with linear elements. Also, the use of linear and higher order conventional elements can be troublesome when dealing with corners.

The governing equation for the advection–diffusion of a scalar potential, φ, can be written as

$$L[\varphi] \equiv \frac{\partial \varphi}{\partial t} + \nabla \cdot (-k\nabla\varphi) + (V \cdot \nabla)\varphi - S\,, \tag{8.32}$$

where V is the velocity vector, k is the diffusion tensor, t is time, and S denotes the source density. Assuming steady state, the steady governing operator $L[\varphi]$ and its adjoint operator $L^*[\psi]$, in which we define ψ as the adjoint potential associated with φ to Green's second identity, can be written as

$$\int_\Omega (L[\varphi]\psi - L^*[\psi]\varphi)d\Omega = \int_\Gamma k(\varphi\frac{\partial \psi}{\partial \hat{n}} - \frac{\partial \varphi}{\partial \hat{n}}\psi)d\Gamma + \int_\Gamma V_n\phi\psi d\Gamma, \tag{8.33}$$

where $\hat{n}$ is the outward normal to φ, and V_n is the normal component of V to Γ. If one introduces the fundamental solution ψ^* of $L^*[\psi] = 0$ instead of φ, Eq. 8.27 can be rewritten as

$$c_i\varphi(r_i) - \int_\Gamma q_n^*\varphi d\Gamma = -\int_\Gamma \psi^* q_n d\Gamma + \int_\Omega S\psi^*, \quad (8.34)$$

where c_i denotes a coefficient that depends on the position vector r_i, $q_n^* = n(-k\nabla\psi^* - V\psi^*)$, $q_n = n(-k\nabla\phi)$ and ψ^* is

$$\psi^*(r;r_i) = \exp\{-(V \bullet r' + |V||r'|)/(2K)\}/(4\pi K|r'|), \quad (8.35)$$

in which $r' = r - r_i$, r is the observation point, and $K_o[\ ]$ is the modified Bessel function of the second kind of order zero. The matrix equivalent form of Eq. 8.35 is

$$[H]\{\Phi\} = [G]\{q\} + \{B\}, \quad (8.36)$$

where [H] and [G] are banded sparse matrices, Φ, q, and B are vectors composed of nodal potentials φ, centroidal q_n and discretized domain integrals, respectively.

Example 8.1.3 BEM calculation of dispersion from a continuous source: A continuous source is located on the floor within a rectangular domain that has a vent located in the ceiling. Flow enters from the left hand side of the room. When the door is opened, the plume bends towards the door. In this problem, a simple 2-D flow is calculated that shows a plume (depicted by particles) exiting through the ceiling vent. When the door is opened, the plume direction is altered as there is more of the room air exhausting through the larger opening on the right-hand side of the domain. A 2-D BEM was used to calculate this dispersion pattern, and was run on a PC (Carrington and Pepper, 1999).

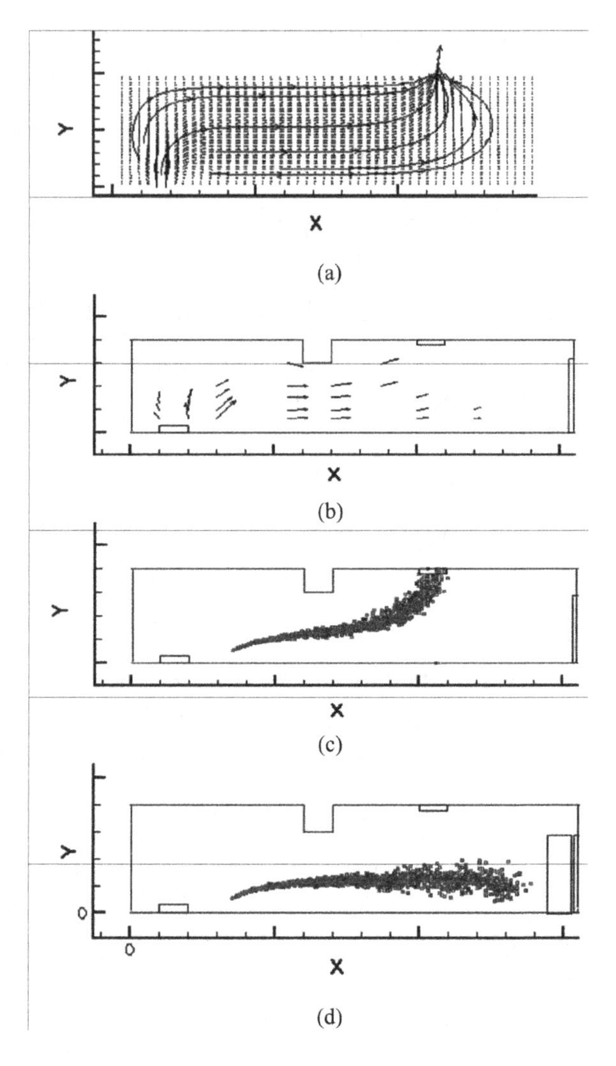

Fig. 8.3 Dispersion within a 2-D room with door closed and opened, (a) typical nodal point distribution in a conventional approach including streamlines, (b) velocity vectors, (c) plume trajectory with door closed, and (d) plume trajectory with door open.

Example 8.1.4 Dispersion within a 2-D Room with Barriers: Flow and dispersion within a room with barriers is shown in Fig. 8.4 and Fig. 8.5(a–c). Figure 8.4 shows the computational domain and grids – in this instance, we need to use only the nodes located along the boundary of the problem domain (denoted by the dark dots). The internal grid is included to show what the mesh would look like if using an FDM, FVM, or FEM model. If one were to employ one of the conventional numerical techniques, the solution matrix would consist of the total number of nodes used to discretize the problem domain.

The velocity vectors are all interior node points (using the BEM to produce values for the internal nodes) is shown in Fig. 8.5(a). Assuming a source is located in the upper right-hand corner, a Lagrangian particle plot is shown in Fig. 8.5(b). Figure 8.5(c) shows contour lines for the dispersion pattern using the BEM model versus an analytical puff model, depicted by the dashed lines.

Clearly there is a great deal of difference in the two solutions, showing the improved accuracy and prediction attributed to using a numerical model versus a simple analytical solution. Much of the improved accuracy is due to the ability of the numerical model to simulate more of the physics of the flow field, i.e., better skill at calculating velocities.

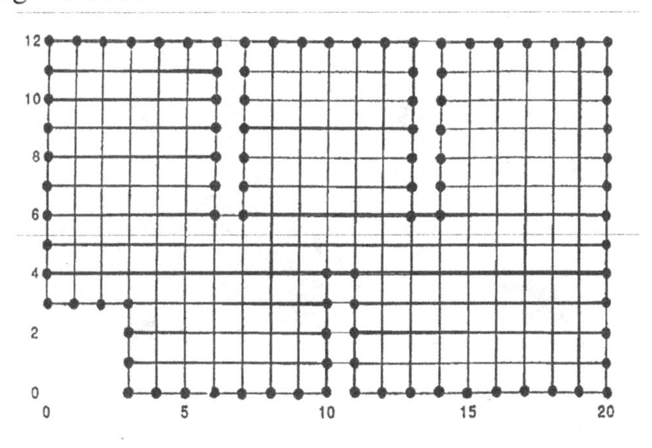

Fig. 8.4 Computational Domain for flow and dispersion within a room with barriers.

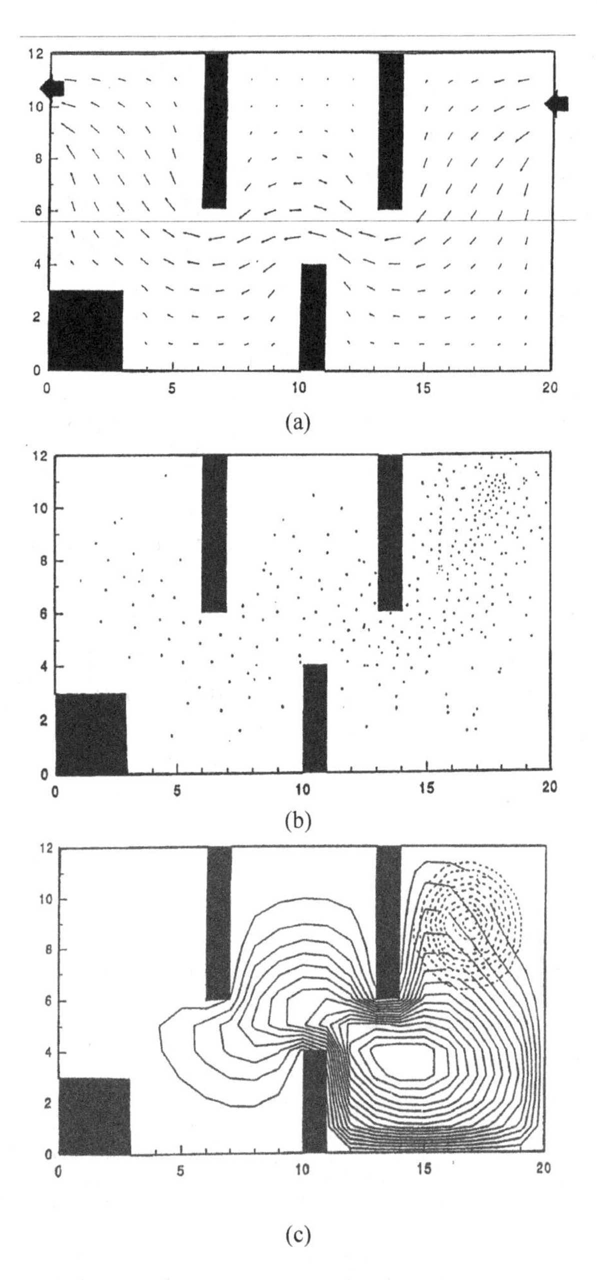

Fig. 8.5 Dispersion within a 2-D room with barriers.

More detailed discussions and implementation of the BEM can be found in many of the textbooks available in the literature, e.g., Ramachandran (1994); Pozrikidis (2002); Kane (1994); Beer and Watson (1992); Beer (2001); Wrobel (2002). A search on the web will list many such references and even computer codes that can be downloaded.

8.2 Lagrangian Particle Technique

Particle positions are calculated to simulate mass transport from both advection and diffusion. The transport equation can be written in the form

$$\frac{\partial C}{\partial t}+\frac{\partial U_i C}{\partial x_i}=0, \tag{8.37}$$

where the velocity vector U_i is expressed in terms of advection and "flux" diffusion as (Runchal, 1980)

$$U_i=\hat{U}_i+U_{f_i}, \tag{8.38}$$

with U_i being the true advection velocity vector and the "flux" velocity defined as

$$U_{f_i}=-\Sigma_j\frac{K_{ij}}{C}\frac{\partial C}{\partial x_j}. \tag{8.39}$$

By combining the advection and diffusion terms together, a total equivalent transport velocity can be obtained. The form of the transport equation becomes identical to the equation of continuity for a general compressible fluid. The original problem of turbulent diffusion is transformed into one describing the advective changes of fluid density in a compressible fluid moving in a velocity field of total equivalent transport velocities. Mass particles are synonymous with density and follow the fluid motion in the velocity field, i.e., they are Lagrangian particles in a non-solenoidal field of total equivalent transport velocity. Their number in any location (volume) determines the concentration of pollutant for the original diffusion problem.

The probability distribution function for a three-dimensional space is (Runchal, 1980)

$$P_{x_i}(x_i, t) = \frac{1}{(4\pi t)^{\frac{3}{4}}(K_1K_2K_3)^{\frac{1}{2}}} \exp\{-\sum_{i=1}^{3} \frac{(x_i - U_i t)^2}{4K_{it}}\}, \quad (8.40)$$

where x_i are the position vectors in the direction of the principal axes and $K_1K_2K_3$ are the diagonal components of the second-order dispersion tensor in the direction of the principal axes.

The transport equation for this distribution can be written as ($P=P(x_i,t)$)

$$\frac{\partial P}{\partial t} + \frac{\partial}{\partial x_i}(U_i P) = \frac{\partial}{\partial x_i}(K_{ij}\frac{\partial P}{\partial x_j}), \quad (8.41)$$

where the tensor summation convention has been employed and K_{ij} is a second-order dispersion tensor. The inclusion of particle decay, settling, and more complex dispersion processes involving specified turbulence correlations, can be included in Eq. 8.41.

The problem of transport of particles by advection and dispersion commonly represented by a deterministic transport equation such as Eq. 8.41 can also be represented simply as a series of random walks. Each of these random walks is composed of a deterministic advection component and a random component.

For example, the increment in the position vector of a particle at any time t can be written as

$$x_t - x_o = \int_{t_o}^{t} U(x_{t'}, t')d't + \int_{t_o}^{t} D(x_{t'}, t')dw_{t'}, \quad (8.42)$$

where D is a deterministic forcing function for the random component of motion. Equation 7.52 can be expressed simply as

$$\delta x(w,t) = \delta x_U + \delta x_D, \quad (8.43)$$

with

$$\delta x_D(w,t) = \int_{t_o}^{t} n_r \sqrt{2K} d't, \quad (8.44)$$

where D is assumed equivalent to K and n_r is a normally distributed random number with a mean value of zero, and a standard deviation of unity. The integral Eq. 8.44 can be further simplified to

$$\delta X_D = n_r \sigma$$
$$\sigma^2 = \int_{t_0}^{t} 2K d't. \tag{8.45}$$

The variance obtained from Eq. 8.44 is the same as that from Eq. 8.45. Thus, Eq. 8.42 can be written as

$$x_t - x_o = \int_{t_0}^{t} U(x_{t'}, t')d't + n_r\{\int_{t_0}^{t}(2K(x_{t'}, t')d't\}^{1/2}. \tag{8.46}$$

For a rigorous application of the random walk method, the net particle displacement must be calculated by integration of Eq. 8.42. However, with U and K as arbitrary functions of space and time, it is not always possible to obtain a closed form solution. It is generally sufficient to assume that the mean velocity and random components can be separately calculated and linearly superimposed.

For steady or quasi-steady flows, the time scale of particle motion is much smaller than the characteristic time scale of change in the mean velocity and the dispersion fields. In such a case, it is often more convenient to express U and K as functions of the position vector x_i, rather than as Lagrangian functions of time.

In the application of the random walk model, the particle displacement in each of the coordinate directions is independently calculated from the displacement algorithm, Eq. 8.46. Before this is performed, however, the mean velocity, U, and the dispersion due to turbulence or other stochastic mechanisms must be specified. The velocity of any particle is obtained from the application of the BEM, which can be used to obtain velocity components anywhere within the problem domain without the need for a nodal mesh or interpolation. A general probability distribution or correlation function for the random component of motion due to dispersion is utilized to account for the dispersivity tensor, K.

The calculation to advance the particle configuration in time proceeds in steps, or cycles, each of which calculates the desired quantities for time $t + \Delta t$ in terms of those at time t. Hence,

$$x_i(t + \Delta t) = x_i(t) + U_i \Delta t. \tag{8.47}$$

The velocity components are the fictitious total velocities determined for the beginning of the time interval and initial particle positions. Every particle is advanced each cycle to a new position using Eq. 8.47. Thus, the particle traces out in time a trajectory for the pollutant mass. Boundary conditions are introduced by modifications of the fictitious total velocities. Solid boundaries are simulated by not allowing particles to be transported across the boundaries. In each cycle, the fictitious total velocity for each cell is calculated as the sum of the advection velocity and the random turbulent flux velocity. The particle positions are updated using an interpolated total velocity. The concentration per unit volume is calculated from the particle masses.

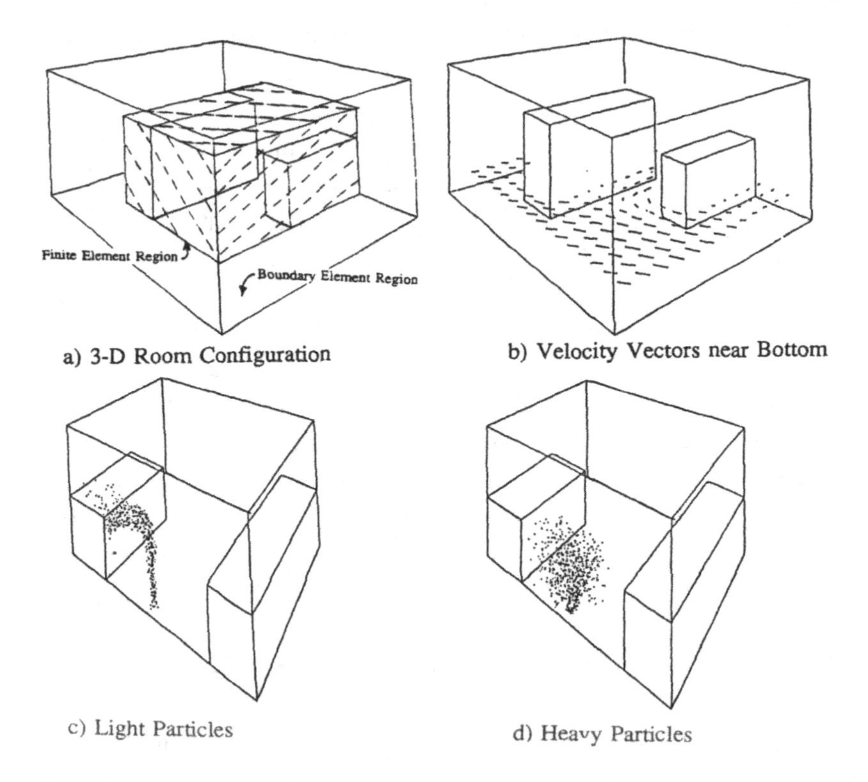

Fig. 8.6 Lagrangian particle transport within a room.

The next method we will examine utilizes the transport of particles within an Eulerian frame of reference, i.e., the mesh. One of the advantages of using particles is their inherent ability to visually display the spread of pollutant, allowing one to quickly grasp the effect of dispersion.

8.3 Particle-in-cell

The Particle-in-Cell (PIC) method is based on the use of Lagrangian particles representing pollutant concentration in an Eulerian mesh. The advection–diffusion equation is solved using a pseudo-velocity technique. The velocities within the problem domain are calculated prior to employing the PIC procedure for concentration. There is also a PIC method (known as the Marker and Cell – MAC – method) for calculating velocities that stems from the work at Los Alamos National Laboratory (LANL) by Welsh *et al.* (1965), Amsden (1966), and Harlow and Amsden (1970). In the MAC method, particles are used as Lagrangian markers to delineate fluid boundaries or free surfaces. Much later, Fogelson (1992) employed a particle method for advection–diffusion equations. However, we will not go into this method for fluid flow but will assume that we can obtain velocities using any number of numerical schemes. The interested reader is referred to the work and numerous articles from LANL.

The application of PIC for species transport comes from the early work by Sklarew *et al.* (1971), and then quickly applied by Lange to develop the ADPIC model (Atmospheric Diffusion Particle-in-Cell) developed for emergency response at Lawrence Livermore National Laboratory (1973). Application of the technique by NASA for pollutant transport was also reported by Spaulding (1976).

The transient advection-diffusion equation for species transport is solved in its flux conservative form using a pseudo-velocity technique. Pollutant concentration is statistically represented by imbedding Lagrangian marker particles in an Eulerian grid. The transport equation can be written in vector form as

$$\frac{\partial C}{\partial t} + V \bullet \nabla C = \nabla \bullet (K \nabla C), \tag{8.48}$$

where C is the concentration, K is the diffusion coefficient, and V is the velocity vector. Using the assumption of incompressibility, we can replace the advection term

$$V \bullet \nabla C,$$

with the expression

$$\nabla \bullet (VC).$$

Combining the advection and diffusion terms into their flux conservative form, Eq. 8.48 becomes

$$\frac{\partial C}{\partial t} + \nabla \bullet \left(C\left(V - \frac{K}{C}\nabla C \right) \right) = \frac{\partial C}{\partial t} + \nabla \bullet \left(CV_P \right) = 0, \qquad (8.49)$$

where $V_P = V - KDC/C$, which is the pseudo transport velocity.

The mesh consists of an Eulerian grid similar to a FDM or FVM mesh. The concentrations are defined at the centers of the cells and the velocities V, V_P, and $-K\nabla C/C$ are defined at the cell corners. The particle locations are defined by their individual Lagrangian coordinates within the mesh structure. A two-step procedure is used to calculate the pollutant transport:

1. Eulerian step:

The concentration, C, is obtained (or defined initially) for each cell; a temporary velocity is calculated based on the gradient of the concentration, i.e.,

$$V_D = -\frac{K}{C}\nabla C,$$

where V_D denotes the diffusion velocity. This velocity is added to the advection velocity to yield V_P, i.e.,

$$V_P = V + V_D.$$

2. Lagrangian step:

Each marker particle is advanced one time step, Dt, with the velocity VP, which is computed from the values at the corners of the cell. A volume weighting scheme is used to determine the particle coordinates, which are calculated using the expression

$$x^{n+1} = x^{n} + V_{P}\Delta t, \tag{8.50}$$

where n + 1 denotes the new value and n is the old value (or location).

A new concentration distribution is then calculated from the new particle positions. The technique is elegantly simple, and eliminates the artificial diffusion inherent in a typical Eulerian scheme. However, the truncation errors are still the same as one would expect in the FDM or FVM.

The basic algorithm on a non-staggered mesh requires three cells, or control volumes, as we saw in the FVM. Figure 8.7 shows transport in the horizontal direction (x) with velocity u (where $\mathbf{V} = u_i + v_j + w_k$).

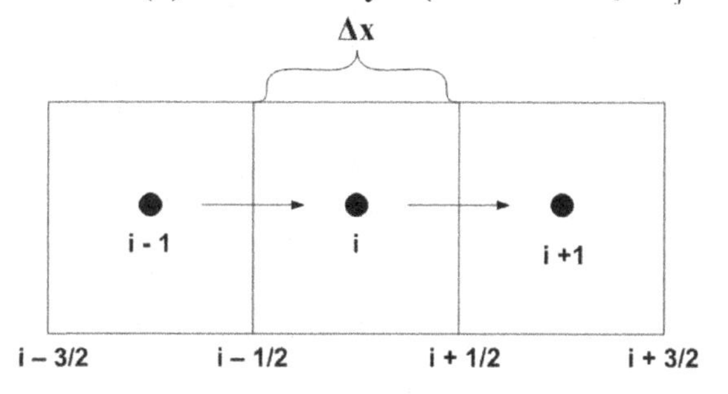

Fig. 8.7 Three-cell PIC mesh.

The horizontal velocity at node, i, is defined as

$$u_i = -\frac{K}{2\Delta x}\frac{C_{i+1} - C_{i-1}}{C_i}. \tag{8.51}$$

Figure 8.8 shows a two-dimensional mesh with the velocity and cell structure. The pseudo-velocities are defined at the cell corners. In this instance, Fig. 8.8 shows the u component of the pseudo-velocity vector.

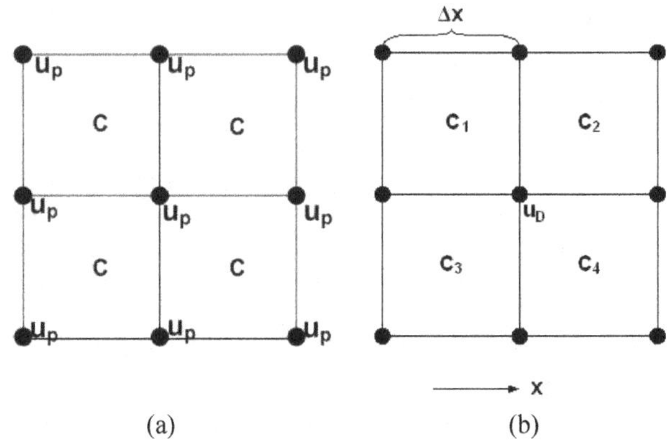

Fig. 8.8 Two-dimensional PIC mesh.

The horizontal velocity u_D is calculated as

$$u_D = -\frac{K_x}{C}\frac{\Delta C}{\Delta x}$$
$$u_D = -2\frac{K_x}{\Delta x}\frac{(C_1 + C_3 - C_2 - C_4)}{(C_1 + C_2 + C_3 + C_4)}, \quad 0 \le u_D \le |2\frac{K_x}{\Delta x}|, \tag{8.52}$$

where the concentration has been averaged over the cells surrounding node i.

Since the particles represent the concentration, it is important to have as many particles as possible. This is principally dictated by the storage limit of the computer. Each particle contributes some fractional component to the mass within a region typically encompassing the eight neighboring cells, i.e., an overlap of its volume with the cell volumes of the neighboring cells. This approach tends to smooth the distribution of concentration. It has been shown that even one particle per cell can yield meaningful results (Lange, 1973).

Concentrations in the PIC method can be obtained using either of two techniques. In the first technique, the cell concentration is just the sum of the masses of all the particles within a cell divided by the cell volume. The second method is equivalent to using an area (or volume) apportionment of a particle's mass among cells utilizing an overlap. The particle mass is considered to be uniformaly spread over an area (or volume) the size of the cell and centered on the particle position. The overlap with adjacent cells is then used to determine the amount of mass apportionment among the cells. The first technique is simple and faster; the second area-averaging technique has the effect of smoothing any artificial gradients caused by relegating the concentration mass into particles. The cell concentrations are associated with the cell centers in both techniques.

A random number generator is used to generate the input particles, which can be input as a Gaussian distribution, constant value, or other choice left to the modeler. For a continuous source (plume), a few hundred particles per time step is generally sufficient.

The basic boundary conditions imposed on the pseudo-velocity field are either a constant mass flux (CV = constant), which corresponds to inflow and outflow of particles, or zero mass flux (CV = 0), which accounts for the reflection of particles from a boundary. Deposition of particles on a surface can also be specified. If a particle leaves the computational domain, it is either eliminated or counted as a deposition or reflection, depending upon the type of boundary.

An example of a Gaussian symmetrical puff using 3896 particles is shown in Fig. 8.9. The particle distribution is shown at t = 0; notice the four cells outlined in the center of the figure.

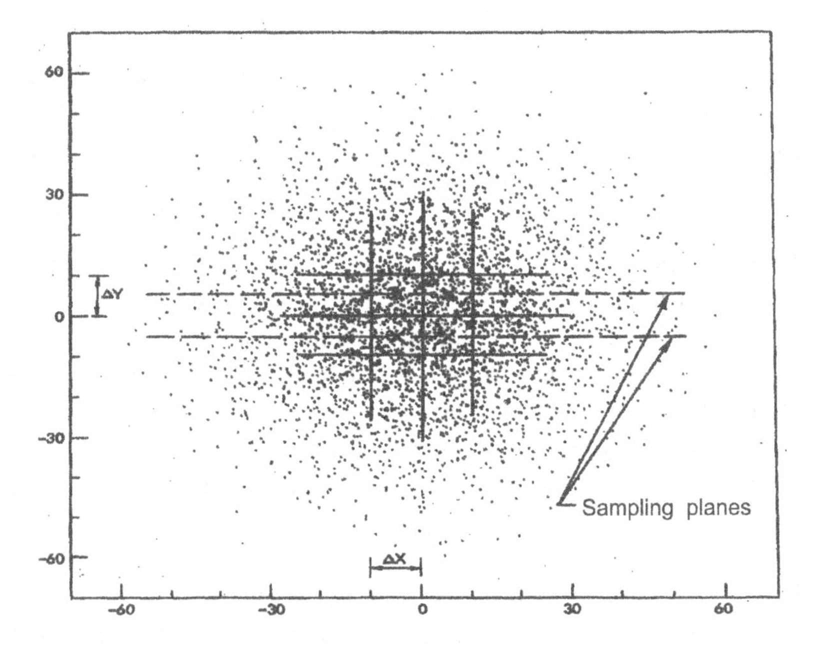

Fig. 8.9 Gaussian puff distribution for PIC (from Lange, 1973).

Assuming a grid spacing of $\Delta x = \Delta y = \Delta z = 10$ m with diffusion coefficients of $K_{xx} = K_{yy} = 1 \times 10^4$ cm^2/s and $K_{zz} = 10$ cm^2/s, the particle distribution after 286,300 s is shown in Fig. 8.10. The grid was allowed to expand with the final cell size reaching $\Delta x = \Delta y = 394$ m and $\Delta z =$ 16.9 m. Simulation results were nearly identical with values obtained from the analytical solution for a Gaussian puff.

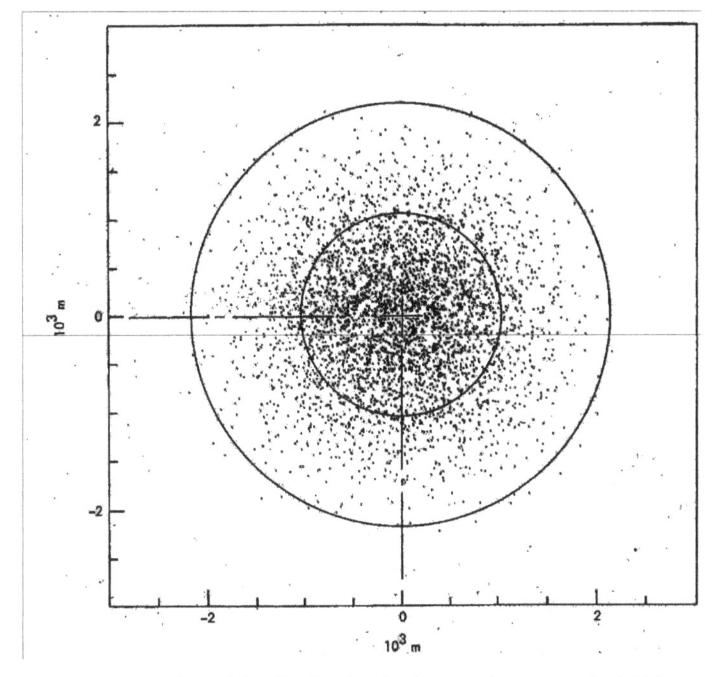

Fig. 8.10 Final particle distribution in the x-y plane at t = 286,300 sec.

An example showing the advection and diffusion of a plume is shown in Figs 8.11 (a,b). The advection velocity is 10 m/s with $K_{xx} = K_{yy} = K_{zz} = 10^7$ cm^2/s. The size of the source is 20 m with $\Delta x = \Delta y = \Delta z = 1000$ m. Figure 8.11b shows the dispersion of particles after 1832 s. Results are within ± 5% of the analytical solution. In this example, the advection term is clearly more dominant than the diffusion terms, and the plume stretches towards the right boundary.

The velocities are the fictitious total velocities which are calculated at the beginning of a time interval and interpolated to initial particle positions. These values are held constant throughout a time step. One must be careful that a particle does not pass through many cells in a time step and out of the computational domain, resulting in large inaccuracies or even instability. An empirical rule used to avoid this problem has been to limit the time step so that a particle does not travel more than 0.4 of a cell in a time step (Sklarew *et al.*, 1971).

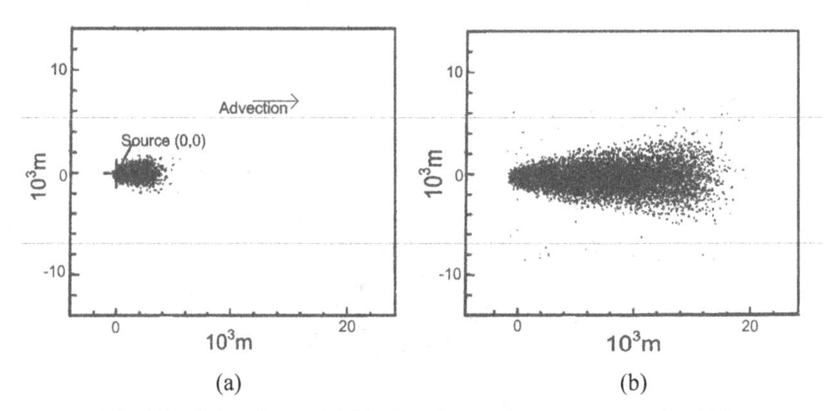

Fig. 8.11 Advection and diffusion of a continuous source using PIC.

8.4 Meshless Method

For decades, the FDM/FVM/FEM have been the dominant numerical schemes employed in most scientific computation. These methods have been used to solve technical problems from aircraft and auto design to medical imaging. Even so, there are often substantial difficulties in applying these techniques, particularly for complicated domain and/or three-dimensional problems.

Common difficulties in the FDM/FVM/FEM include considerable amounts of time and effort required to discretize and index domain elements. This is often the most time-consuming part of the solution process and is far from being fully automated, particularly in 3-D. One method for alleviating this difficulty is to use the boundary element method (BEM), as noted previously. The major advantage of the BEM is that only boundary discretization is required rather than domain. Efficiency is significantly improved over these more traditional methods. However, the BEM involves sophisticated mathematics beyond the FEM and FDM/FVM and some difficult numerical integration of singular functions. Furthermore, the discretization of surfaces in 3-D can still be a complex process even for simple shapes, such as spheres (Fig. 8.12). In addition, all these traditional methods are often slowly convergent, frequently requiring the solution of 10s–100s of thousands of equations in order to get acceptable accuracy.

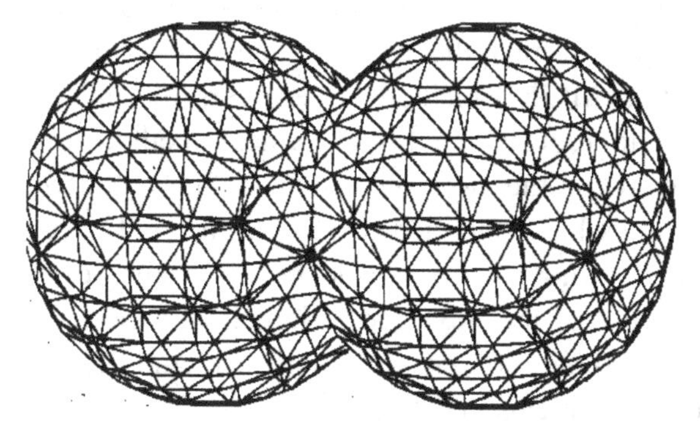

Fig. 8.12 Discretization of spherical shapes.

In recent years, a novel numerical technique called "meshless methods" (or "mesh-free methods") has been undergoing strong development and has attracted considerable attention from both scientific and engineering communities. Currently, meshless methods are now being developed in many research institutions all over the world. Various methods belonging to this family include:

1. Diffuse Element Methods (Nayroles, *et al.*, 1992)
2. Smooth Particle Hydrodynamics Methods (Lucy, 1977)
3. Element-Free Galerkin Methods (Belytschko, *et al.*, 1994)
4. Partition of Unity Methods (Melenk and Babuska, 1996)
5. Hp-Cloud Methods (Duarte and Oden, 1996)
6. Moving Least Squares Methods (Atluri and Zhu, 1999)
7. Local Petrov–Galerkin Methods (Atluri and Zhu, 1999)
8. Reproducing Kernel Particle Methods (Liu, *et al.*, 1998)
9. Radial Basis Functions (Kansa, 1990).

A common feature of meshless methods is that neither domain nor surface meshing is required during the solution process. These methods are designed to handle problems with large deformation, moving boundaries, and complicated geometry. Recently, advances in the development and application of meshless techniques show they can be strong competitors to the more classical finite difference/volume and finite element approaches (Lewis *et al.*, 1996; Huang and Usmani, 1994). A quote from Alturi *et al.* (1999) alludes to the promise of meshless methods: "We show that the basic framework of the meshless local Petrov–Galerkin (MLPG) method is very versatile indeed, and holds a great promise to replace the finite element method, as a method of choice, someday in the not too distant future." Indeed, research in meshless methods has continued to grow at a rapid pace over the past few years, and is now being considered by some researchers as *the numerical method* of the next generation. It is expected that meshless methods will become a dominant numerical method for solving science and engineering problems in the 21st century.

Liu (2002) discusses mesh-free methods, implementation, algorithms, and coding issues for stress-strain problems, and includes Mfree2D, an adaptive stress analysis software package available for free from the web. Atluri and Shen (2002) also recently produced a research monograph that describes the meshless method in detail, including much in-depth mathematical basis.

A flow chart of the procedures for numerically solving a problem using FVM/FEM versus the meshless approach is shown in Fig. 8.13. The first step in any numerical procedure is to define the problem, and establish the governing equations. Once this preliminary step has been done, the next task is to create the geometry. This is now routinely done using various CAD packages. After generating the geometry, a mesh must be created when using either a finite volume or finite element technique. This step can be the most time consuming, especially if one is using a combination of hybrid and unstructured elements (or volumes). On the other hand, the meshless method requires that one only place nodes throughout the physical geometry, i.e., the boundary is represented (and not discretized) by a set of nodes. However, some meshless methods may require a background mesh for integration of the system matrices (but any element or volume may be accepatable since it is only needed for sufficient accuracy in the integrations). Likewise, inputting material properties are generally defined for subdomains of a problems, whereby FVM/FEM methods permit individual definitions per volume or element. When dealing with FEM methods, inputting initial and boundary conditions are not difficult. In meshless methods, special techniques are usually required to impose Dirichlet or Neumann conditions since the shape functions do not satisfy Kronecker delta conditions.

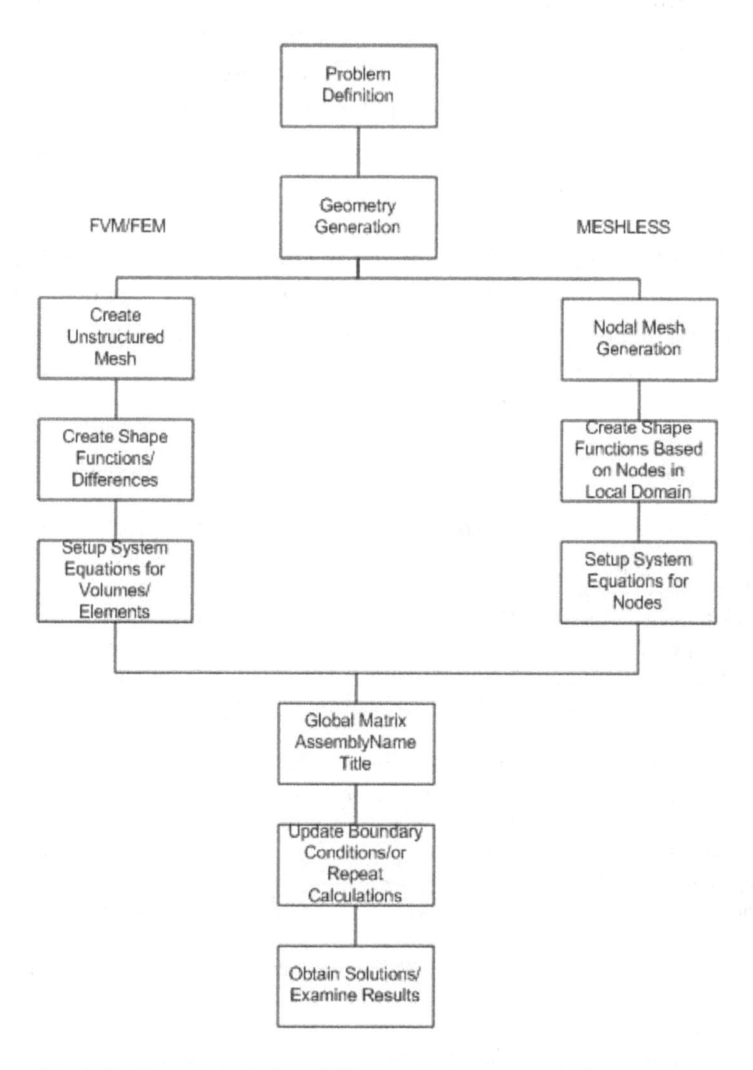

Fig. 8.13 Flow chart for FVM/FEM methods versus meshless methods.

Similar to FEM techniques, meshless methods produce banded system matrices that can be handled in similar fashion. Either set of methods can utilize either direct methods based on Gauss elimination or matrix decomposition methods or iterative methods, e.g., Gauss–Seidel or SOR techniques. When dealing with nonlinear problems, additional iterative loops are needed. Meshless methods generally require more CPU time since the creation of shape functions are more time consuming and are performed during the computation. However, since less time is spent in setting up meshes, and that results using meshless methods are typically more accurate, the ratio of accuracy to CPU is still greater for meshless methods.

In Galerkin-based meshless methods (Belytschko *et al.*, 1994), the highest order derivatives are lowered using weak forms; however, proper evaluation of integrals (generally using a mesh) or a nodal integration scheme is required. Collocation-based methods are attributed to SPH, least-square, and RBF techniques, and are easier to program.

Table 8.1 Differences Between FVM/FEM and Meshless Methods

Items	**FVM/FEM**	Meshless
Element mesh or grid	Required	Not required
Mesh creation	Can be difficult	Relatively easy
Mesh automation/adaptation	Difficult for 3-D	Easy
Create shape functions	Element based	Node based
Shape function properties	Satisfy Kronecker delta; valid for all elements	May or may not satisfy Kronkecker delta conditions
Discretized system stiffness matrix	Sparse or symmetrical depending on problem (fluid flow is sparse)	Dense; may or may not be symmetrical based on the method used
Boundary conditions	Easy to implement	Special methods required
Computational speed	Generally fast	Up to 50X slower
Retrieval of results	Special techniques needed; post processing	Generally easier to extract; post processing standard
Accuracy	Generally $0(2^{nd})$; varies on shape function choice	Can be more accurate than FEM
Stage of development	Mature; well established	Beginning
Commercial packages	Many packages	Essentially none

The similarities and differences between FVM/FEM and meshless methods are listed in Table 8.1 (after Liu *et al.*, 2002). While there are advantages and disadvantages to any numerical method, the meshless approach holds promise for becoming a fast and convenient technique for the near future.

8.4.1 Application of meshless methods

As previously mentioned, there exists various types of meshless methods and each method has its advantages and disadvantages. Intensive research efforts conducted in many major research institutions all over the world are now underway to improve the performance of these approaches.

Meshless methods hold some promising alternative approaches for problems involving fluid flow, heat transfer, and species transport analyses. The most attractive feature is the lack of a mesh that is required in the more conventional numerical approaches. This becomes particularly interesting in that one can begin to conduct adaptive analyses for CFD problems.

There are essentially four meshless-related methods that are typically used for fluid flow and transport related problems:

8.4.1.1 *Smoothed particle hydrodynamics (SPH) techniques including finite integral methods (e.g., Kernel Particle Methods – RKPM – and general kernel reproduction methods – GKR)*

SPH methods use integral representations of a function. A function is approximated by a finite integral form and a kernel or weight (known as a smoothing function) is employed, as shown in the following relation (Liu *et al.*, 2002)

$$u(x) = \int_{-\infty}^{+\infty} u(\xi)\delta(x-\xi)d\xi, \tag{8.60}$$

where $\delta(x)$ is the Dirac delta function. In SPH, $u(x)$ is approximated in the following form:

$$u^h(x) = \int_{\Omega} u(\xi)\hat{W}(x-\xi,h)d\xi, \tag{8.61}$$

where $u^h(x)$ is the approximation of the function u(x) and W(x-ξ,h) is the smoothing weight. The smoothing length, h, controls the size of the domain (known as a smoothing domain). Weight functions include such choices as cubic splines, quartic splines, or exponential functions. In an SPH simulation, the system state is represented as a collection of arbitrarily distributed particles with forces calculated through interparticle interactions in a smoothed manner. The particles are free to move in space and carry all necessary computational information – allowing them to be regarded as interpolation points or field nodes. There is no direct boundary condition in SPH simulations, i.e., for particles near a solid surface, only those particles adjacent to the boundary contribute to the particle interaction – which can lead to incorrect solutions. Hence, some specialized action must be performed, such as using virtual particles. While the SPH method has some limitations, it is very effective for problems that are difficult to simulate using the more conventional approaches, e.g., explosions and free surface flows.

8.4.1.2 *Meshless Petrov–Galerkin (MLPG) methods including finite series representations (e.g., moving least squares (MLS), point interpolation methods (PIM), and hp-clouds)*

The MLPG approach, originally proposed by Atluri and Zhu (1999), has been used to solve incompressible flow problems. The MLPG technique uses a local weak statement integrated over a local quadrature domain which can be of any simple geometry. The field variables within the problem domain are approximated using MLS. A quartic spline function is used to compute the MLS shape functions.

Since the MLPG method creates nodal equations, interior nodes can be treated separately from the boundary nodes. Numerical integration is achieved using subdivision of quadrature domains, coupled with coordinate transformations and Gaussian quadrature. The procedure follows closely the conventional FEM approach using Galerkin's method but with the weight function centered about each node. However, for arbitrary node distributions, large domains containing too many nodes have been found to be troublesome, leading to divergence. The MLPG can become computationally expensive due to the lack of diagonal dominance in the linear system matrix.

An approach first introduced by Pepper and Baker (1980) in the mid-1970s was based on a point discretization of the PG-FEM weak statement, resulting in a tridiagonal system of equations in 1-D (also known as Chapeau functions). However, a conventional mesh with orthogonal nodal arrays was needed to establish the recursive form of the equations (due to the inherent implicit nature of the FEM method). Employing time-splitting, multidimensional equations could be solved quickly. Further elaboration of the method was made by Fletcher (1982) in the 1980s.

8.4.1.3 *Local radial point interpolation Methods (LRPIM) using finite difference representations*

LRPIM methods have been used successfully in solid mechanics problems. Application of the method for incompressible fluid flow has been used to much lesser extent. The PIM approach, proposed by Liu and Gu (2001), is used to replace the MLS approximation for creating shape functions. The PIM maintains superb accuracy in function fitting with the shape functions possessing the Kronecker delta property. This permits simple imposition of the boundary conditions as in the FEM approach. However, efforts are still underway to overcome problems dealing with singular moment matrices, and to make the algorithm numerically stable for arbitrarily distributed points. The LRPIM requires a large number of numerical integrations that generally consume a great deal of CPU time. When coupled with radial basis functions, the method can be made computationally efficient. Examples of the use of LRPIM in 2-D natural convection studies is given in Wu and Liu (2003).

8.4.1.4 *Radial basis functions (RBFs)*

Radial basis functions (RBFs) are simple to implement, and easy to follow. We will discuss this method in more detail. Currently, there are two major approaches in this direction: (i) a domain-type meshless method that was developed by Kansa (1990); (ii) a boundary-type meshless method that has evolved from the BEM. Before we proceed to introduce these meshless methods, it is important to understand exactly what we mean by RBFs. RBFs are the natural generalization of univariate polynomial splines to a multivariate setting. The main advantage of this type of approximation is that it works for arbitrary geometry with high dimensions and it does not require a mesh at all. A RBF is a function whose value depends only on the distance from some center point. Using distance functions, RBFs can be easily implemented to reconstruct a plane or surface using scattered data in 2-D, 3-D or higher dimensional spaces. Due to the uses of the distance functions, the RBFs can be easily implemented to reconstruct the surface using scattered data in 2D, 3D or higher dimensional spaces.

To be more specific, let $\Omega \subset \Re^2$ be a bounded, sufficiently smooth domain. Let $S = \{x_1, x_2, \ldots, x_N\} \subset \Omega$ be a given finite set of distinct points (referred to as interpolation points). We are interested in the following problem:

Let $\{y_1, y_2, \ldots, y_N\}$ be given values. Find a function $f : \Omega \rightarrow \Re$ such that the interpolation equations $f(x_i) = y_i$, for $i=1,2,\ldots,N$, are satisfied. From the theory of radial basis functions, the given function is approximated by a linear combination of radial functions centered in points scattered throughout the domain of interest; i.e.,

$$f(x) \approx s(x) = \sum_{j=1}^{N} c_j \phi\left(\left|x - x_j\right|\right), \quad x \in \Omega, \tag{8.62}$$

where $\{c_1, c_2, \ldots, c_j\}$ is the unknown coefficient to be determined, ϕ the trial function and $|\bullet|$ the Euclidean distance. For convenience, we denote $r = |\bullet|$. Some popular choices of trial function ϕ include:

1. linear (r), cubic (r^3)
2. multiquadrics (MQ) ($(r^2+c^2)^{1/2}$)
3. polyharmonic splines ($r^{2n+1}\log r$ in 2-D, r^{2n+1} in 3-D)
4. Gaussian ($\exp(-cr^2)$)

The unknown coefficients can be computed by a collocation method, which means the $s(x)$ reproduces the original given data set; i.e.,

$$f(x_i) = s(x_i) = \sum_{j=1}^{N} c_j \phi\left(\left|x_i - x_j\right|\right), \quad i = 1, 2, \ldots N. \tag{8.63}$$

The above expression implies a linear system whose size is equal to the number of scattered data points. Once the unknown coefficients are obtained by solving the above linear system of equations, one can approximate *f(x)* by *s(x)* at any point *x* in . For further details, we refer readers to the theory of RBFs discussed in Powell (1992).

In the early 1980s, Franke (1982) published a review paper testing 29 interpolation methods in 2-D and ranked RBFs as the best (MQ followed by Thin Plate Splines ($r^2 \log r$)) based on its accuracy, speed, storage requirements, and ease of implementation. The following surface *f(x,y)* is one of the benchmark problems tested by Franke. In Fig. 8.14, 100 scattered points is used in the domain of a unit square [0,1] X [0,1]. These sample points are randomly chosen and there is no connectivity among these points. Thin Plate Splines is selected as the trial function. Using the collocation method, we have reconstructed the surface *f(x,y)* as shown in Fig. 8.15. The reader is referred to the following website for some interesting applications of surface reconstruction using RBFs: http://www.aranz.com/research/modelling/theory/rbffaq.html.

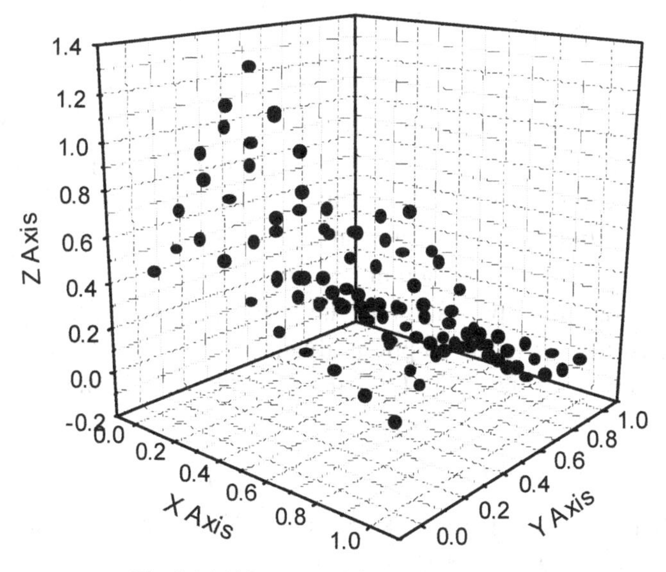

Fig. 8.14 100 scattered data points of *f(x,y)*.

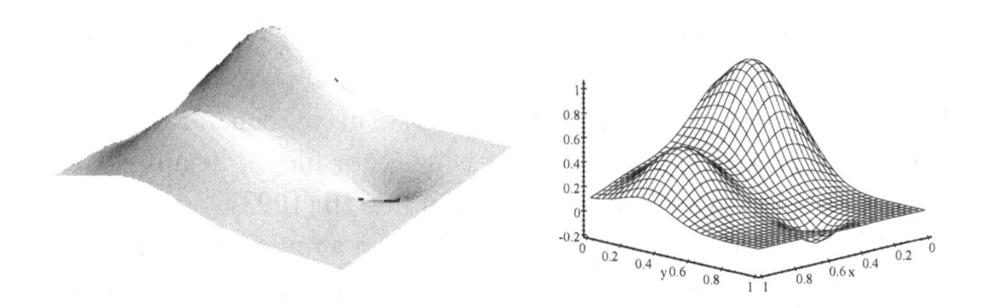

Fig. 8.15 Reconstructed surface using RBF.

The analytical solution for the surface can be expressed as

$$\begin{aligned} f(x,y) = {} & \frac{3}{4}\exp\left(\frac{-1}{4}\left((9x-2)^2+(9y-2)^2\right)\right)+ \\ & \frac{3}{4}\exp\left(\frac{-1}{49}(9x+1)^2-\frac{1}{10}(9y+1)^2\right)+ \\ & \frac{1}{2}\exp\left(\frac{-1}{4}(9x-7)^2-\frac{1}{4}(9y-3)^2\right)- \\ & \frac{1}{5}\exp\left(-(9x-4)^2-(9y-7)^2\right). \end{aligned} \tag{8.64}$$

In 1990, Kansa (1990) extended the idea of interpolation scheme using RBFs to solving various types of engineering problems. The method is simple and direct and is becoming very popular in the engineering community. The boundary type meshless methods indicated in the last section is rather technical and we will only focus on a brief introduction of Kansa's method in this section.

To illustrate the application of the meshless method using Kansa's method, we first consider elliptic problems. For simplicity, we consider the 2-D Poisson problem with Dirichlet boundary condition

$$\begin{aligned} &\nabla^2 \mathrm{T} = \mathrm{f}(\mathrm{x},\mathrm{y}), \quad && (\mathrm{x},\mathrm{y}) \in \Omega, \\ &\mathrm{T} = \mathrm{g}(\mathrm{x},\mathrm{y}), && (\mathrm{x},\mathrm{y}) \in \Gamma. \end{aligned} \tag{8.65}$$

Notice that the solution of Eq. (8.65) is in fact nothing but a surface. The technique in surface interpolation shown in the last section can be applied to solve Eq. (8.65).

To approximate T, Kansa (1990) proposed to assume the approximate solution can be approximated by a linear combination of RBFs

$$\hat{T}(x,y)=\sum_{j=1}^{N} T_j \phi(r_j), \tag{8.66}$$

where $\{T_1, T_2, \ldots, T_N\}$ are the unknown coefficients to be determined, $\phi(r_j)$ is some form of RBF (trial function), and r is defined as

$$r_j=\sqrt{(x-x_j)^2+(y-y_j)^2}. \tag{8.67}$$

Since MQ is an infinitely smooth function, it is often chosen as the trial function for ϕ, i.e.,

$$\phi(r_j)=\sqrt{r_j^2+c^2}=\sqrt{(x-x_j)^2+(y-y_j)^2+c^2}, \tag{8.68}$$

where *c* is a shape parameter provided by the user. The optimal value of *c* is still a subject of outstanding research. We will not further elaborate it here. Other trial functions such as polyharmonic splines can also be chosen as the trial function.

By direct differentiation of Eq. (8.68), the first and second derivatives of with respect to x and y can be expressed as

$$\begin{aligned} &\frac{\partial \phi}{\partial x}=\frac{x-x_j}{\sqrt{r_j^2+c^2}}, \quad \frac{\partial \phi}{\partial y}=\frac{y-y_j}{\sqrt{r_j^2+c^2}} \\ &\frac{\partial^2 \phi}{\partial x^2}=\frac{(y-y_j)^2+c^2}{\sqrt{r_j^2+c^2}}, \quad \frac{\partial^2 \phi}{\partial y^2}=\frac{(x-x_j)^2+c^2}{\sqrt{r_j^2+c^2}}. \end{aligned} \tag{8.69}$$

Substituting Eq. 8.69 into 8.63 and by collocation method, one obtains

$$\begin{aligned} &\sum_{j=1}^{N} T_j \left(\frac{(x_i-x_j)^2+(y_i-y_j)^2+2c^2}{\left((x_i-x_j)^2+(y_i-y_j)^2+c^2\right)^{3/2}} \right)=f(x_i,y_i),\ i=1,2,\cdots,N_I \\ &\sum_{j=1}^{N} T_j \sqrt{(x_i-x_j)^2+(y_i-y_j)^2+c^2}=g(x_i,y_i),\ i=N_I+1, N_I+2,\cdots,N, \end{aligned} \tag{8.70}$$

where N_I denotes the total number of interior points and N_I+1, …, N are the boundary points. Figure 8.16 shows two sets of interpolation points: interior and boundary points. Equation 8.70 is a linear system of N X N equations and can be solved by direct Gaussian elimination. Once the unknown coefficients $\{T_1, T_2, \ldots, T_N\}$ are found, the solution of T in Eq. 8.62 can be approximated by Eq. 8.63 at any point in the domain.

For time-dependent problems, we consider the following heat equation as an example:

$$\frac{\partial T}{\partial t} - \alpha \nabla^2 T = f(x, y, T, \frac{\partial T}{\partial x}, \frac{\partial T}{\partial y}). \tag{8.71}$$

An implicit time-marching scheme can be used and Eq. 8.71 becomes

$$\frac{T^{n+1} - T^n}{\Delta t} - \alpha \left(\frac{\partial^2 T^{n+1}}{\partial x^2} + \frac{\partial^2 T^{n+1}}{\partial y^2} \right) = f(x, y, T^n, \frac{\partial T^n}{\partial x}, \frac{\partial T^n}{\partial y}), \tag{8.72}$$

where Δt denotes the time step and superscript n+1 is the unknown (or next time-step) value to be solved and superscript n is the current known value. The approximate solution can be expressed as

$$\hat{T}(x, y, t^{n+1}) = \sum_{j=1}^{N} T_j^{n+1} \phi_j(x, y) \,. \tag{8.73}$$

Substituting Eq. 8.73 into Eq. 8.72, one obtains

$$\begin{aligned} &\sum_{j=1}^{N} T_j^{n+1} \left(\frac{\phi_j}{\Delta t} - \alpha \left(\frac{\partial^2 \phi}{\partial x^2} + \frac{\partial^2 \phi}{\partial y^2} \right) \right) (x_i, y_i) = \frac{1}{\Delta t} T^n (x_i, y_i) + \\ &f(x_i, y_i, t^n, T^n(x_i, y_i), T_x^n(x_i, y_i), T_y^n(x_i, y_i)) \quad i = 1, 2, \ldots, N_I \\ &\sum_{j=1}^{N} T_j^{n+1} \phi(x_i, y_i) = g(x_i, y_i, t^{n+1}) \quad i = N_I + 1, \ldots, N, \end{aligned} \tag{8.74}$$

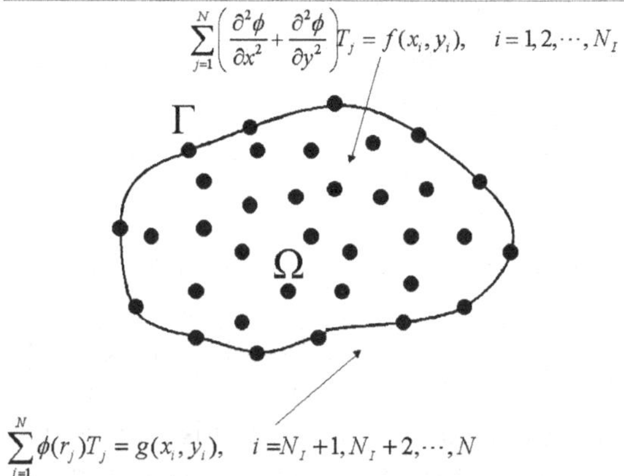

Fig. 8.16 Interior points and boundary points using Kansa's method.

which produces a N x N linear system of equations for the unknown T_j^{n+1}. Note that the right hand side of the first equation in Eq. 8.74 can be updated before the next time step, i.e.,

$$T^n(x_i,y_i)=\sum_{j=1}^{N}T_j^n\phi_j(x_i,y_i),\;\; T_x^n(x_i,y_i)=\sum_{j=1}^{N}T_j^n\frac{\partial\phi_j}{\partial x}(x_i,y_i)$$
$$T_y^n(x_i,y_i)=\sum_{j=1}^{N}T_j^n\frac{\partial\phi_j}{\partial y}(x_i,y_i). \tag{8.75}$$

Figure 8.17 shows an arbitrary domain discretized using three-noded triangular elements, boundary elements, and a meshless method. An internal mesh is required in the FEM (Fig. 8.17, a) and linear elements are needed along the boundary in the BEM (Fig. 8.17, b), as noted by the dotted lines. Both methods require the use of efficient matrix solvers to obtain values at the prescribed nodes, which can become resource limiting and time consuming. The meshless method, with arbitrarily distributed interior and boundary points, requires no mesh as illustrated in Fig. 8.17 (c).

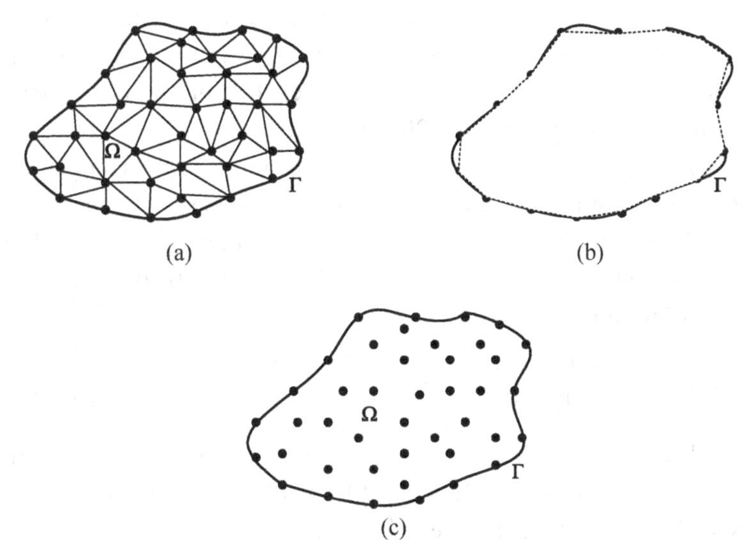

Fig. 8.17 Irregular domain discretized using (a) three-noded triangular finite elements, b) boundary element, and (c) arbitrary interior and boundary points meshless method.

8.4.2 Example cases – Heat transfer

8.4.2.1 *Heat transfer in a 2-D plate*

To illustrate the use of meshless methods, let us begin with a simple heat transfer problem. The governing equation for temperature transport can be written as

$$\frac{\partial T}{\partial t}+\mathbf{V}\cdot\nabla T=\alpha\nabla^2 T+Q \tag{8.76}$$

$$q+k\nabla T-h(T-T_\infty)-\varepsilon\sigma(T^4-T_\infty^4)=0 \tag{8.77}$$

$$T(\mathbf{x},0)=T_o, \tag{8.78}$$

where $\mathbf{V}$ is the vector velocity, $\mathbf{x}$ is vector space, $T(\mathbf{x},t)$ is temperature, T_∞ is ambient temperature, T_0 is initial temperature, α is thermal diffusivity ($\kappa/\rho c_p$), ε is emissivity, σ is the Stefan–Boltzmann constant, h is the convective film coefficient, q is heat flux, and Q is heat source/sink. Velocities are assumed to be known and typically obtained from solution of the equations of motion (a separate program is generally used for fluid flow (Pepper *et al.*, 2000).

In this first example problem, a two-dimensional plate is subjected to prescribed temperatures applied along each boundary (Pepper *et al.*, 2000), as shown in Fig. 8.18. The temperature at the mid-point (1,0.5) is used to compare the numerical solutions with the analytical solution. The analytical solution is given as

$$\theta(x,y)\equiv\frac{T-T_1}{T_2-T_1}=\frac{2}{\pi}\sum_{n=1}^{\theta}\frac{(-1)^{n+1}+1}{n}\sin\left(\frac{n\pi x}{L}\right)\frac{\sinh(n\pi y/L)}{\sinh(n\pi W/L)}, \tag{8.79}$$

which yields $\theta(1,0.5)$ = 0.445, or $T(1,0.5)$ = 94.5°C. The analytical solution simply denotes the diffusion of heat within a rectangular domain with fixed (Dirichlet) temperatures along each face.

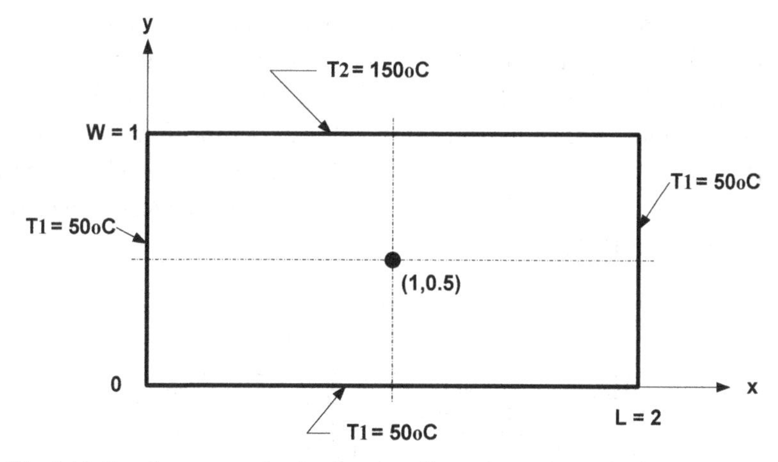

Fig. 8.18 Steady-state conduction in a two-dimensional plate (from Incropera and DeWitt, 2002).

Table 8.2 lists the final temperatures at the mid-point using a finite element method, a boundary element method, and a meshless method.

Table 8.2 Comparison of Results for Poblem 1 from Exact, FEM, BEM, and Meshless Methods (from Pepper and Chen, 2002)

Method	mid-point (°C)	Elements	Nodes
Exact	94.512	0	0
FEM	94.605	256	289
BEM	94.471	64	65
Meshless	94.514	0	325

8.4.2.2 *Singular point in a 2-D domain*

As a second example problem, a two-dimensional domain is prescribed with Dirichlet and Neumann boundary conditions applied along the boundaries, as shown in Fig. 8.19. This problem, described in Huang and Usmani (1994), was used to assess an h-adaptive FEM technique.

A fixed temperature of 100°C is set along side AB; a surface convection of 0°C acts along edge BC and DC with h = 750 W/m°C and k = 52 W/m°C. The temperature at point E is used for comparative purposes.

The severe discontinuity in boundary conditions at point B creates a steep temperature gradient between points B and E. Figures 8.20(a,b) show the initial and final FEM meshes after two adaptations using bilinear triangles. The analytical solution for the temperature at point B is T = 18.2535°C. Table 8.3 lists the results for the three methods compared with the exact solution. The initial 3-noded triangular mesh began with 25 elements and 19 nodes.

Table 8.3 Comparison of Results for Problem 2 from Exact, FEM, BEM, and Meshless Methods (from Pepper and Chen, 2002).

Method	Point E (oC)	Elements	Nodes
Exact	18.2535	0	0
FEM	18.1141	256	155
BEM	18.2335	32	32
Meshless	18.2531	0	83

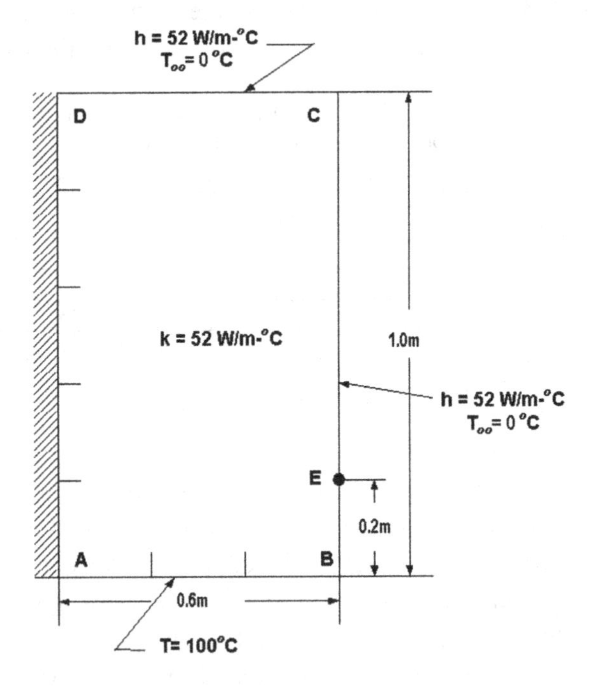

Fig. 8.19 2-D domain with prescribed Dirichlet and Neumann boundaries.

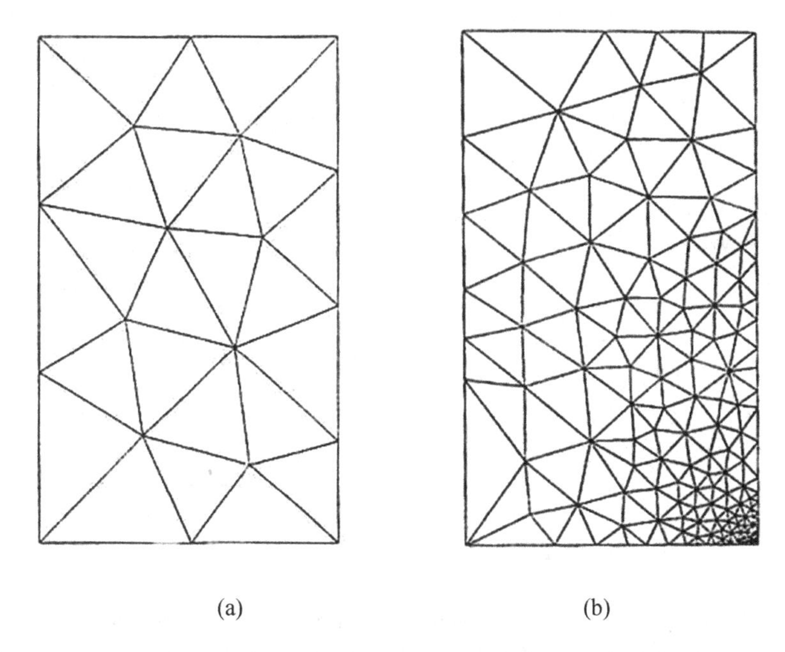

(a) (b)

Fig. 8.20 Problem for assessing an h-adaptive FEM technique, (a) initial FEM mesh and (b) final FEM adapted mesh (from Huang and Usmani, 1994).

8.4.2.3 *Heat transfer within an irregular domain*

A simple irregular domain is used for the third example problem and results compared from the three methods. Results from a fine mesh FEM technique (without adaptation) are used as a reference benchmark (Pepper *et al.*, 2000). The discretized domain and accompanying boundary conditions set along each surface are shown in Fig. 8.21. The FEM results are displayed as contour intervals. Figure 8.22 (a,b) shows meshless results (using FEM fine mesh nodes for contouring) versus FEM solutions using adapted quadrilateral elements. Heat conduction occurs as a result of constant temperatures set on the top and bottom surfaces, adiabatic faces in the upper right cutout and lower cutout portions, and convective heating along the right and left vertical walls. Adaptive meshing occurs in the corners as a result of steep temperature gradients; this is not evident when using meshless methods.

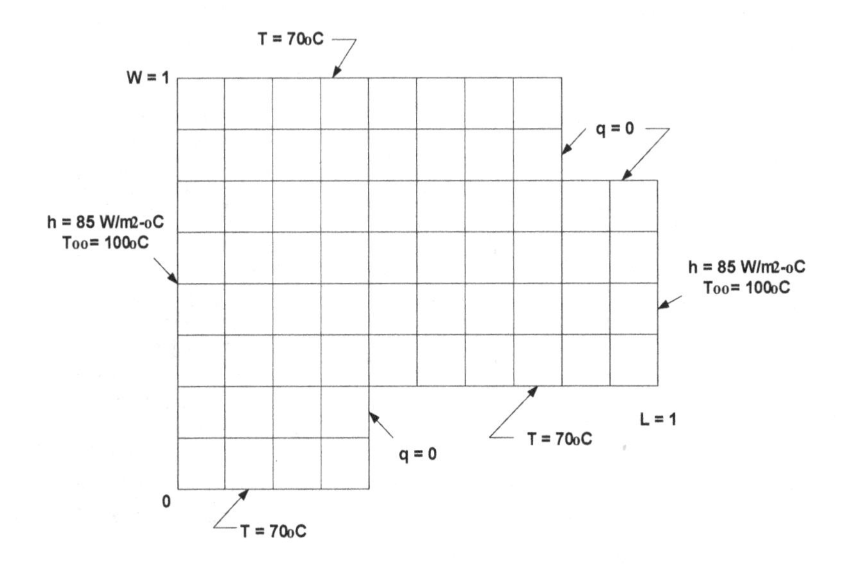

Fig. 8.21 Problem specification for heat transfer in a user-defined domain.

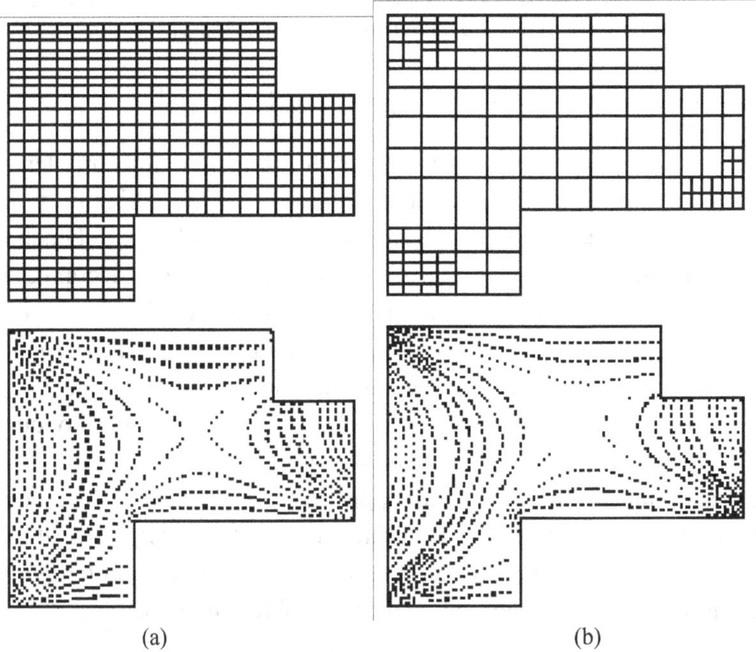

Fig. 8.22 FEM solutions (a) meshless (on FEM fine mesh) and (b) adapted mesh.

The FEM, BEM, and meshless mid-point values at (0.5,0.5) are listed in Table 8.4.

Table 8.4. Comparison of Results for Problem 3 from FEM, BEM, and Meshless Methods (from Pepper and Chen, 2002).

Method	mid-point (°C)	Elements	Nodes
FEM	75.899	138	178
BEM	75.885	36	37
Meshless	75.893	0	96

All three techniques provide accurate results for the three example cases. The meshless method was clearly the fastest, simplest, and least storage demanding method to employ. Advances being made in meshless methods will eventually enable the scheme to compete with the FEM and BEM on a much broader range of problems (Wu and Liu, 2003; Atluri and Shen, 2002; Sarler *et al.*, 2002). Hon has also done much work in engineering modeling using Kansa's method. We refer the readers to his website: http://www.cityu.edu.hk/ma/staff/ychon.html.

8.4.2.4 *Natural convection*

Natural convection within a 2-D rectangular enclosure is a well-known problem commonly used to test the ability of a numerical algorithm to solve for both fluid flow and heat transfer. The equations are strongly coupled through the buoyancy term in the momentum equations and the temperature. There are various forms of dimensionless equations, and numerous references can be found in the literature and on the web regarding these various forms. The solution to the problem generally splits between solving either the primitive equations for velocity or the vorticity equation, coupled with the transport equation for temperature.

The issue in this early development of the meshless approach was not to dwell on various schemes dealing with pressure (e.g., projection methods or the SIMPLE scheme both of which are well known). Hence, most researchers developing meshless approaches deferred to using the streamfunction–vorticity–temperature equations. These equations are the well-known set generally formulated as follows:

$$\frac{\partial\omega}{\partial t}+u\frac{\partial\omega}{\partial x}+v\frac{\partial\omega}{\partial y}=\Pr\left(\frac{\partial^2\omega}{\partial x^2}+\frac{\partial^2\omega}{\partial y^2}\right)-\Pr\cdot Ra\cdot\frac{\partial T}{\partial x}, \quad (8.80)$$

$$\frac{\partial T}{\partial t}+u\frac{\partial T}{\partial x}+v\frac{\partial T}{\partial y}=\frac{\partial^2 T}{\partial x^2}+\frac{\partial^2 T}{\partial y^2}, \quad (8.81)$$

$$\frac{\partial^2\psi}{\partial x^2}+\frac{\partial^2\psi}{\partial y^2}=-\omega, \quad (8.82)$$

with the conventional definitions for velocity in terms of the streamfunction gradients. Figure 8.23 shows the physical and computational domain with accompanying boundary conditions. Two types of nodal configurations are shown in Fig. 8.24 (a,b) utilizing 256 nodes. Results are in excellent agreement with well-known results in the literature for $10^3 \le Ra \le 10^5$. Figure 8.25 (a,b) shows streamlines and isotherms for the differentially heated enclosure for Ra = 10^5. A convergence plot showing the difference in rates between a conventional FDM and applications of the MLPG and LRPIM techniques is shown in Fig. 8.26. Notice the more rapid rate of convergence of the two meshless methods versus the finite difference scheme.

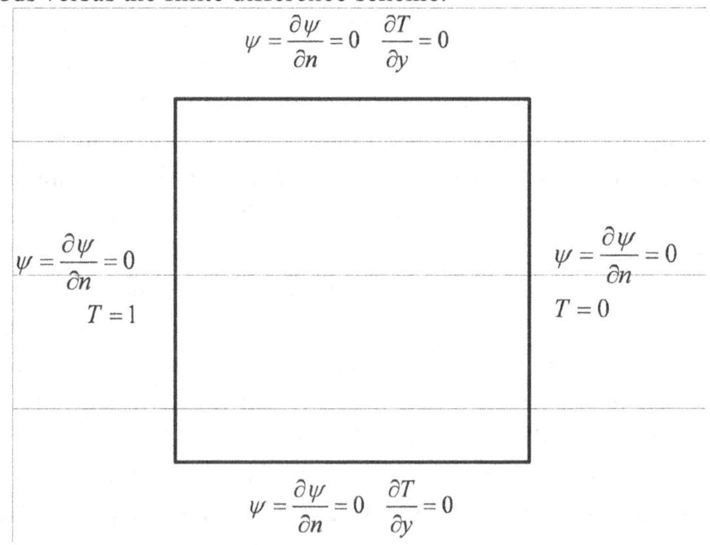

Fig. 8.23 Boundary conditions for natural convection within a rectangular enclosure.

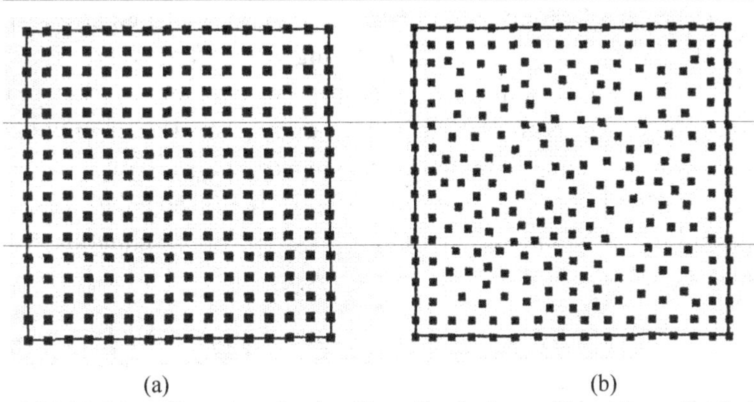

Fig. 8.24 Nodal configurations for a) uniform distribution and b) arbitrary distribution for 256 nodes.

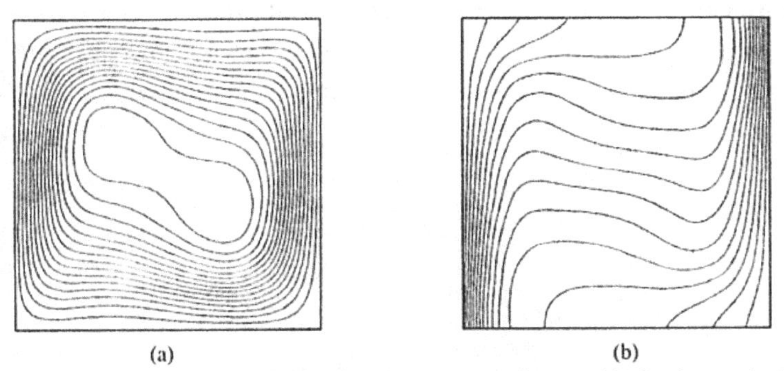

Fig. 8.25 Natural convection results showing a) streamlines and b) isotherms for Ra = 10^5 using the MLPG method (from Wu and Liu, 2003).

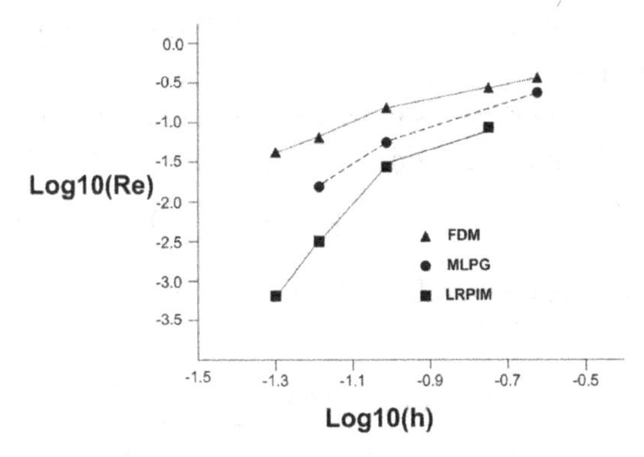

Fig. 8.26 Convergence rates for FDM, MLPG, and LRPIM meshless methods for natural convection within a 2-D enclosure (adapted from Wu and Liu, 2003).

A similar study of natural convection within a rectangular enclosure was conducted by Sarler *et al.* (2002) using the RBF approach of Kansa (1990). Solving the nonlinear Poisson reformulation of the general transport equation representing mass, energy, and momentum, the problem was solved by dividing the physical domain into two parts consisting of an internal array of nodes and a set of boundary nodes for the Dirichlet and Neumann conditions. The governing equation for the transport variable is of the form

$$\frac{\partial}{\partial t}\left(\rho C(\phi)\right)+\nabla\cdot\left(\rho VC(\phi)\right)=-\nabla\cdot\left(-D\nabla\phi\right)+S, \tag{8.83}$$

where ρ, ϕ, V, t, D, and S denote density, transport variable, velocity, time, diffusion matrix, and source. The transport variable C consisted of enthalpy C(h(ϕ = T)), velocity C(ϕ = u,v), and pressure C(ϕ = p), with a pressure correction Poisson equation used to resolve the pressure. The nonlinear equations solved with the meshless technique were of the form

$$\nabla^2\phi=\theta+\nabla\cdot\Theta, \tag{8.84}$$

$$\theta=\left[\frac{\partial}{\partial t}\left(\rho C(\phi)-S\right)\right]/D, \tag{8.85}$$

$$\Theta=\left[\rho VC(\phi)-D'\nabla\phi\right]/D, \tag{8.86}$$

where and D′ denote density, transport variable, time, velocity, and D is the diffusion matrix with D′ being the nonlinear anisotropic part. The variable C denotes the relation between the transported and the diffused variable. The solution requires the use of an iterative technique. The final form of the transformed Poisson equation is

$$\nabla^2\phi=\theta+\theta_{,\phi}\left(\phi-\bar{\phi}\right)+\nabla\cdot\Theta+\nabla\cdot\Theta_{,\phi}\left(\phi-\bar{\phi}\right), \tag{8.87}$$

where the bar denotes values from the previous iteration. Time discretization utilizes the relation

$$\theta\approx\left[\frac{\rho C(\phi)-\rho C(\phi_o)}{\Delta t}-S\right]/D, \tag{8.88}$$

with the unknown field ϕ approximated by the N global approximation functions $\psi_n(p)$ and their coefficients ς_n, i.e.,

$$\phi(p) \approx \psi_n(p)\varsigma_n; \quad n = 1, 2, \cdots N_\Gamma \,. \tag{8.89}$$

The global radial basis function approximation was based on multiquadrics with the free parameter r_o:

$$\psi_n = \left(r_n^2 + r_o^2\right)^{1/2} . \tag{8.90}$$

The coefficients were calculated from the N collocation equations of which N_Γ were equally distributed over boundary Γ and N_Ω over the domain Ω. Separate relations were established for the boundary condition indicators.

The computational domain was discretized into 80 boundary nodes and 361 domain nodes. The multiquadrics constant r_o was set to 0.2. Steady-state results were achieved after 34 iterations for Ra = 10^3, 187 iterations for Ra = 10^4, and 293 iterations for Ra = 10^5. The calculated values for temperature and velocity were in excellent agreement with results obtained using a fine grid FDM.

Kalla and Pepper (2008) demonstrated the application of a meshless approach to solve the primitive equations of motion and energy using radial basis functions. A projection scheme was employed to account for pressure, similar to the techniques used by Pepper and Carrington (1997) and later by Wang and Pepper (2007) to simulate fluid flow with heat transfer. Kalla and Pepper (2008) describe the technique in more detail in the paper from her thesis.

Further application of the LRPIM is shown in Fig. 8.27 for natural convection within a concentric annulus. In this instance, the nodal distribution is 967 with the inner cylinder heated and the outer cylinder cooled. The early work of Kuehn and Goldstein (1976) utilized a second-order FDM technique (which supplemented their earlier experimental work) to simulate the flow and heat transfer within the annulus. Their results have served as reference values for many years. Figure 8.28 (a,b) shows streamlines and isotherms for Ra = 10^4, and the results agree closely with those of Kuehn and Goldstein (1976) and others in the literature.

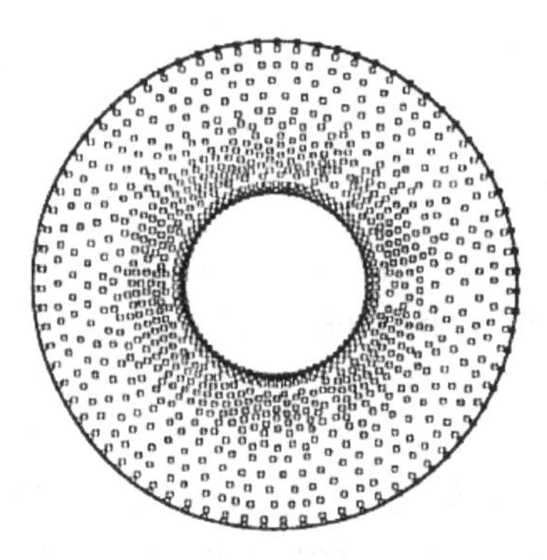

Fig. 8.27 Nodal distribution for natural convection within a concentric annulus with r_i = 0.625 and r_o = 1.625.

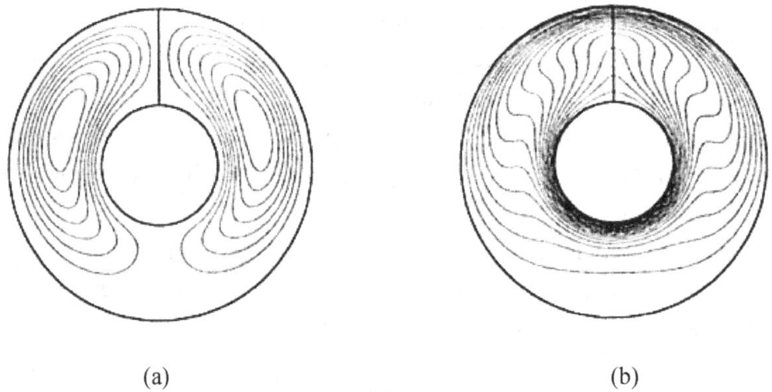

(a) (b)

Fig. 8.28 Natural convection within a concentric annulus a) streamlines and b) isotherms for Ra = 10^4.

Application of the meshless approach to porous media flow is discussed in Li *et al.* (2003). Utilizing RBF collocation, the transient dispersion of contaminant and pressure head were calculated for various flow parameters. A regional groundwater system was simulated for a cross section of x = (0,200 m) by z = (0,100 m). Results were obtained using multiquadric functions. The governing equations consisted of head and concentration expressed in the forms

$$\nabla^2 h = 0, \tag{8.91}$$

$$\frac{\partial C}{\partial t} + V \cdot \nabla C = \nabla \left(D \cdot \nabla C \right) - \lambda C \,, \qquad (8.92)$$

where h denotes the pressure head, C is the contaminant concentration, V is the seepage velocity, and λ is the rate of decay. The top boundary is set with h = 0.05x + 100 with zero flux conditions on the two side boundaries and no-flow bottom boundary denoting impermeable bedrock. The coefficients were all set equal to 1. Following similar procedures employed when using Kansa's method (1990), results were obtained quickly using a PC.

For the first example case, a 41 x 21 mesh was used to calculate 2-D steady-state water head within the computational domain. Figure 8.29 shows contours of the head and velocity vectors (obtained from the gradients of the head), and are in agreement with expected results.

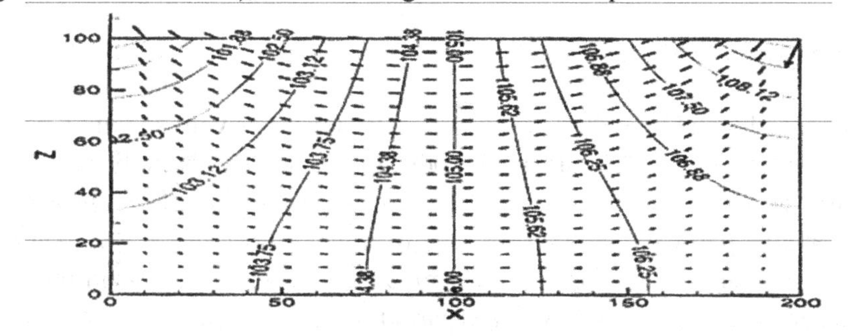

Fig. 8.29 Contour of water head and velocity vectors.

The transport of a scalar quantity is illustrated in a second case as shown in Fig. 8.30 (a,b). In this example, an 11 x 11 x 11 mesh was used with uniformly distributed set of collocation points for $\Omega = [0,1]^3$, and follows from the 3-D heat transfer model analyzed by Zerroukat *et al.* (1998). The maximum number of time steps was set to 800 with Δt = 0.01. Contours of the numerical solution and relative error are shown in Fig. 8.30 (a,b) after 200 time steps for the z = 0.5 plane. Both fully implicit marching scheme and Crank–Nicolson time-marching schemes were compared with the Crank–Nicolson scheme producing a slightly faster convergence.

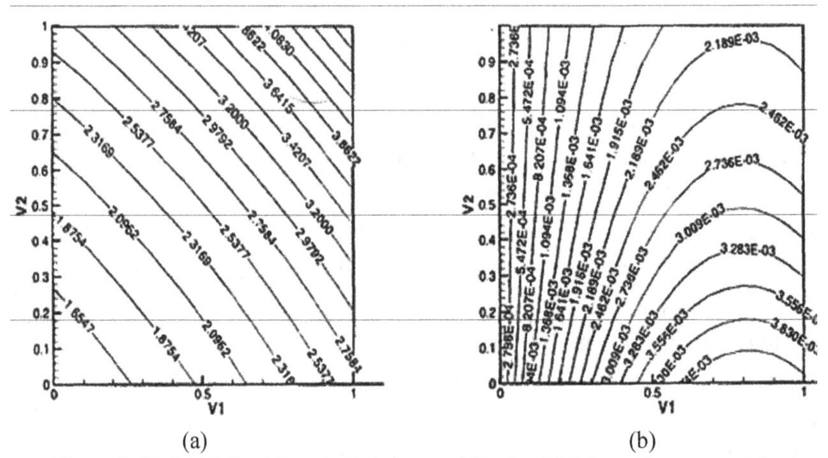

(a) (b)

Figure 8.30 Isopleths (a) and relative error (b) after 200 time steps at z = 0.5.

8.5 Molecular Modeling

The idea behind modeling at the molecular level lies with solving a non-continuum problem using an N-body approach. The N-body problem refers to solving the forces on N particles as a function of the distances among the particles. In other words, given initial positions and velocities of all the particles, we seek to determine the motion of the system if each particle interacts with all the other particles in the system.

The N-body problem was formulated around 1900, dealing with the solar system and forces due to gravitational attraction. In this instance, we are interested in the interactions of molecules and particles, with forces more complex than just gravitation.

Feynman *et al.* (1963) discuss classical molecular forces. Basically, when two molecules that are close are pulled apart, they attract. When they are pushed together, they repel, with the force of repulsion being an order of magnitude greater than the force of attraction (an exception to this rule is liquid water). There are many classical molecular potentials for the interactions of molecules. The most popular model is the Lennard–Jones potential, which can be written as

$$\phi(r_{i,j}) = 4\varepsilon\left(\frac{\sigma^{12}}{r_{i,j}^{12}} - \frac{\sigma^{6}}{r_{i,j}^{6}}\right) \quad \text{erg}, \tag{8.93}$$

where $r_{i,j}$ is measured in angstroms (Å). To obtain the force, $F_{i,j}$, that is exerted on molecule P_i by P_j, we can use the chain rule on Eq. 8.93, i.e.,

$$F_{i,j} = -\frac{d\phi(r_{i,j})}{dR} = -\frac{d\phi(r_{i,j})}{dr_{i,j}}\frac{dr_{i,j}}{dR}, \tag{8.94}$$

where R is the distance between molecules with R cm = 10^8 R Å, or $r_{i,j} = 10^8$ R. In other words, if $R > 1$, F acts toward the origin, corresponding to attraction; if $R < 1$, F acts away from the origin and indicates repulsion.

Example 8.5.1 Force acting on argon vapor: The Lennard–Jones potential for argon can be obtained using Eq. 8.93 using values for ε and σ found in most chemistry textbooks,

$$\phi(r_{i,j}) = (6.848)10^{-14}\left(\frac{3.418^{12}}{r_{i,j}^{12}} - \frac{3.418^{6}}{r_{i,j}^{6}}\right) \quad \text{erg}, \tag{8.95}$$

with $r_{i,j}$ measured in angstroms. The force, $F_{i,j}$, exerted on P_i by P_j is

$$F_{i,j} = (6.848)10^{-6}\left(\frac{12(3.418)^{12}}{r_{i,j}^{13}} - \frac{6(3.418)^{6}}{r_{i,j}^{7}}\right)\frac{\overline{r}_{i,j}}{r_{i,j}} \quad \text{dynes}, \tag{8.96}$$

where the overbar on $r_{i,j}$ denotes the vector. Equation 8.96 can be simplified to the form

$$F_{i,j} = \left\|\overline{F}_{i,j}\right\| = \left(\frac{209}{r_{i,j}^{13}} - \frac{0.06551}{r_{i,j}^{7}}\right),$$

which produces $F_{i,j}(r_{i,j}) = 0$ and that $r_{i,j} = 3.837$Å, which is the equilibrium distance.

The equation of motion for a single argon vapor atom P_i that is acted upon by a single argon vapor atom P_j can be expressed as (remember from Newton's second law, F = ma)

$$ma_i = \left(\frac{209}{r_{i,j}^{13}} - \frac{0.06551}{r_{i,j}^{7}}\right)\frac{\overline{R}_{i,j}}{R_{i,j}}. \tag{8.97}$$

The mass of an argon atom is 6.63 x10-23 gm. Since r = 108 R, then Eq. 8.97 can be rewritten as

$$a_i = \frac{10^{23}}{6.63}\left(\frac{209}{r_{i,j}^{13}} - \frac{0.06551}{r_{i,j}^{7}}\right)\frac{\overline{r}_{i,j}}{r_{i,j}} \quad \left(\frac{cm}{sec^2}\right). \tag{8.98}$$

If we now replace centimeters by angstroms and seconds by picoseconds, with $a_i = dv/dt = d^2r_i/dt^2$, Eq. 8.98 can be rewritten as

$$\frac{d^2\overline{r}_i}{dt^2} = 98810\left(\frac{3190}{r_{i,j}^{13}} - \frac{1}{r_{i,j}^{7}}\right)\frac{\overline{r}_{i,j}}{r_{i,j}} \quad \left(\frac{\dot{A}}{ps^2}\right). \tag{8.99}$$

Assuming a local interaction distance of D = 2.5σ = 2.5(3.418) = 8.545 Å, which represents the distance for a local force on molecule P_i, the equations of motion for a system of N argon vapor atoms would be

$$\frac{d^2\overline{r}_i}{dt^2} = 98810\sum_{\substack{j \\ j\neq i}}\left(\frac{3190}{r_{i,j}^{13}} - \frac{1}{r_{i,j}^{7}}\right)\frac{\overline{r}_{i,j}}{r_{i,j}} \quad i = 1,2,3,\ldots,N; r_{i,j} < D. \tag{8.100}$$

We can neglect the effect of gravitational settling since 980 cm/s^2 = $(980)10^{-16}$ Å/ps^2.

An example of molecular motion of argon vapor within a 2-D cavity is shown in Fig. 8.31 (a–c) (from Greenspan, 2005). The top wall, CD, moves in the X direction at a constant speed V = –50 Å/ps. The length of the cavity is 230.22 Å. The initial grid consists of 4235 molecules, as shown in Fig. 8.31 (b). We also define initial velocities of the atoms at v = 3.58 Å/ps, determined with random direction.

Figure 8.31 (c) shows a primary vortex developed at t = 3.5 with Δt = 0.00002 s. This is quite amazing as we are only solving Eq. 8.100 for the motion of the argon atoms, yet we see an effect analogous to the classical continuum flow in a cavity with a moving top wall (Burgraff, 1966). Although the distances are quite small, we can readily see the interaction of the molecules upon one another, and the resulting evolution of a larger motion.

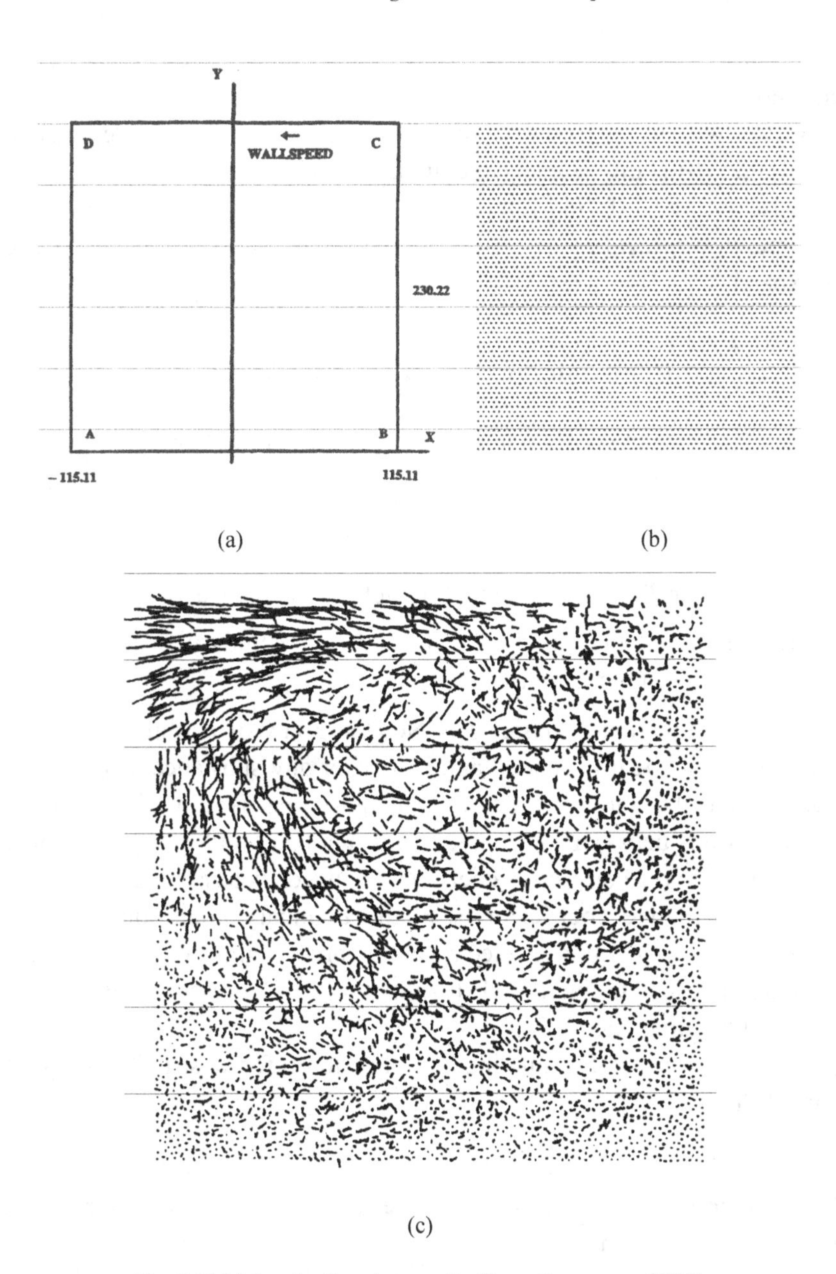

Fig. 8.31 Molecular flow in a cavity (from Greenspan, 2005).

From a practical point of view, one would have to model an enormous number of atoms to begin to represent a continuum fluid. However, we can see the effect of molecular forces at the atomic level on the fluid structure, which eventually escalates into motion on a macroscopic level. In the future, modeling at the microscopic level will become more prevalent as computational power and resources continue to expand.

8.6 Boundary Conditions for Mass Transport Analysis

Gas and particulate flux boundary conditions are of the form

$$q_{p_j} = h_{p_j}\left(C_{j\,\text{at boundary}} - C_{\text{bluk}_j}\right), \tag{8.101}$$

where q_p is the flux rate of the jth particle or substance, C_{bluk_j} is a bulk concentration in the fluid stream and, a boundary concentration, $C_{j\,\text{at boundary}}$ is just at the boundary.

Such an equation is basic, it is Newton's law of cooling applied to mass. The statement is general and is true for any substance. It is merely stating that the rate of transfer per unit area is the difference in concentration between one place and another multiplied by some constant. Only the convective coefficient h_p needs determining.

The convective coefficient h_p in the boundary condition above is determined by geometry, electrostatic forces, gravity, other forces affecting particulate inertia, diffusivities, partial pressures of vapors, chemical bonding, etc.

Another way of formulating the flux term for a strictly diffusion related flux is

$$J_j = D_j \nabla C_j. \tag{8.102}$$

This is Fick's first law of diffusion (Reist, 1993) where J_i is mass flux and in one dimension.

Both forms of boundary conditions have units of mass per unit time per unit area. Different formulations are required as the significance of the forces acting on the mass changes and the type of mass in consideration. For an evaporating liquid, the first form is typically employed. For particulate depositing to a surface, or gases migrating without phase change at the boundary, the second form is more appropriate as it directly incorporates the diffusion coefficient. Particles with a one micron aerodynamic diameter or less do not experience gravitational settling or significant relaxation times. Therefore, a division at one-micron makes a natural delineation between the behavior of larger and smaller particles.

Diffusion, thermophoretic forces, and particle agglomeration (with associated increase in settling velocity) are responsible for deposition. For small particles, a settling velocity is essentially nonexistent. In the absence of inertial, thermophoretic forces and other forces, molecular diffusion through a boundary layer is responsible for most deposition of particles smaller than one micron.

If an analytic boundary layer solution were to be obtained a deposition velocity could be calculated. The deposition velocity is given by the following diffusion velocity

$$V_d = \frac{\text{Rate of Deposition}}{C_{bluk} - C_{wall+}}, \tag{8.103}$$

and for a concentration at the boundary or wall which is not affecting the rate of deposition (Davies, 1966)

$$V_d = \frac{J_j}{C_{bulk_j}}. \tag{8.104}$$

For purely diffusive deposition, substituting for

$$\frac{dC_j}{dx} = \frac{C_{boundary} - C_{bulk_j}}{\delta_{p_j}},$$

where dx is a boundary thickness over which there is a change in concentration then,

$$V_d = \frac{D_j}{\delta_{p_j}}, \tag{8.105}$$

where V_d is deposition velocity, δ_p is a boundary layer thickness for the jth particulate. This equation is for the case of sedimentation where the surface concentration is unimportant to the flow (Davies, 1966).

Rate of deposition is not affected by the flow. The difficulty with this equation is the determination of the boundary layer thickness, δ_p. It is essentially defined as the distance at which the gradient of concentration is zero. The bulk concentration is a function of distance, changing in time as material is deposited from the flow.

Deposition of particulate in the numerical model is treated as a flux boundary. The mass flux to the wall has a value of the deposition equal to the deposition velocity. The mechanism for deposition is Fickian diffusion in the absence of other influencing forces. Deposition of inert gases onto a surface is zero. Other gases deposit via some reaction mechanisms for which rates must be specified. For deposition other than by diffusion, rates are generally determined experimentally.

Turbulence provides good mixing therefore, if a homogenous concentration everywhere beyond the diffusive boundary layer is assumed, the deposition velocity becomes

$$V_d = \frac{\text{Rate of Deposition}}{C_\infty - C_0} = \frac{D_j}{\delta_p}. \tag{8.106}$$

The difficulty again lies with the determination of δ_p.

The rate of transport in turbulent flow of a substance towards a surface in non-imensional form is (Davies, 1966)

$$\frac{R}{u_* c_o} = \left(\frac{D_j + \mu_{turb}}{\nu_{fluid}}\right)\frac{dc_+}{dy_+}, \tag{8.107}$$

where:

$c_+ = c / c_o$ non-imensional concentration at any given time.

$y_+ = y u_* / \nu_{fluid}$ non-imensional distance normal to a surface.

$V_d = R / c_o$ *deposition velocity.*

$u_* = \sqrt{\frac{\tau_w}{\rho_f}}$ *is the friction velocity.*

Substituting these terms, an expression for the non-dimensional V+ is found as

$$\frac{V_+}{u_*} = \left(\frac{D_j + \mu_{turb}}{\nu_f}\right)\frac{dc_+}{dy_+}. \tag{8.108}$$

The diffusive term is a linear combination of the turbulent eddy diffusivity μ_{turb}, and Fickian diffusivity, D_j. The deposition velocity is derived from the non-dimensional deposition velocity by $V_+ = V_d / u_*$ where u_* is the friction velocity.

As mentioned previously, a numerical model for deposition of particulate is essentially treated as a flux boundary. The mechanism for deposition is Fickian diffusion only through the laminar sublayer. The mechanism for distribution into the turbulent sublayers is by turbulent diffusion. Unless a concentration is specified on the surface, the law of the wall is not necessary for calculation of the mass gradient. A listing of diffusion coefficients for various gas mixtures is given in Appendix A.

8.7 Comments

We have examined a set of interesting alternative numerical methods to the more conventional FDM, FVM, and FEM techniques. These schemes are now being used for a variety of problems, including dispersion as well as heat transfer and fluid flow. Applications of these advanced numerical techniques are showing up more frequently in the literature and will likely become even more widespread over time. The interested reader can easily scan the web using any of these schemes as the principal subject and access an extensive list of articles and discussions.

Meshless methods are a unique and novel numerical technique now making inroads into various fields. Their advantages in solving problems associated with crack propagation and stress/strain including deformation over more conventional numerical schemes have been demonstrated repeatedly in the literature. The application of meshless methods for heat transfer is equally advantageous; such methods have become very competitive with both finite volume and finite element methods for problems involving irregular geometries. The requirements for creating grids as well as the detailed input necessary for establishing volume or element properties is greatly reduced or eliminated. However, much has yet to be done before meshless methods can handle a wide range of fluid flow problems and produce results with confidence and surety. This is true of the other numerical schemes mentioned in this chapter as well. Only a small portion of incompressible flow problems have been addressed; advances are just now being made in the area of porous media flows.

While BEM and meshless methods may be as accurate as FDM/FVM/FEM techniques, they can be much slower with regards to computational time to achieve convergence. This is due in part to some of the effort needed for numerical integration and subsequent use of a direct matrix solver. However, meshless methods do not need any preknowledge of their nodal arrangement, as in conventional numerical schemes. This makes the method particularly attractive for developing adaptive capabilities. Since much of a modeler's efforts are generally spent on developing a good mesh that will lead to a converged solution, the overall time for obtaining problem solutions using meshless methods can be significantly less.

Details regarding the development and use of meshless methods can be obtained by accessing website http://rbf-pde.uah.edu/. Additional information regarding FDM, FVM, and FEM algorithms and some of the meshless techniques can be obtained from the website developed by the authors. The site address is: http://www.ncacm.unlv.edu.

Chapter 9

Turbulence Modeling

Turbulent flow is a very difficult subject to master and much work is underway to fully understand the myriad intricacies and interactions. There is much that is yet to be discovered and formulated. In this chapter, we introduce the fundamental equations and relations associated with turbulence, including various forms of closure. The equations can be intimidating, but can be grasped with a little patience and perseverance.

9.1 Brief History of Turbulence Formulation

Understanding turbulent flow for problems has been of interest for over a century, dating back to Boussinesq (1877) when he introduced the idea of an eddy viscosity in addition to molecular viscosity, and to Reynolds (1895) who developed what is now called Reynolds time averaging. In 1925, Prandtl introduced the idea of a mixing length for determining the eddy viscosity. Since then various ideas have been developed to achieve a closure to the momentum equations, that is, to determine ways to solve the Reynolds stress terms. In 1945 Prandtl theorized an eddy viscosity which is dependent on turbulent kinetic energy. In 1942 Kolmogorov developed the k-ω concept which provides the turbulent length scale, $k^{1/2}/\omega$ where $1/\omega$ is the turbulent time scale. This was a two-equation model, which is a complete model because it doesn't require *a priori* knowledge of the turbulent flow to solve the equations. Wilcox (2006) presents a very good and concise history of the subject of turbulence.

Development of supercomputing and advanced high-speed desktop workstations has greatly enhanced our ability to study the evolution of the model equations for turbulence in whatever form they are theorized and prepared. There are numerous ways to model turbulence, from the simplest ideas of Prandtl mixing length models (algebraic equations), to one and two equation closure, two-equation with renormalization, and second-order closure for solving the Reynolds (time) averaged equations. With today's even larger and faster computers, Direct Numerical Simulation (DNS) is possible, to a limited extent. A relative to DNS is Large eddy Simulation (LES).

Some of the first work in second-order closure methods or stress transport methods was developed by Harlow in the late 60s (see Harlow and Hirt, 1969) and early 70s. Second-order methods introduce six additional equations, and new closure terms or models, often the dissipation rate or the specific dissipation rate, that add yet another equation to the mix. Because of these complexities, the models remain in little commercial use.

Work using the DNS method by Rai and Moin (1991) used finite difference to solve duct flow. They were among the first researchers to compare finite difference based DNS to spectral methods. Hung *et al.* (1997) studied three-dimensional turbulent flow over a backward-facing step by DNS. Time advancing was performed using a semi-implicit method. The advancement scheme for the velocity components was a compact-storage third-order Runge–Kutta scheme. The convective terms were treated explicitly, whereas the viscous term was handled implicitly. The calculated span-wise averaged reattachment length showed quasi-periodic behavior as observed in the experiments by Jovic and Driver (1994).

Since the solution to the second-order closure equations are computationally very expensive, Large eddy Simulation (LES) is a better alternative. LES resolves the dynamics of the large-scale flow while modeling the effects of the small-scale fluctuations. The LES equations are developed by spatially averaging the governing equations where all but the smallest turbulent scales are solved. The small turbulent scales are modeled. Smagorinsky (1963) developed a Subgrid-Scale (SGS) model which was used in modeling general atmospheric circulation. The SGS model incorporates the familiar eddy viscosity formulation and in many formulations requires a tuned parameter for each flow. Others have developed dynamic SGS models, where filtering techniques are used for the subgrid parameter as the flow develops (Wilcox, 2006).

Koutmos and Mavridis (1997) formulated a hybrid time-dependent Navier–Stokes model that blended elements from both the LES methodology and the standard eddy-viscosity approach. They aimed at using this hybrid model to consistently analyze unsteady separated flows including periodic and quasi-periodic flows. A finite-volume scheme based on a staggered mesh was used. The pressure and velocity fields were coupled by the Semi-Implicit Method for Pressure-Linked Equations (SIMPLE) algorithm (Patankar, 1980). The model was tested by application to a range of unsteady separated flows such as flow over square cylinders and backward-facing steps. For the backward-facing step flows, this formulation improved on the results obtained by steady-state standard k-ε closures. However, the two-equation k-ε closure scheme is known to perform poorly for flows with adverse pressure gradients and detached/reattachment of the flow.

There are limitations associated with LES because of isotropic eddy-viscosity models. Turbulent flows of practical importance are inherently three-dimensional, unsteady, and often subjected to strong inhomogeneous effects that cannot be captured by isotropic models. Persson *et al.* (2002) have developed a homogenization-based LES model using a multiple-scales expansion technique and taking advantage of the scaling properties of the Navier–Stokes equations. From the homogenization-based LES model they obtained better agreement with DNS data than the ordinary LES model. However, their approach is mathematically intensive, and they have proposed further modifications to their model.

Fureby (1999), Fuerby and Grinstein (1999), and Grinstein and Fuerby (2002) used an alternative approach to the large eddy simulation, called Monotone Integrated LES (MILES). In conventional large eddy simulation (LES) models, the filtered Navier–Stokes equations (NSE) are supplemented by subgrid-scale (SGS) models that emulate the energy transfer from large scales to the subgrid scales. In MILES, the NSE are solved by high-resolution monotone methods with embedded nonlinear filters providing implicit closure models. Governing equations were solved by finite difference technique. Time differencing was done using the Crank–Nicholson scheme. When simulations were carried out corresponding to the experiments of Pitz and Daily (1981), the LES models were found reasonably accurate and well behaved.

Kolmogorov's k-ω wasn't explored in great depth until the development of moderately quick and larger computers, that is, for computers at that time. Some of the first work was performed by Wilcox and his collaborators starting around 1972 continuing through today (Wilcox, 2006). Work by Ilegbusi (1983) revised the model to correct the near-wall treatment of the flow. Ilinca, Hetu, and Pelletier performed the development of natural logarithmic based k-ω method with a FEM formulation employing h-adaptive grid and grid remeshing (Ilinca *et al.*, 1998). They have made numerous comparisons to the k-ε closure system, using the same formulations. The benefits of the logarithmic based system are related to maintaining the positivity of the eddy dissipation without the need for cut-off or clipping limiters which can be problematic.

We introduce the formulation of a two-equation closure method in this chapter as this type of closure is considered *complete*, i.e., no advance knowledge other than initial and boundary conditions are needed. We first discuss the k-ε approach, primarily because of its popularity and wide use in many commercial CFD packages. We then introduce the k-ω model and its implementation. There is much in the literature regarding mixing length (which also includes algebraic or zero-equation) methods, along with one-equation models of turbulence. These are rather dated closure schemes, and are not popular today. Such schemes are considered to be *incomplete*, i.e., you must know something about the flow other than initial and boundary conditions. Extended closure schemes include stress-transport, or second-order closure, techniques, which represent the individual Reynolds stress terms, but these are more complicated and difficult to resolve.

The solution method for the Reynolds averaged equations is performed by the fractional step method (Carrington and Pepper, 2002). The advantage of the fractional step scheme is found in the projection of the velocity field onto the solenoid space, rendering it a mass-consistent method without the need for iteration on the velocity-pressure coupling that is found with the SIMPLE-type methods. Also, the method employs an upwinding scheme (Kelly *et al.*, 1980). These schemes are easy to implement, as discussed in Chapter 7.

In this chapter, we explain in detail a fractional-step method used to solve the equations of momentum, heat and mass transfer. The method is quite capable of providing engineers and environmental scientists, robust and reasonable solutions in the areas of combustion, solidification, heat exchanger design and environmental flow in the regimes of laminar, transitional, and turbulent. The k-ω method enhances the capabilities for modeling flow through the regimes of transition, low Reynolds number (Re) flows, to high-speed incompressible flows.

The basic postulates about flow in an Eulerian reference frame are repeated from Chapter 2 with the addition of Reynolds or time-averaged relations to obtain the two-equation closure model. Since we are interested in coupled energy and momentum transport, we formulate the basic equations, and present the turbulent transport equation for energy. Following the development of the equations, we formulate the basics of a numerical approach in the context of the fractional step method.

Among the best numerical models for simulating turbulence are FLUENT (FVM-based), STAR-CD (FVM-based), CFX (FVM-based), ANSWER (FVM-based), and COMSOL (FEM-based). These FVM codes are essentially CFD models; COMSOL is a very versatile FEM code that permits a great deal of flexibility in solving a wide range of problems.

9.2 Physical Model

Fluids are governed by mass conservation (continuity) and the instantaneous conservation of momentum and energy equations. By averaging these equations in time (Reynolds time averaging) the instantaneous equations result in additional terms, the Reynolds stresses or turbulent stresses.

Invoking mass conservation for incompressible flows,

$$\frac{\partial u}{\partial x} + \frac{\partial v}{\partial y} = 0, \tag{9.1}$$

the final form (in vector notation) for the conservation of instantaneous momentum is

$$\rho \frac{\partial \vec{V}}{\partial t} + \rho \vec{V}(\nabla \bullet \vec{V}) = \rho \vec{F} - \nabla p + \nabla \cdot \mu \nabla \vec{V}, \tag{9.2}$$

where $V\left(\nabla \bullet \vec{V}\right) = V_j \partial V_i / \partial x_j$. Using the material derivative notation, the conservation of instantaneous momentum in an incompressible fluid is

$$\rho \frac{D\vec{V}}{Dt} = \rho \vec{F} - \nabla p + \nabla \cdot \mu \nabla \vec{V}. \tag{9.3}$$

9.2.1 Turbulent flow

Turbulent flows are characterized by eddies with a wide range of length and time scales. The largest eddies are typically comparable in size to the characteristic length of the mean flow. The smallest scales are responsible for the dissipation of turbulence kinetic energy. Creating time average Navier–Stokes equations by Reynolds averaging (Tennekes and Lumley, 1972) one obtains

$$\frac{\partial \rho}{\partial t} + \frac{\partial \left(\rho \bar{u}_i\right)}{\partial x_i} = 0, \tag{9.4}$$

$$\frac{\partial \left(\rho \bar{u}_i\right)}{\partial t} + \frac{\partial \left(\rho \bar{u}_i \bar{u}_j\right)}{\partial x_j} = \frac{\partial}{\partial x_j}\left(\mu \frac{\partial s_{ij}}{\partial x_j}\right) - \frac{\partial \bar{p}}{\partial x_j} - \frac{\partial \tau_{ij}}{\partial x_j}. \tag{9.5}$$

In familiar incompressible form the momentum equation is

$$\rho \frac{\partial \bar{u}_i}{\partial t} + \rho \bar{u}_j \frac{\partial \bar{u}_i}{\partial x_j} = -\frac{\partial P}{\partial x_i} + \frac{\partial}{\partial x_j}\left(\mu s_{ij} - \rho \overline{u_i' u_j'}\right), \tag{9.6}$$

where s_{ij} is the strain rate tensor due to molecular viscosity defined as

$$s_{ij} \equiv \left(\frac{\partial \bar{u}_i}{\partial x_j} + \frac{\partial \bar{u}_j}{\partial x_i}\right), \tag{9.7}$$

and τ_{ij} is the Reynolds stress tensor given by

$$\tau_{ij} \equiv \rho \overline{u_i u_j} . \tag{9.8}$$

The momentum equation has six new unknowns related to the Reynolds stress tensor, for a total of ten unknowns in three-dimensional flows. Transport equations for the Reynolds stress tensor are derived from a moment of the momentum equation for fluctuations. Modeling the equations directly is Reynolds stress modeling. The derivation is lengthy, and is well presented in many texts (Wilcox, 2006). These six unknowns in the Reynolds stress tensor create a closure problem which can be modeled with higher moments, e.g., a two-equation model with closure coefficients. The system is not fully closed without some form of analysis to determine the closure coefficients.

Simplifying the equation, that is, creating a model that is less complicated and gives an estimate to the Reynolds stress can be performed using the average velocity terms. For time averaging, the Reynolds stress tensor can be modeled with higher moments, e.g., a two-equation model and closure coefficients. The burden in computational effort is reduced by considering turbulent kinetic energy, the trace of the Reynolds stress equation, $k = \tau_{ij}\delta_{ij} / 2$, i.e., $k = \overline{u}_i^{'}\overline{u}_i^{'} / 2$. The Reynolds stresses can be written in the form

$$\tau_{ij} = \mu_t \left(\frac{\partial \overline{u}_i}{\partial x_j} + \frac{\partial \overline{u}_j}{\partial x_i} \right) - \frac{2}{3}\delta_{ij} k, \tag{9.9}$$

by introducing a turbulent viscosity, $\mu_t = \rho c_u k^2 / \varepsilon$ where ε is the turbulent dissipation rate. The equation is meticulously derived by time averaging the velocity moment of the Navier–Stokes equations and subtracting out the kinetic energy of the mean flow, leaving only the kinetic energy of the turbulent flow.

Turbulent kinetic energy is transported having generation and dissipation terms. Turbulent kinetic energy, when using the Boussinesq assumption

$$\tau_{ij} = {}^{2}\!/_{3} K\delta_{ij} - 2\mu S_{ij}, \tag{9.10}$$

(see Hoffman and Chiang, 2000, Wilcox, 2006) is given by

$$\rho \frac{\partial \vec{k}}{\partial t} + \rho(\vec{u} \cdot \nabla)\vec{k} = c_2 \mu_t \nabla^2 \vec{k} + \vec{P}_k - \nabla \vec{D}_k - \vec{\varepsilon}, \tag{9.11}$$

where we now assume all velocities are averaged, except where noted, and where c_2 is a closure coefficient. Turbulent production, P_k, is given by,

$$P_k = \tau_{ij} \frac{\partial u_i}{\partial x_j} = \left[\mu_t \left(\frac{\partial u_i}{\partial x_j} + \frac{\partial u_j}{\partial x_i} - \frac{2}{3} \frac{\partial u_k}{\partial x_k} \delta_{ij} \right) - \delta_{ij} \frac{2}{3} \rho k \right] \frac{\partial u_i}{\partial x_j}, \tag{9.12}$$

where the turbulent diffusion, D_k, is

$$D_k = -\frac{\rho \mu_t}{\sigma_k} \frac{\partial k}{\partial x_j}, \tag{9.13}$$

and turbulent dissipation rate, ε, (noting the averaged fluctuating components) is

$$\varepsilon = \mu \overline{\frac{\partial u_i'}{\partial x_j} \frac{\partial u_i'}{\partial x_j}}. \tag{9.14}$$

The unit for turbulent kinetic energy is length^2/time^2, and ε has units of length^2/time^3.

9.2.2 Two-equation turbulence closure models

The two most popular methods that employ two-equation closure are the k-ε and k-ω techniques. These two models fall under the general class of closure schemes known as Reynolds Averaged Navier–Stokes, or RANS, models. Launder and Spalding (1972) did the most extensive work on formulating the k-ε approach, and until the late 1990s this method was the most widely used two-equation model.

The k-ω model, independently formulated by Saffman (1970), but originally developed by Kolmogorov (1942) and then essentially forgotten until the recent computer age, enjoys advantages over the k-ε model in predicting effects of adverse pressure gradients.

Two-equation models provide for computation of k as well as the turbulence length scale, i.e., they are complete since no prior knowledge of the turbulence structure is needed. An extensive discussion of turbulence and the application of various closure schemes, with particular emphasis on k-ω, is given by Wilcox (2006). This is a variation on the two-equation models.

9.2.2.1 *Two-equation k-ε*

The basic k-ε model most commonly used today was created by Launder and Sharma (1974) as an outgrowth of the earlier work by Jones and Launder (1972). The basic idea is to develop an exact equation for ε including suitable approximations for the embedded coefficients, and to solve the equation along with a similar equation for k. The equations are typically expressed as follows:

$$\frac{\partial(\rho u_i)}{\partial t}+\frac{\partial(\rho u_i u_j)}{\partial x_j}=\frac{\partial}{\partial x_i}\left((\mu+\mu_t)\left(\frac{\partial u_i}{\partial x_j}+\frac{\partial u_j}{\partial x_i}\right)\right)-\frac{\partial p}{\partial x_j}, \quad (9.15)$$

where

$$\mu_t = C_\mu k^2/\varepsilon. \quad (9.16)$$

Turbulent kinetic energy and dissipation rate are written as

$$\rho\frac{\partial k}{\partial t}+\rho u_j\frac{\partial k}{\partial x_j}=\frac{\partial}{\partial x_j}\left[(\mu+\mu_t/\sigma_k)\frac{\partial k}{\partial x_j}\right]+P_k-\varepsilon, \quad (9.17)$$

$$\rho\frac{\partial \varepsilon}{\partial t}+\rho u_j\frac{\partial \varepsilon}{\partial x_j}=\frac{\partial}{\partial x_j}\left[(\mu+\mu_T/\sigma_\varepsilon)\frac{\partial \varepsilon}{\partial x_j}\right]+C_{\varepsilon 1}\frac{\varepsilon}{k}P_k-C_{\varepsilon 2}\frac{\varepsilon^2}{k}, \quad (9.18)$$

where the closure coefficients and added relations are defined as

$$C_{\varepsilon 1}=1.44,\quad C_{\varepsilon 2}=1.92,\quad C_\mu=0.09,\quad \sigma_k=1.0,\quad \sigma_\varepsilon=1.3.$$

More recent versions of the k-ε model involve the use of renormalization group theory, or RNG. The eddy viscosity, k and ε are still used as defined above, but the other coefficients are modified (see Yakhot *et al.*, 1992).

The k-ε model has been applied to many types of flow problems. Its main drawback is its inability to respond to adverse pressure gradients when dealing with separated flows. The method also requires corrections to reproduce the law of the wall for incompressible flows over flat surfaces. Fine-tuning is generally required for an application. This scheme is closely coupled with the need to employ law-of-the-wall and wall blending functions.

9.2.2.2 *Two-equation k-w*

The two-equation mode is incorporated into an effective viscosity formulation for the Navier–Stokes equation, stated as

$$\frac{\partial(\rho u_i)}{\partial t}+\frac{\partial(\rho u_i u_j)}{\partial x_j}=\frac{\partial}{\partial x_i}\left((\mu+\mu_t)\left(\frac{\partial u_i}{\partial x_j}+\frac{\partial u_j}{\partial x_i}\right)\right)-\frac{\partial p}{\partial x_j}, \quad (9.19)$$

where

$$\mu_t=\bar{\rho}k/\omega. \quad (9.20)$$

Specific dissipation rates are related by

$$\varepsilon=\beta^{*}\omega k, \quad (9.21)$$

and the mixing length is

$$l=k^{1/2}/\varpi. \quad (9.22)$$

Turbulent kinetic energy (as above with the addition of the closure coefficient for *k)*, and dissipation rate, ω, are, respectively,

$$\rho\frac{\partial k}{\partial t}+\rho u_j\frac{\partial k}{\partial x_j}=\frac{\partial}{\partial x_j}\left[(\mu+\sigma^{*}\mu_t)\frac{\partial k}{\partial x_j}\right]+P_k-\beta^{*}\rho k\omega, \quad (9.23)$$

$$\rho\frac{\partial \omega}{\partial t}+\rho u_j\frac{\partial k}{\partial x_j}=\frac{\partial}{\partial x_j}\left[(\mu+\sigma\mu_\tau)\frac{\partial \omega}{\partial x_j}\right]+\alpha\frac{\omega}{k}P_k-\beta\rho\omega^2, \quad (9.24)$$

where the closure coefficients are defined as
$\alpha=5/9$, $\beta=3/40$, $\beta^{*}=C_\mu=0.09$, $\sigma=0.5$, $\sigma^{*}=0.5$ and $\mu_t=\rho k/\omega$.

The specific dissipation rate is a measure of the root mean square of the fluctuating vorticity with units of inverse time, 1/time. This is a measure of the enstrophy (or RMS fluctuating vorticity) in the system. The advantages of this formulation become more apparent in wall-bounded flows because the equation can be integrated to the wall when sufficient discretization of the boundary layer is employed (Carrington and Pepper, 2002).

The k-ω model is more accurate than the k-ε approach in dealing with two-dimensional boundary layers with adverse and favorable pressure gradients. The method is also effective when dealing with free shear flows and separated flows. However, both two-equation models can be inaccurate when predicting turbulent flows over curved surfaces and may not predict secondary motions in non-circular duct flows.

9.2.3 Large eddy Simulation (LES)

Large eddy Simulation, or LES, involves the computation of large eddies with the smallest, or subgrid-scale eddies (SGS) modeled. The largest eddies, which include the Reynolds stress terms, are affected by the boundary conditions and must be calculated. The less significant small-scale turbulence is more suitable for modeling. A major difficulty for LES is when the flow is near a solid surface (since the eddies are small), typically requiring a much finer grid and smaller time steps than the two-equation models.

An important part of LES modeling is the use of filters. A filter function is generally employed to eliminate scales smaller than the mesh size. There are various forms of this filter function, e.g., volume-average box filter, Fourier cutoff filter, and Gaussian filters have been popular (see Ferziger, 1977). The filter essentially introduces a scale (usually denoted as Δ where $\Delta = (\Delta x \Delta y \Delta z)^{1/3}$) that represents the smallest turbulence scale that is permitted by the filter. The filter serves to separate the resolvable scales from the subgrid scales.

For incompressible flow, the equation for fluid motion can be expressed as

$$\frac{\partial(u_i)}{\partial t}+\frac{\partial(u_i u_j)}{\partial x_j}=-\frac{1}{\rho}\frac{\partial P}{\partial x_i}+\frac{\partial}{\partial x_j}\left(\nu\frac{\partial u_i}{\partial x_j}+\tau_{ij}\right), \tag{9.25}$$

where

$$\begin{aligned}\tau_{ij} &= -\left(Q_{ij}-\frac{1}{3}Q_{kk}\delta_{ij}\right)\\ P &= p+\frac{1}{3}\rho Q_{kk}\delta_{ij}\\ Q_{ij} &= R_{ij}+C_{ij},\end{aligned} \tag{9.26}$$

with R_{ij} and C_{ij} being the SGS Reynolds stress and cross-term stress terms, respectively (see Wilcox, 2006). The problem is apparent in that one must create a model for the SGS stresses as denoted by the tensor Q_{ij}. Ferziger (1977) states that the subgrid scales account for a significant amount of the energy spectrum.

Models have been formulated that vary from a simple gradient–diffusion scheme (see Smagorinsky, 1963) to a second-order closure model (Deardorff, 1973). More recent efforts are based on a dynamic SGS model (Carati and Eijnden, 1997). Our efforts in modeling turbulence using LES have centered more on the Smagorinsky (1963) approach where

$$\begin{aligned}\tau_{ij} &= 2\nu_T S_{ij}, \quad S_{ij}=\frac{1}{2}\left(\frac{\partial u_i}{\partial x_j}+\frac{\partial u_j}{\partial x_i}\right)\\ \nu_T &= (C_S\Delta)^2\sqrt{S_{ij}S_{ij}},\end{aligned} \tag{9.27}$$

with C_s being the Smagorinsky coefficient ($0.10 < C_s < 0.24$).

As one can see, it becomes apparent that the SGS model needs to incorporate more and more of the Reynolds stresses as the flow approaches a wall. Research efforts continue to improve and modify the LES approach to modeling turbulence.

9.2.4 Direct Numerical Simulation (DNS)

Since turbulent flows are transient, three-dimensional processes, one can essentially solve the full 3-D Navier–Stokes equations to simulate all aspects of flow. In other words, the N–S equations are complete and permit one to exactly calculate laminar or turbulent flows. Solving these governing equations is referred to as Direct Numerical Simulation, or DNS. However, implementing a DNS approach is exceedingly time consuming and essentially impractical for applied problems, especially when dealing with indoor pollutant dispersion. A massive number of node points must be employed to accurately capture the detailed microstructure and eddy cascade inherent in turbulence. On the other hand, DNS solutions can provide insight and understanding of turbulence structure when viewed as a means to investigate fundamental physical processes.

The principal issue involving DNS modeling is determining the number of node points and time steps needed to conduct a valid simulation. The grid must be fine enough to resolve eddies at the Kolmogorov length scale. Likewise, the time step must be of the same order as the Kolmogorov time scale, $t = (\nu/\varepsilon)^{1/2}$. For example, assume that you wish to solve for flow in a device with a length scale of 1 m. If the flow is in a fully developed turbulent condition with $Re_\tau = 500$, then a rule of thumb is that the largest eddies are about 1/10 the characteristic scale length, or 0.1 m. The characteristic length scale to model the smallest dissipating eddies is about 1 mm. Hence, grid spacings of around 1 mm should be used, i.e., $O(10^9)$ grid points/m^3. The total computational effort is proportional to Re_τ^5. It is easy to see that a magnitude change in the order of Re_τ would require an increase of five orders of magnitude in the computational process. A simple formula (Wilcox, 2006) is

$$N_{DNS} = (3\,Re_\tau)^{9/4}, \quad Re_\tau = \frac{u_\tau L}{\nu},$$

where u_τ is the shear velocity adjacent to a surface. Such requirements can easily exhaust even the largest of present-day supercomputers. While the development of massively parallel computers over the past few years has greatly improved execution times, the problem of storage is still troublesome.

Typically high order numerical schemes are employed to achieve desired accuracy, e.g., spectral methods (Fourier series in each spatial direction), in order to limit numerical dispersion. The more conventional techniques using FDM, FVM, or FEM are not appropriate since they are typically low order schemes. One must ensure that grid convergence is also achieved. In addition, one must be careful to avoid *rolloff* (where the energy spectrum rapidly decays near the Kolmogorov length scale) and *aliasing*, which occurs as a result of nonlinear interactions among the resolved wavenumbers producing spurious waves. While spectral methods are accurate when computing derivatives at the smallest scales, they do not work well with unstructured or nonuniform grids.

9.2.5 Turbulent transport of energy or enthalpy

Generally, it is of interest to solve energy transport and its influence on the momentum equation, since often a tight coupling between fluid momentum and energy exist via changes in density and effects on pressure. The first law of thermodynamics as applied to a small volume allows for a straight forward development of a governing equation. By balancing the energy generated in the volume with the sum of energy fluxes entering and leaving the volume must equal the increase (decrease) of energy in the volume. Once the time rate of change of energy in the system is determined, it can be averaged in time to provide both mean and fluctuating components, just as was done for the momentum equation, yielding the turbulent enthalpy transport equation. This energy balance, or conservation of energy of a system, is often written as

$$d\dot{W} + d\dot{E}_c + d\dot{E}_d = d\dot{E}_i, \tag{9.28}$$

where $d\dot{W}$ is the rate of work done on the system, rate $d\dot{E}_c$ is energy which is advected through the boundaries of the fluid control volume, $d\dot{E}_d$ is the rate that energy is diffused into the volume either my conduction of molecular diffusion, and $d\dot{E}_i$ is the time rate of increase (decrease) of the change of energy in the volume.

9.2.6 Derivation of enthalpy transport

Using the small control volume shown in Fig. 9.1 we can easily develop a transport equation for energy and subsequently the equation for enthalpy transport. Figure 9.1 shows only the contributions to rate of change in energy for the control volume in the x direction by internal and kinetic energy, body, pressure, heat flux, advection, and stress forces. We have neglected heat sources, such as chemical energy or radiation heat fluxes, or radiation energy density.

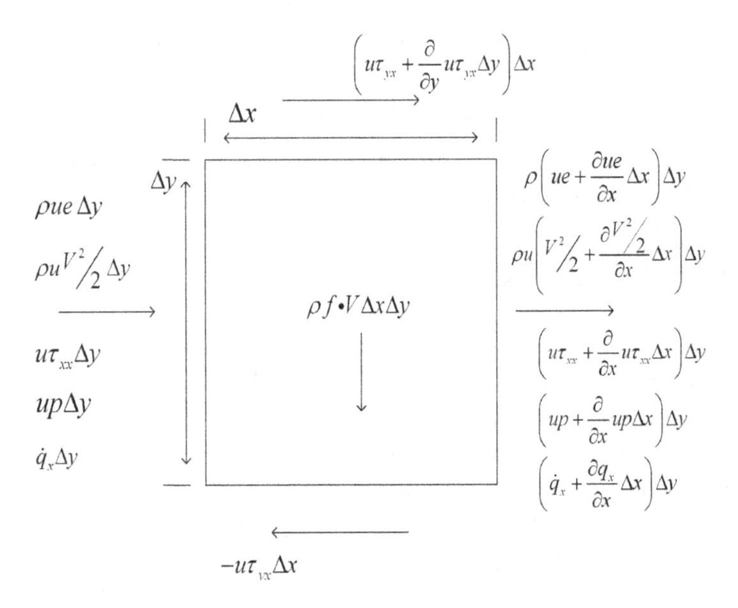

Fig. 9.1 Conservation of energy – x-component in 2-D Cartesian coordinate system.

Again, the rate of change of total energy in time is equal to the sum of the rate of kinetic and internal energy entering and leaving the control volume (change in energy flux), the rate of work done to the volume, the amount of heat transported by diffusion (conduction), and the amount of heat generation in an incremental volume of fluid. From Fig. 9.1, the net rate of internal energy flowing into the volume is

$$\left(u\frac{\partial \rho e}{\partial x} + v\frac{\partial \rho e}{\partial y}\right), \tag{9.29}$$

with kinetic energy $(u^2 + v^2)/2 \equiv V^2/2$, the total change in energy flux in the volume is

$$\left(\frac{\partial u\rho\left(e+\frac{V^2}{2}\right)}{\partial x}+\frac{\partial v\rho\left(e+\frac{V^2}{2}\right)}{\partial y}\right)=\frac{\partial u_i\rho\left(e+\frac{V^2}{2}\right)}{\partial x_i}=$$
$$\frac{\partial u_i\rho e}{\partial x_i}+\frac{\partial u_i\rho\frac{V^2}{2}}{\partial x_i}. \tag{9.30}$$

Rate of work done on the fluid volume includes action by pressure, body and stress forces. Rate of work done by pressure forces acting on a small volume is given as

$$-\left(\frac{\partial}{\partial x}(pu)+\frac{\partial}{\partial y}(pv)\right)=-\left(u\frac{\partial p}{\partial x}+v\frac{\partial p}{\partial y}\right)-\left(p\frac{\partial u}{\partial x}+p\frac{\partial v}{\partial y}\right). \tag{9.31}$$

Rate of stress work acting normal to the volume's surface is

$$-\frac{\partial}{\partial x}\left(u_i\sigma_{ij}\delta_{ij}\right)=-\frac{\partial}{\partial x}\left(u\sigma_{xx}\right)-\frac{\partial}{\partial y}\left(v\sigma_{yy}\right), \tag{9.32}$$

and that acting tangentially is

$$-\frac{\partial}{\partial y}\left(u\sigma_{xy}\right)-\frac{\partial}{\partial x}\left(v\sigma_{yx}\right). \tag{9.33}$$

The rate of work by stress is incorporated into the dissipation term

$$\phi=\sigma_{ij}\frac{\partial u_i}{\partial x_j}=\sigma_{xx}\frac{\partial u}{\partial x}+\sigma_{xy}\frac{\partial u}{\partial y}+\sigma_{yx}\frac{\partial v}{\partial x}+\sigma_{yy}\frac{\partial v}{\partial y}, \tag{9.34}$$

which becomes, for a Newtonian fluid,

$$\phi=2\mu\left(\frac{\partial u}{\partial x}\right)^2+2\mu\left(\frac{\partial v}{\partial y}\right)^2+\mu\left(\frac{\partial u}{\partial y}+\frac{\partial v}{\partial x}\right)^2$$
$$+\left(\beta-\frac{2}{3}\mu\right)\left(\frac{\partial u}{\partial x}+\frac{\partial v}{\partial y}\right)^2. \tag{9.35}$$

The rate of heat flux per unit area by diffusion into the incremental volume is

$$-\left(\frac{\partial \dot{q}_x}{\partial x}+\frac{\partial \dot{q}_y}{\partial y}\right). \tag{9.36}$$

Rate of work done by gravitational body force is

$$\rho f \bullet V dxdy = \rho u g_x + \rho v g_y. \tag{9.37}$$

Neglecting radiation energy density, and the divergence of radiation flux, and adding the above terms, the transport of total energy is

$$\begin{aligned}\frac{d\rho E_t}{dt} &= -\left(u\frac{\partial \rho e}{\partial x}+v\frac{\partial \rho e}{\partial y}\right)-\frac{\partial u\rho\frac{V^2}{2}}{\partial x}-\frac{\partial v\rho\frac{V^2}{2}}{\partial y}\\ &+\frac{\partial}{\partial x}\left(\kappa\frac{\partial T}{\partial x}\right)+\frac{\partial}{\partial y}\left(\kappa\frac{\partial T}{\partial y}\right)-p\left(\frac{\partial u}{\partial x}+\frac{\partial v}{\partial y}\right)\\ &-\left(u\frac{\partial p}{\partial x}+v\frac{\partial p}{\partial y}\right)+\phi+\rho u g_x+\rho v g_y,\end{aligned} \tag{9.38}$$

where we have invoked Fourier's Law for heat flux, $\dot{q}_x = \kappa\partial T/\partial x$ and κ the conductance.

After performing the chain rule on the second term of the right-hand side of Eq. 9.38 and using mass conservation the total energy equation becomes

$$\begin{aligned}&\rho\frac{DE_t}{Dt}+p\left(\frac{\partial u}{\partial x}+\frac{\partial v}{\partial y}\right)+\left(u\frac{\partial p}{\partial x}+v\frac{\partial p}{\partial y}\right)=\\ &\frac{\partial}{\partial x}\left(\kappa\frac{\partial T}{\partial x}\right)+\frac{\partial}{\partial y}\left(\kappa\frac{\partial T}{\partial y}\right)+\phi+\rho u g_x+\rho v g_y.\end{aligned} \tag{9.39}$$

With enthalpy defined as

$$h = e + \frac{p}{\rho}, \tag{9.40}$$

we see that

$$\frac{Dh}{Dt} = \frac{De}{Dt} - \frac{Dp}{Dt} + \frac{p}{\rho^2}\frac{D\rho}{Dt}. \tag{9.41}$$

From mass consistency,

$$\frac{D\rho}{Dt} = -\rho\frac{\partial v_i}{\partial x_i}.$$

The enthalpy equation is then

$$\rho\frac{Dh}{dt} = \left[\frac{\partial}{\partial x}\left(k\frac{\partial T}{\partial x}\right) + \frac{\partial}{\partial y}\left(k\frac{\partial T}{\partial y}\right)\right] + \frac{\partial p}{\partial t} + \phi + \rho u g_x + \rho v g_y. \tag{9.42}$$

From Maxwell's relations (Callen, 1985), we have

$$dh = c_p dT + \frac{1}{\rho}(1 - \beta T)dp, \tag{9.43}$$

then

$$\frac{dc_p dT}{dt} + \frac{\beta T}{\rho}\frac{dp}{dt} =$$

$$\left(u\frac{\partial e}{\partial x} + v\frac{\partial e}{\partial y}\right) + \frac{1}{\rho}\left[\frac{\partial}{\partial x}\left(k\frac{\partial T}{\partial x}\right) + \frac{\partial}{\partial y}\left(k\frac{\partial T}{\partial y}\right)\right] \tag{9.44}$$

$$+\frac{1}{\rho}\left(u\frac{\partial p}{\partial x} + v\frac{\partial p}{\partial y}\right) + \frac{1}{\rho}\phi,$$

where β is the coefficient of thermal expansion. Enthalpy transport is further simplified using dyadic notation

$$\rho c_p\frac{\partial T}{\partial t} + \rho c_p u_j\frac{\partial T}{\partial x_j} + (1 - \beta T)\left[\frac{\partial P}{\partial t} + u_j\frac{\partial P}{\partial x_j}\right] = \frac{\partial}{\partial x_j}\left(\kappa\frac{\partial T}{\partial x_i}\right). \tag{9.45}$$

Assuming an ideal gas law ($h = \rho c_p T$) and incompressible flow, the conservation of energy becomes under the first law

$$\rho c_p\frac{\partial T}{\partial t} + \rho c_p u_j\frac{\partial T}{\partial x_j} = \frac{\partial}{\partial x_j}\left(\kappa\frac{\partial T}{\partial x_i}\right). \tag{9.46}$$

9.2.7 Turbulent energy transport

Equation (9.46) is the instantaneous change in thermal energy for an incompressible fluid. After separating the instantaneous into its mean and fluctuating components, just as was done to derive the momentum equation and averaging, thermal transport is given as,

$$\rho c_p \frac{\partial \bar{T}}{\partial t} + \rho c_p \bar{u}_j \frac{\partial \bar{T}}{\partial x_j} = \frac{\partial}{\partial x_j}\left(\kappa \frac{\partial \bar{T}}{\partial x_i} \right) - \frac{\partial \bar{T}' \bar{u}'}{\partial x_i}. \tag{9.47}$$

Representing the turbulent heat flux we introduce in the same fashion as with turbulent inertia term,

$$\frac{\partial \bar{T}' \bar{u}_j'}{\partial x_j} = \varepsilon_H \frac{\partial \bar{T}}{\partial x_j} = \frac{\mu_t}{\mathrm{Pr}_t} \frac{\partial \bar{T}}{\partial x_j}, \tag{9.48}$$

where ε_H is the turbulent eddy viscosity for enthalpy transport, and is shown as a scale of the fluid's eddy viscosity, i.e., a turbulent Prandtl number, Pr_t (Bejan, 1984). Now the turbulent heat transport becomes

$$\rho c_p \frac{\partial T}{\partial t} + \rho c_p u_j \frac{\partial T}{\partial x_j} = \frac{\partial}{\partial x_j}\left(\left(\frac{\mu_t}{\mathrm{Pr}_t} + \kappa \right) \frac{\partial T}{\partial x_i} \right), \tag{9.49}$$

where we have discarded the average notation bar, that is, assumed average values for temperature and velocity. The eddy viscosity scaling for general engineering solutions is equal to one for fluids with moderate to high Prandtl numbers, Pr, and therefore, $\varepsilon_H = \mu_t$. Actually, Pr_t varies as a function of the distance away from a wall in the boundary layer but averages out to one, explaining why the value of $\mathrm{Pr}_t \approx 1$ produces reasonably accurate engineering results. For liquid metals however, fluids having $\mathrm{Pr} << 1$; correlations have been developed for various liquid metals (Thomas, 1999). In terms of internal energy alone the energy equation becomes

$$\frac{de}{dt} = -\left(u\frac{\partial e}{\partial x} + v\frac{\partial e}{\partial y}\right)$$
$$+\frac{1}{\rho}\left[\frac{\partial}{\partial x}\left(\left(\frac{\mu_t}{\Pr_t} + \kappa\right)\frac{\partial T}{\partial x}\right) + \frac{\partial}{\partial y}\left(\left(\frac{\mu_t}{\Pr_t} + \kappa\right)\frac{\partial T}{\partial y}\right)\right] \quad (9.50)$$
$$-\frac{1}{\rho}\left(u\frac{\partial p}{\partial x} + v\frac{\partial p}{\partial y}\right) + \frac{1}{\rho}\phi.$$

9.2.8 Turbulent transport species

If the concentration of a species is not influencing the momentum equations, it is possible to advect the species with the mean fluid flow. Under these situations, typically found in pollution transport, a scalar equation is appropriate and has the same form as the thermal energy transport Eq. 9.49, given by

$$\rho\frac{\partial C_s}{\partial t} + \rho u_j\frac{\partial C_s}{\partial x_j} = \frac{\partial}{\partial x_j}\left(\left(\frac{\mu_\tau}{Sc_t} + \kappa_s\right)\frac{\partial C_s}{\partial x_i}\right). \quad (9.51)$$

9.2.9 Coupled fluid-thermal flow

Most fluids experience expansion when absorbing heat. This expansion changes the fluid's density. The difference in density between the warmer and cooler fluid, a fluid that is differentially heated, produces a buoyant body force within the fluid. This is a body force represented by $(\rho_o - \rho)g_{x_i} / \rho$, where ρ_o is the reference density of the fluid (the reservoir's hydrostatic density). In this case the momentum is given by

$$\frac{\partial(\rho u_i)}{\partial t} + \frac{\partial(\rho u_i u_j)}{\partial x_j} =$$
$$\frac{\partial}{\partial x_i}\left((\mu + \mu_t)\left(\frac{\partial u_i}{\partial x_j} + \frac{\partial u_j}{\partial x_i}\right)\right) - \frac{\partial p}{\partial x_i} + (\rho_o - \rho)g_{x_i}. \quad (9.52)$$

Determining the density is by equation of state (EOS), which for an ideal gas is

$$\rho = P_H / RT, \tag{9.53}$$

where P_H is the hydrostatic pressure and $-\rho_o g_{x_i} = dP_H / dx_i$ is the gradient in hydrostatic pressure at some point. Then $(\rho_o - \rho) g_{x_i}$ is the total difference in the pressure gradient for the x_i component caused by thermal expansion in the direction of gravitational force. The mechanical pressure gradient $\partial p / \partial x_i = 0$ for natural convection can be quite large for forced convective flow and is therefore still present in Eq. 9.52.

Since we are assuming incompressible flow, we can make use of the thermal expansion coefficient for the fluid,

$$\beta = \frac{1}{V}\frac{\partial V}{\partial T}\bigg|_P = -\frac{1}{\rho}\frac{\partial \rho}{\partial T}\bigg|_P . \tag{9.54}$$

Solving for density in terms of temperature, we get (by the mean value theorem of calculus)

$$\int_{\rho_o}^{\rho} d\rho = \rho - \rho_o = -\int_{T_o}^{T} \rho\beta dT = -\overline{\rho\beta}\left(T - T_o\right), \tag{9.55}$$

where $\overline{\rho\beta} \approx \rho\beta$ for moderate differences in temperature. Using this result, Eq. 9.52 becomes

$$\frac{\partial(\rho u_i)}{\partial t} + \frac{\partial(\rho u_i u_j)}{\partial x_j} = \frac{\partial}{\partial x_i}\left((\mu + \mu_t)\left(\frac{\partial u_i}{\partial x_j} + \frac{\partial u_j}{\partial x_i}\right)\right) - \frac{\partial p}{\partial x_i} + g_{x_i}\beta(T - T_o). \tag{9.56}$$

9.3 Numerical Modeling

Numerous methods exist for the solution of these nonlinear equations. The most popular numerical approach is the FVM, as seen in the number of commercial CFD codes now available. However, the FEM is becoming more popular, especially with regards to modeling turbulence. The absorption of unstructured meshing capabilities into the FVM commercial codes comes from the FEM. The merging of these two methods is evolving – over time, the blending may become rather transparent. The elegance and enhanced accuracy of the FEM make it the preferred choice of numerical schemes for the authors.

A method for velocity, pressure and vorticity formulation employing Newton's method to linearize the momentum equation (Bochev and Gunzburger, 1998), and using the least squares finite element method with a conjugate gradient technique, was demonstrated to work well (Jiang and Lin, 1993). However, without some form of projection in the finite element formulation, mixed methods are required, i.e., different approximations or finite dimensional spaces are necessary for the velocity (Lebesgue or $\mathbf{L_2}$) and pressure (Sobolev or $\mathbf{H}$) to satisfy the Div-Stability condition (Gunzburger, 1989) also known as the LBB conditions. With the use of a projection method, where the pressure is being estimated from the flow as in well-known SIMPLE, semi-implicit or self-adjoint projections (Chorin, 1968, Lohner, 1990), the LBB conditions are satisfied.

A self-adjoint projection scheme as developed by various researchers including Gresho and Chan (1990), Lohner (1990), and Ramaswamy *et al.* (1992), also provides a solution for nonlinear problems. This semi-implicit scheme ideally has an advantage over iterative methods that may not have good convergence rates, that is, when the method is not supplied with a reasonable first guess.

The projection method for the solution of the Navier–Stokes equations is a self-adjoint system created by decomposing the momentum into gradient-driven or curl-free portions and divergence-free portions. A divergence-free velocity field is maintained by the projection of the predicted velocity onto the divergence-free space. An Euler–Lagrangian variational seeks to minimize the functional (Gresho, 1985)

$$E(\mathbf{V},\lambda)=\frac{1}{2\,dt}\int_{\Omega}\left(\left[\mathbf{V}-\mathbf{V}^{*}\right]^{2}+[\lambda][C]\{\mathbf{V}\}\right)d\Omega\,. \qquad (9.57)$$

In the incompressible case, pressure is recovered from the divergence of the momentum equation using some initial or recently calculated velocity. The velocities are then updated from the pressure, which enforces continuity. This splitting method is discussed below as an Euler–Lagrange Variational projection into divergence-free space.

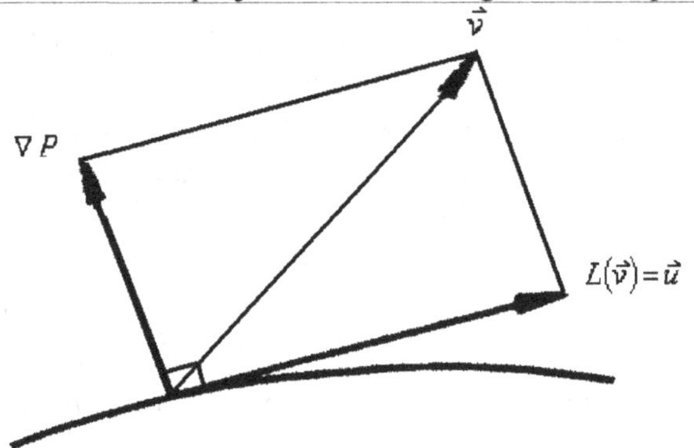

Fig. 9.2 Decomposition of vector.

9.3.1 Projection algorithm

The projection-step algorithm incorporated in this computer model is based on the method initially developed by Chorin (Fletcher, 1984). Using the Helmholtz–Hodge decomposition theorem which states that any vector field in domain, Ω, can be uniquely decomposed as

$$\mathbf{V}=\mathbf{U}+\nabla \mathrm{P}, \qquad (9.58)$$

where $\mathbf{U}$ is the divergence-free velocity vector, i.e., $\nabla\bullet\mathbf{U}=0$ in Ω, and $\mathbf{U}\bullet\mathbf{n}=0$ on the boundary Γ, i.e., parallel to the boundary. This portion of the decomposition is a projection onto a divergence-free field (Chorin and Manderson, 1993, Marchioro and Pulvirenti, 1994).

The projection is shown is Fig. 9.2 for the velocity field,$\mathbf{V}$. Notice the gradient portion has zero curl under the decomposition since the vector identity $\nabla \times \nabla P=0$. The curl of a vector field that is a function of the gradient of a scalar is irrotational or curl free.

Under the projection we seek the proper P such that

$$\nabla P = \mathbf{V} + \mathbf{U}, \tag{9.59}$$

taking the divergence of each side yields

$$\begin{gathered}\nabla \bullet \nabla P = \nabla^2 P = \nabla \bullet (\mathbf{V} + \mathbf{U}) = \\ \nabla \bullet (\mathbf{V}) + \nabla \bullet (\mathbf{U}) = \nabla \bullet (\mathbf{V})\end{gathered} . \tag{9.60}$$

The linear orthogonal projection operator L, applied to the incompressible Navier–Stokes vector field yields

$$L\left(\frac{\partial \mathbf{U}}{\partial t} + \nabla P\right) = L\left(-(\mathbf{U}\bullet\nabla)\rho\mathbf{U} + \mu\nabla^2\mathbf{U}\right), \tag{9.61}$$

and since L is a linear operator and $L(\nabla P) = 0$ as shown previously, pressure is removed from the equation set. The projection under L is given as

$$\frac{\partial \mathbf{U}}{\partial t} = -(\mathbf{U}\bullet\nabla)\rho\mathbf{U} + \mu\nabla^2\mathbf{U}, \tag{9.62}$$

where U is divergence-free, the averaged time velocity. During the time advancement of the averaged velocity field a projection onto the divergence-free space is performed, maintaining a divergence-free velocity field. This is accomplished with the proper choice of P, and splitting velocity into the divergence-free field, and the perturbed field or predictor.

The fractional split is explained as follows: splitting the velocity into two averaged components, **V*** and **V**, the momentum equations under the linear orthogonal projection operator L described above become

$$\rho\frac{\mathbf{V}^*_{n+1} - \mathbf{V}_n}{dt} + \rho\mathbf{V}_n \bullet \nabla \mathbf{V}_n = \mu\nabla^2\mathbf{V}_n, \tag{9.63}$$

where the velocity components of $\mathbf{V}$ are either from the initial guess or from the previously calculated time step which is the divergence-free velocity attained through the proper choice of grad (P).

Given the approximate velocity just advanced from the previous explicit marching, the goal is to find some velocity $\mathbf{V}$ that satisfies continuity. We seek the projection of $\mathbf{V}^*$, a perturbed velocity, onto the divergence-free space to complete the calculation of the velocities subject to incompressibility. Under the decomposition of the vector field $L(\mathbf{V}^*)$, we make the projection,

$$\mathbf{V}^* = \mathbf{V} + dt\ \nabla P \text{ where } \nabla \bullet \mathbf{V} = 0. \tag{9.64}$$

Taking the gradients of both sides, a Poisson equation for P is obtained in the form

$$\nabla^2 P = -\nabla \bullet \mathbf{V}^* / dt. \tag{9.65}$$

In discretized finite element representation we have

$$M(\mathbf{V} - \mathbf{V}^*) / dt + \nabla P = 0, \tag{9.66}$$

where M is the mass matrix. Eqs. (9.65) and (9.66) are essentially the Euler–Lagrange equation

$$\frac{M}{dt}\left(\mathbf{V} - \mathbf{V}^*\right) + C\,P = 0, \tag{9.67}$$

where C is the gradient operator. The equation is subject to the constraint of continuity

$$C^T \mathbf{V} = 0. \tag{9.68}$$

The system is solved sequentially by creating a diagonal form of the mass matrix (a lumped matrix), multiplying by its inverse, and by taking the gradients of both sides and also enforcing continuity, that is,

$$C^T M^{-1} C P = C^T \mathbf{V}^*, \tag{9.69}$$

$$\mathbf{V} = \mathbf{V}^* - dt\, M^{-1}\, C\, P. \tag{9.70}$$

9.3.2 Finite volume approach

By far the most popular approach in discretizing the governing equations of fluid motion, including turbulence, is the FVM. Most commercial CFD codes utilize the FVM approach. There are many books and references with regards to formalizing the computational method, along with free codes that can be downloaded from the web (this is evident when conducting a Google search on CFD and Finite Volume Method). We shall give a basic overview of the method here as it would apply to modeling turbulence.

We begin with the general form of the transport equation (representing momentum, temperature, or species transport)

$$\frac{\partial(\rho\phi)}{\partial t}+\nabla\bullet(\rho\mathbf{V}\phi)=\nabla\bullet(\Gamma\nabla\phi)+S_{\phi}, \tag{9.71}$$

where φ = [u, v, w, T, C, k, ε], Γ is the diffusion coefficient (or viscous term), and S is the source or sink term. This equation is integrated over the control volume (CV) and time, i.e.,

$$\begin{aligned}&\int_{CV}\left(\int_{t}^{t+\Delta t}\frac{\partial(\rho\phi)}{\partial t}\right)d\forall+\int_{t}^{t+\Delta t}\left(\int_{CS}\hat{n}\bullet(\rho\mathbf{V}\phi)dA\right)dt\\&=\int_{t}^{t+\Delta t}\left(\int_{CS}\hat{n}\bullet(\Gamma\nabla\phi)dA\right)dt+\int_{t}^{t+\Delta t}\int_{CV}S_{\phi}d\forall dt,\end{aligned} \tag{9.72}$$

where φ denotes the variables for velocity, temperature, species transport, turbulent kinetic energy, and turbulent energy dissipation (if using k-ε closure) and CS denotes the control surface (area). A staggered grid configuration is the most common meshing technique for the FVM. A fully implicit discretization of Eq. 9.72 using a finite volume approach produces the following relation (Versteeg and Malalasekera, 1995), where a hybrid differencing scheme is employed to implement upstream weighting (to insure stability and minimize dispersion errors associated with advection)

$$\begin{aligned}a_P\phi_P&=a_W\phi_W+a_E\phi_E+a_S\phi_S+a_N\phi_N+\\&\quad a_B\phi_B+a_T\phi_T+a_P^0\phi_P^0+S_u,\end{aligned} \tag{9.73}$$

where

$$a_P = a_W + a_E + a_S + a_N + a_B + a_T + a_P^0 + \Delta F - S_P,$$

with

$$a_P^0 = \frac{\rho_P^0 \Delta V}{\Delta t}, \quad \underline{S}\Delta V = S_u + S_P \phi_P,$$

where $\underline{S}$ is the average source term obtained from linear approximation (another popular method is to integrate this term using Simpson's rule). The coefficients are defined as

$$a_W = \max\left[F_w, \left(D_w + \frac{F_w}{2}\right), 0\right], a_E = \max\left[-F_e, \left(D_e - \frac{F_e}{2}\right), 0\right]$$
$$a_S = \max\left[F_s, \left(D_s + \frac{F_s}{2}\right), 0\right], a_N = \max\left[-F_n, \left(D_n - \frac{F_n}{2}\right), 0\right] .$$
$$a_B = \max\left[F_b, \left(D_b + \frac{F_b}{2}\right), 0\right], a_T = \max\left[-F_t, \left(D_t - \frac{F_t}{2}\right), 0\right]$$
$$\Delta F = F_e - F_w + F_n - F_s + F_t - F_b$$

The expressions for F and D are given as

Face	W	E	S	N	B	T
F	$(\rho u)_w A_w$	$(\rho u)_e A_e$	$(\rho v)_s A_s$	$(\rho v)_n A_n$	$(\rho w)_b A_b$	$(\rho w)_t A_t$
D	$\frac{\Gamma_w}{\Delta x_{WP}} A_w$	$\frac{\Gamma_e}{\Delta x_{PE}} A_e$	$\frac{\Gamma_s}{\Delta x_{SP}} A_s$	$\frac{\Gamma_n}{\Delta x_{PN}} A_n$	$\frac{\Gamma_b}{\Delta x_{BP}} A_b$	$\frac{\Gamma_t}{\Delta x_{PT}} A_t$

with

$$\Delta V = \Delta x \Delta y \Delta z$$
$$A_E = A_W = \Delta y \Delta z$$
$$A_N = A_S = \Delta x \Delta z$$
$$A_B = A_T = \Delta x \Delta y$$

The control volume is illustrated in Fig. 9.3 below.

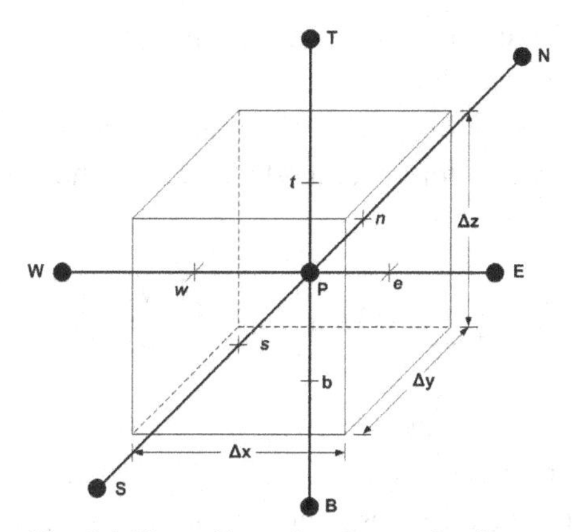

Fig. 9.3 Three-dimensional control volume.

The nodes are labeled N, S, E, W, T, B and stand for north, south, east, west, top and bottom; the faces of the control volume are identified by the lower-case letters.

Further modifications to the technique include modifying the advection terms using a quadratic upwinding, known as QUICK (Leonard, 1979), and Total Variation Diminishing, or TVD (Osher, 1984), schemes to reduce numerical oscillations. The general procedure is to solve the primitive equations of fluid motion using the SIMPLE algorithm (Patankar and Spalding, 1972), which stands for Semi-Implicit Method for Pressure-Linked Equations (essentially a guess and correct technique for calculating pressure), and then solve the remaining set of equations sequentially. Other variations of the SIMPLE procedure include SIMPLER (Patankar, 1980) and SIMPLEC (Van Doormal *et al.*, 1986). Another popular procedure employs a splitting operator technique known as PISO (Issa, 1986).

9.3.3 Finite element approach

The weak statement for the projection, or Euler-Lagrange variational, is

$$\int_{\Omega}\left[\frac{\partial N_j}{\partial x_j}\right]\{N_k\}\sum_{l=1}^{n}[N_l]\left\{\frac{\partial N_i}{\partial x_i}\right\}\{P_i\}\,d\Omega = \int_{\Omega}\left[\frac{\partial N_j}{\partial x_j}\right]\left\{\mathbf{V}_i^*\right\}d\Omega/dt, \quad (9.74)$$

where N denotes the shape (or basis) functions in the FEM and the subscripts refer to coordinate directions (or rows and columns when performing FEM assembly); the summation creates the diagonalized mass matrix. Solving for averaged **V** from the weighted residual statement

$$\{\mathbf{V}_i\} = \{\mathbf{V}_i^*\} - dt\left[\{N_k\}\sum_{l=1}^{n}[N_l]d\Omega\right]^{-1}\left[\int_{\Omega_e}\left\{\frac{\partial N_i}{\partial x}\right\}[N_j]\{P_i\}d\Omega\right], \quad (9.75)$$

produces the divergence-free velocity.

A time-explicit advancement of velocity is made using the weakened momentum equations and an assumed initial pressure at time n = 0. The projection onto a divergent free field is made to ensure mass continuity. Pressure can be determined if desired. The whole process is repeated, i.e., marched forward in time. Once the inverse matrix is established for the solution of the pressure, the most time-consuming part of the process is the solution of the Euler–Lagrange equation enforcing mass consistency.

9.3.3.1 *Weak forms of the governing equations*

To apply the finite element method to the solution of the governing equations, the weak statements of the equations must be established and then coded. The energy and mass transport equations are included below in the development. These will be needed to solve convective and species transport problems.

The Method of Weighted Residuals is applied to the weak statements resulting in the following representation of the governing equations. The dependent variables are replaced with their trial functions, that is,

$$Z_i = \sum_{i=1}^{N}\phi_n(x)z_i^n(t) = [N_j]\{Z_i\}, \quad (9.76)$$

where Z_i are the dependent variables, $[N_j]$ is the basis (shape) function notation for the element $[\ ]$ is a row vector (row matrix), the transpose of a column vector $\{\ \}$. Other terms in the following weak statements include the integral over the domain or element $\int_{\Omega} d\Omega$ and over the boundary surface $\int_{\Gamma} d\Gamma$.

The specifics of evaluating the boundary term is presented after combining the following into matrix equations, where the boundary terms become the load vector in the matrix statement. Substituting the dependent variable trail functions into governing equations, or model equations, and second order terms as described above, produces the following set of integrated ordinary differential equations.

Weighted residual statement of velocity under decomposition

$$
\begin{aligned}
&\left(\rho\int_{\Omega}\{N_k\}\sum_{l=1}^{n}[N_l]d\Omega\right)\left\{\dot{\mathbf{V}}_i\right\}+\rho\left(\int_{\Omega}\{N_i\}(N_k\mathbf{V}_k)\left[\frac{\partial N_j}{\partial x_i}\right]d\Omega\right)\{\mathbf{V}_i\}\\
&+\int_{\Omega}\{N_i\}\left[\frac{\partial N_j}{\partial x_i}\right]\{P_i\}d\Omega+\int_{\Omega}\{N_i\}\frac{2}{3}\left[\frac{\partial N_j}{\partial x_i}\right]d\Omega\{k_i\}\\
&-\left(\int_{\Omega}\{N_i\}\left[\frac{\partial N_j}{\partial x_i}\right]\{\mu_t\}\left[\frac{\partial N_j}{\partial x_i}\right]d\Omega\right)\{\mathbf{V}_i\}\\
&+\left(\int_{\Omega}[\mu+\mu_t]\left\{\frac{\partial N_j}{\partial x_i}\right\}\left[\frac{\partial N_j}{\partial x_i}\right]d\Omega\right)\{\mathbf{V}_i\}\\
&+\rho\int_{\Omega}f(x_i)\{N_i\}d\Omega-\int_{\Gamma}\{N_i\}[\mu+\mu_t]\left[\frac{\partial N_j}{\partial x_i}\right]\{\mathbf{V}_i\}\hat{n}_i d\Gamma=0,
\end{aligned}
\tag{9.77}
$$

where $f(x_i)$ is the body force per unit mass, typically supplied by gravity. As mentioned previously, for slightly compressible fluids, i.e., those subject to the Boussinesq approximation for density change as a function of temperature, this body force is the difference in gravity forces and buoyant forces, $(\rho_o - \rho)g_x$.

Weighted residual statement of thermal energy

The thermal energy equation can be written in the form

$$\left(\rho c_p \int_\Omega \{N_k\} \sum_{l=1}^{n} [N_l] d\Omega\right)\{\dot{T}_i\}$$
$$+\left(\rho c_p \int_\Omega \{N_i\}(N_k \mathbf{V}_k)\left[\frac{\partial N_j}{\partial x_i}\right] d\Omega\right)\{T_i\}$$
$$-\int_\Omega \{N_i\}\left(\left[\frac{\partial N_j}{\partial x_i}\right]\left\{\frac{\mu_t}{Pr_t}\right\}\left[\frac{\partial N_j}{\partial x_i}\right]\right) d\Omega \{T_i\} \tag{9.78}$$
$$+\left(\int_\Omega \left[\kappa + \frac{\mu_t}{Pr_t}\right]\left\{\frac{\partial N_j}{\partial x_i}\right\}\left[\frac{\partial N_j}{\partial x_i}\right] d\Omega\right)\{T_i\}$$
$$-\left(\int_\Omega \{N_i\}\{Q_i\} d\Omega\right) - \left(\int_\Gamma \{N_i\}\{q_i\}\Gamma d\Gamma\right) = 0.$$

Weighted residual statement of turbulent kinetic energy

$$\left(\rho \int_\Omega \{N_k\} \sum_{l=1}^{n} [N_l] d\Omega\right)\{\dot{k}_i\} + \left(\rho \int_\Omega \{N_i\}(N_k \mathbf{V}_k)\left[\frac{\partial N_j}{\partial x_i}\right] d\Omega\right)\{k_i\}$$
$$-\left(\int_\Omega \{N_i\}\left[\frac{\partial N_j}{\partial x_i}\right]\{\sigma^* \mu_t\}\left[\frac{\partial N_j}{\partial x_i}\right] d\Omega\right)\{k_i\}$$
$$+\left(\int_\Omega \left[\mu + \sigma^* \mu_{t_i}\right]\left\{\frac{\partial N_j}{\partial x_i}\right\}\left[\frac{\partial N_j}{\partial x_i}\right] d\Omega\right)\{k_i\} \tag{9.79}$$
$$+\left(\rho \beta^* \int_\Omega \{N_i\}[N_j]\{k_i\}[N_j]\{\varpi_i\} d\Omega\right)$$
$$-\{P_k\} - \left(\int_\Gamma \{N_i\} q_k \Gamma d\Gamma\right) = 0,$$

where

$$\{P_k\} = \int_\Omega \begin{bmatrix} \{N_j\}[\mu_{t_i}] \begin{pmatrix} \left[\dfrac{\partial N_j}{\partial x_i}\right]\{u_i\} + \left[\dfrac{\partial N_i}{\partial x_j}\right]\{u_i\} \\ -\dfrac{2}{3}\dfrac{\partial N_k}{\partial x_k}\{u_k\}\delta_{ij} \end{pmatrix} \\ -\delta_{ij}\dfrac{2}{3}\rho[N_j]\{k_i\} \end{bmatrix} d\Omega. \quad (9.80)$$

Weighted residual statement of specific dissipation rate

The specific dissipation rate equation can be expressed as

$$\begin{aligned} &\left(\rho \int_\Omega \{N_k\}\sum_{l=1}^{n}[N_l]d\Omega\right)\{\dot{\varpi}_i\} + \left(\rho \int_\Omega \{N_i\}(N_k\mathbf{V}_k)\left[\frac{\partial N_j}{\partial x_i}\right]d\Omega\right)\{\omega_i\} \\ &\quad -\left(\int_\Omega \{N_i\}\left[\frac{\partial N_j}{\partial x_i}\right]\{\sigma\mu_t\}\left[\frac{\partial N_j}{\partial x_i}\right]d\Omega\right)\{\varpi_i\} \\ &\quad +\left(\int_\Omega [\mu+\sigma\mu_t]\left\{\frac{\partial N_j}{\partial x_i}\right\}\left[\frac{\partial N_j}{\partial x_i}\right]d\Omega\right)\{\varpi_i\} \qquad (9.81) \\ &\quad +\left(\beta\rho\int_\Omega \{N_i\}\left([N_j]\{\omega_i\}\right)^2 d\Omega\right) \\ &\quad +P_\omega d\Omega - \left(\int_\Gamma \{N_i\} q_\varpi \Gamma d\Gamma\right) = 0, \end{aligned}$$

where

$$\{P_\varpi\} = \alpha \int_\Omega \begin{bmatrix} \{N_j\}[\mu_{t_i}] \begin{pmatrix} \left[\dfrac{\partial N_j}{\partial x_i}\right]\{u_i\} + \left[\dfrac{\partial N_i}{\partial x_j}\right]\{u_i\} \\ -\dfrac{2}{3}\dfrac{\partial N_k}{\partial x_k}\{u_k\}\delta_{ij} \end{pmatrix} \\ -\delta_{ij}\dfrac{2}{3}\rho[N_j]\{k_i\} \end{bmatrix} \left(\frac{[N_j]\{\omega_i\}}{[N_j]\{k_i\}}\right) d\Omega \cdot \quad (9.82)$$

9.3.3.2 *Matrix equations*

By integrating over each element and combining the contributions from each element to nodes in common to those elements, a matrix equation is formed that will be solved for these nodal values. It is important to note that when integrating over each element and summing, the contributions of the surface flux, $\int_{d\Gamma} d\Gamma$ cancels everywhere except at the boundaries.

This is an important distinction between the finite volume methods (FVM) and the classical finite element method. It also leads some to the idea that the FEM statements are not locally conservative. The fact is, an FEM is precisely conservative, where as FVM is not because of truncation error associated with evaluating the surface fluxes everywhere within the domain.

The matrix equations for the explicit time advancement of momentum, heat and mass transport can be written as

$$[\mathbf{M}]\left\{\dot{\mathbf{V}}\right\}+[\mathbf{A}(\mathbf{V})]\{\mathbf{V}\}-\mathbf{C}\{\mathbf{P}\}+[\mathbf{K}_v]\{\mathbf{V}\}=\{\mathbf{F}_v\}, \tag{9.83}$$

$$[\mathbf{M}_T]\left\{\dot{T}\right\}+[\mathbf{A}(\mathbf{V})]\{T\}+[\mathbf{K}_T]\{T\}=\{\mathbf{F}_T\}, \tag{9.84}$$

$$[\mathbf{M}]\left\{\dot{k}\right\}+[\mathbf{A}(\mathbf{V})]\{k\}+[\mathbf{K}_k]\{k\}=\{\mathbf{P}_k\}+\left\{\beta^*\mathbf{k}\,\mathbf{w}\right\}+\{\mathbf{F}_k\}, \tag{9.85}$$

$$[\mathbf{M}]\left\{\dot{\varpi}\right\}+[\mathbf{A}(\mathbf{V})]\{\varpi\}+[\mathbf{K}_\varpi]\{\omega\}=\left\{\alpha \mathbf{k}/\boldsymbol{\omega}\right\}\{\mathbf{P}_k\}+\left\{\beta\mathbf{w}^2\right\}\{\mathbf{F}_\varpi\}. \tag{9.86}$$

The individual matrices for these equations are defined as

$$\mathbf{M}=\rho\int_\Omega\{N_k\}\sum_{l=1}^{n}[N_l]\,d\Omega, \tag{9.87}$$

$$\mathbf{M}_T=\rho c_p\int_\Omega\{N_k\}\sum_{l=1}^{n}[N_l]\,d\Omega, \tag{9.88}$$

$$\mathbf{A}(\mathbf{V})=\rho\int_\Omega N_i\left(N_k\mathbf{V}_k\right)\frac{\partial N_i}{\partial x_j}\,d\Omega, \tag{9.89}$$

$$\mathbf{K}_v = -\left(\int_\Omega \{N_i\} \left[\frac{\partial N_j}{\partial x_i} \right] \{\mu_t\} \left[\frac{\partial N_j}{\partial x_i} \right] d\Omega \right) + \int_\Omega \left([\mu + \mu_t] \frac{\partial N_j}{\partial x_i} \frac{\partial N_j}{\partial x_i} \right) d\Omega, \tag{9.90}$$

$$\mathbf{C} = \int_\Omega \frac{\partial N_j}{\partial x_i} N_i d\Omega, \tag{9.91}$$

$$\mathbf{F}_v = \int_\Omega N_j f(x_i) d\Omega + \int_\Gamma [\mu + \mu_t] N_j \mathbf{n}_j \frac{\partial \mathbf{V}_j}{\partial x_j} d\Gamma, \tag{9.92}$$

$$\mathbf{K}_T = -\left(\int_\Omega \{N_i\} \left[\frac{\partial N_j}{\partial x_i} \right] \left\{ \frac{\mu_t}{\mathrm{Pr}_t} \right\} \left[\frac{\partial N_j}{\partial x_i} \right] d\Omega \right) + \int_\Omega \left(\left\{ \kappa + \frac{\mu_t}{\mathrm{Pr}_t} \right\} \frac{\partial N_i}{\partial x_j} \frac{\partial N_i}{\partial x_j} \right) d\Omega, \tag{9.93}$$

$$\mathbf{K}_k = -\left(\int_\Omega \{N_i\} \left[\frac{\partial N_j}{\partial x_i} \right] \{\mu_t\} \left[\frac{\partial N_j}{\partial x_i} \right] d\Omega \right) + \int_\Omega \left(\left[\mu + \sigma^* \mu_t \right] \frac{\partial N_i}{\partial x_j} \frac{\partial N_i}{\partial x_j} \right) d\Omega, \tag{9.94}$$

$$\mathbf{K}_\varpi = -\left(\int_\Omega \{N_i\} \left[\frac{\partial N_j}{\partial x_i} \right] \{\sigma \mu_t\} \left[\frac{\partial N_j}{\partial x_i} \right] d\Omega \right) + \int_\Omega \left([\mu + \sigma \mu_t] \frac{\partial N_i}{\partial x_j} \frac{\partial N_i}{\partial x_j} \right) d\Omega, \tag{9.95}$$

$$\mathbf{F}_T = \left(\rho \int_\Omega \{N_i\} \{Q_i\} d\Omega \right) + \left(\int_\Gamma \{N_i\} \left[N_j \right] \{q_i\} d\Gamma \right), \tag{9.96}$$

$$\mathbf{P}_k = \int_{\Omega} \left[\begin{array}{c} \{N_j\}\left[\mu_{t_i}\right]\left(\begin{array}{c} \left[\frac{\partial N_j}{\partial x_i}\right]\{u_i\} + \left[\frac{\partial N_i}{\partial x_j}\right]\{u_i\} \\ -\frac{2}{3}\frac{\partial N_k}{\partial x_k}\{u_k\}\delta_{ij} \end{array} \right) \\ -\delta_{ij}\frac{2}{3}\rho\left[N_j\right]\{k_i\} \end{array} \right] d\Omega \cdot \tag{9.97}$$

9.3.3.3 *Time advancement of the explicit/implicit matrix equations*

The initial guess of velocity is marched in time explicitly by

$$\left\{\mathbf{V}_i^{n+1}\right\} = \left\{\mathbf{V}_i^{n}\right\} + \Delta t\left[\mathbf{M}^{-1}\right]\left[\begin{array}{c} \left\{\mathbf{F}_{v_i}\right\} - \left[\mathbf{K}_v\right]\left\{\mathbf{V}_i^n\right\} - \\ \left[\mathbf{A}(V)\right]\left\{\mathbf{V}_i^n\right\} + \left[\mathbf{C}\right]\left\{\mathbf{P}_i^n\right\} \end{array} \right]. \tag{9.98}$$

This explicit marching also applies equally to the scalar quantities of temperature, turbulent kinetic energy, specific dissipation rate, and species transport. Before marching these quantities forward in time, then the velocities need to be projected onto the divergence-free field. The velocities are updated from the components of P

$$\mathbf{V}^{n+1} = \mathbf{V}^{*} + dt\,\mathbf{M}^{-1}\mathbf{C}\,P. \tag{9.99}$$

The pressure is calculated from either the discretized Poisson equation or is extracted directly from the projection algorithm by dividing λ with *dt*. This pressure is associated with the projection, the time-advanced divergent velocity. To calculate the pressure experienced in the momentum equations, the gradient is taken of the divergent free Navier–Stokes equations, resulting in the Poisson equation. For a better estimate of the pseudo-velocity this “dynamic” pressure can be supplied into the time advancement as a gradient.

Scalar transport for enthalpy (or internal energy) and species are performed as per the scalar transport equation

$$\left\{\mathbf{T}_i^{n+1}\right\} = \left\{\mathbf{T}_i^{n}\right\} + \Delta t\left[\mathbf{M}^{-1}\right]\left[\left\{\mathbf{F}_{v_i}\right\} - \left[\mathbf{K}_T\right]\left\{\mathbf{T}_i^n\right\} - \left[\mathbf{A}(\mathbf{V})\right]\left\{\mathbf{T}_i^n\right\}\right]. \tag{9.100}$$

Time-step size should be a consideration on this explicit statement. The time scale for most engineering and environmental problems governed by the faster time scales of turbulence and momentum transport.

Scalar transport for turbulent kinetic energy and species is performed as per equation the scalar transport equation

$$\left\{K_i^{n+1}\right\} = \left\{K_i^{n}\right\} + \Delta t\left[\mathbf{M}^{-1}\right]\left[\begin{matrix}\{\mathbf{F}_k\}+\{\mathbf{P}_k\}+\{\beta^{*}\mathbf{k}\,\mathbf{w}\}- \\ [\mathbf{K}_k]\{K^{n}\}-[\mathbf{A(V)}]\{K^{n}\}\end{matrix}\right], \quad (9.101)$$

$$[\mathbf{M}]\left\{\dot{\varpi}\right\} = \left\{\omega_i^{n}\right\} + \Delta t\left[\mathbf{M}^{-1}\right]\left[\begin{matrix}\{\mathbf{F}_\varpi\}+\{\alpha\,\mathbf{k}/\omega\}\{\mathbf{P}_k\}+\{\beta\,\mathbf{w}^2\}- \\ [\mathbf{K}_\varpi]\{\omega\}-[\mathbf{A(V)}]\{\varpi\}\end{matrix}\right]. \quad (9.102)$$

The explicit and implicit equations for velocity and pressure are always solved to the boundaries, based on the latest update to the boundary conditions, i.e., those boundaries which are changing with the flow. These are the turbulent closure model boundary conditions k, ω and μ_t, which are discussed next along with boundaries for velocity, energy and pressure. When using the law of the wall, the $k-\varpi$ equations are only solved to the point next to the solid boundary because the boundary for these points is determined by the wall function. Otherwise the model can be solved to all boundaries provided the grid resolution is sufficient enough to provide for accurate solution in the boundary layer.

9.3.3.4 *Mass lumping*

Mass lumping is the combining of the time-dependent terms in the mass matrix, row by row, into a diagonal matrix. This is done by simply adding the terms of each row. Lumping creates a matrix that has as its inverse

$$\mathbf{M}_L^{-1} \equiv \frac{1}{\mathbf{M}_L}, \quad (9.103)$$

where

$$\mathbf{M}_L = \int_\Omega N_i \sum_{j=1}^{n} N_j d\Omega . \quad (9.104)$$

Mass lumping makes the time dependent equation an explicit equation (Pepper, 1987, Carrington and Pepper, 2002). Mass lumping speeds the transient solution significantly; multiplication by the inverted mass matrix is not required at each time step. The size of the time increment for explicit advancement is governed by stability requirements based on the Courant and Reynolds cell numbers.

9.3.3.5 *General numerical solution*

A Petrov–Galerkin (P–G) scheme is used to weight the advection terms, i.e.,

$$W_i = N_i + \frac{\alpha h_e}{2V}(\mathbf{V} \cdot \nabla N_i), \quad (9.105)$$

where h_e is the element size, $\alpha = \coth \beta/2 - 2/\beta$ with $\beta = h_e|V|/2K_e$, and K_e is an effective diffusion in the direction of the local velocity vector (Kelly, *et al.*, 1980, Heinrich and Yu, 1988, Brooks and Hughes, 1982, Hughes, 1987). This weighting introduces *selective* artificial diffusion into the numerical scheme that acts along the local streamline. This method is effective at removing numerical dispersion, leaving perhaps about 1 to 2% dispersive noise in the solution in very steep gradient areas. The dispersive error associated with modeling advection is precisely measured prior to the time advancement and is removed during integration. It is important to note, that this P–G method is also a good shock capturing scheme, even in the absence of molecular viscosity, i.e., in the absence of a Peclet number (Brueckner and Heinrich, 1991).

For non-hydrostatic calculations, the pressure is obtained from solution of a Poisson equation based on the discrete momentum equations, i.e.,

$$[K]\{p\} = C^T [M]^{-1} [\{F_V\} - ([K_v] + [A(\mathbf{V})])\{\mathbf{V}\}]. \quad (9.106)$$

Sparse Cholesky and Krylov solvers can be used to solve the Poisson pressure equation. A time-dependent form of the continuity equation is used to correct the velocities. A forward-in-time Euler scheme is employed to advance the discretized equations in time (Carrington and Pepper, 2002).

9.4 Stability and Time-Dependent Solution

The explicit Euler time-integration scheme has time-advancement restrictions that are met by the requirements on the Courant and Reynolds (Peclet) cell numbers. The determining equations for a forward-in-time, centered-in-space finite difference scheme (FTCS) can be found using a Von Neumann stability analysis (Fletcher, 1991). In fact, only the stability of linear equations can be analyzed with this type of analysis. Linearizing a nonlinear equation can be performed and the stability analyzed although it is applicable locally only (Hoffman and Chiang, 1993). The stability analysis produces guidelines to constrain the time increments.

A von Neumann stability analysis is based on Fourier mode analysis. For example, velocity can be expressed in its Fourier modes as

$$u_j^n = U^n \, e^{i\kappa\Delta x\, j}, \tag{9.107}$$

where the amplitude at time n, κ is wave number in the x direction, $\kappa\Delta x$ is the phase angle, i $=\sqrt{-1}$, and j are the discretized coordinate indices. These components are substituted into the discretization and reduced. An amplification factor 'G' is introduced such that

$$U^{n+1} = GU^n. \tag{9.108}$$

Stability requires the absolute value of 'G' be bounded for all values, or $\kappa\,\Delta x$, i.e., bounded for all phase angles.

If it is assumed that the fluid motion is wave-like in nature and a discretization is made to represent the motion over a length, Δx, the highest frequency in the interval that can be approximated is $2\,\Delta x$. That is, it requires at least three points to approximately determine a sine wave between 0 and 2π.

Hindmarsh *et al.* (1984) determined necessary and sufficient conditions for stability of the advection–diffusion equation. This analysis as applied to the explicit Euler forward scheme produces the time-increment limits

$$\Delta t \leq \frac{1}{\sum_{i=1}^{3} \frac{2K_i}{\Delta x_i^2}}, \tag{9.109}$$

and

$$\Delta t \leq \frac{1}{\sum_{j=1}^{3} \frac{u_j^2}{2K_j}}, \tag{9.110}$$

The terms K_j, refer to the jth directional component of the diffusion matrix. Since the Galerkin method utilizing linear interpolating polynomials has a centered-in-space type architecture, these stability constraints certainly give some idea as to what time increments might be allowed.

Numerical experimentation with various types of problems has shown the following stability conditions are usually satisfactory

$$\Delta t \leq \frac{2}{\mathrm{Re}} \frac{1}{\|U\|^2}, \tag{9.111}$$

$$\Delta t \leq \frac{\mathrm{Re}/2}{1/\Delta x^2 + 1/\Delta y^2 + 1/\Delta z^2}. \tag{9.112}$$

Construction of 'Δx' or 'h' in three dimensions is performed by finding the average value for the coordinates of each face and then taking the difference between opposing faces. The entire grid is searched for the constraining values in order to optimize the time step.

9.5 Boundary Conditions

Evaluating the boundary integral for the second-order bilinear equation, given as

$$\int_{\Gamma} W_i \left(-k \frac{\partial \hat{\mathbf{T}}}{\partial \mathbf{n}} \right) d\Gamma, \tag{9.113}$$

over the surface Γ requires simply noticing that

$$\frac{\partial \hat{\mathbf{T}}}{\partial \mathbf{n}} = \hat{n} \bullet \nabla \hat{\mathbf{T}} = n_x \frac{\partial \hat{\mathbf{T}}}{\partial x} + n_y \frac{\partial \hat{\mathbf{T}}}{\partial y}. \tag{9.114}$$

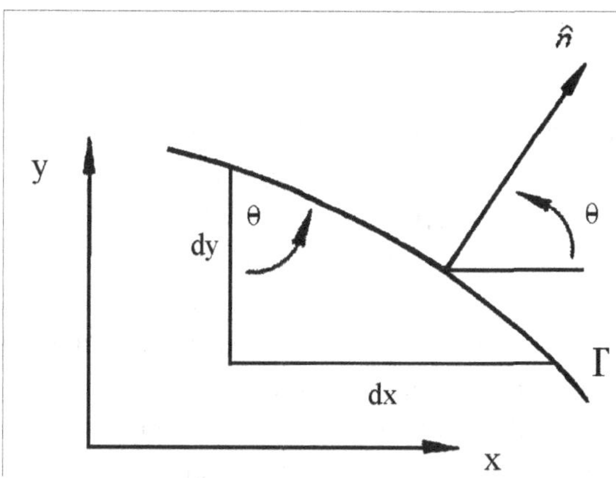

Fig. 9.4 Normal to surface and direction cosines.

Determining the value of the direction cosines, n_x and n_y are obtained from noticing that

$$n_x = \cos\theta = \frac{dy}{d\Gamma} \quad \text{and} \quad n_y = \sin\theta = -\frac{dx}{d\Gamma}, \tag{9.115}$$

as defined in Fig. 9.4. Therefore the equation for the surface integral in two-dimensions (actually a line integral) becomes

$$-\int_\Gamma k\, W_i \left(n_x \frac{\partial \hat{\mathbf{T}}}{\partial x} + n_y \frac{\partial \hat{\mathbf{T}}}{\partial y} \right) d\Gamma. \tag{9.116}$$

9.5.1 Boundary conditions for velocity under decomposition

Dirichlet boundary conditions for average velocity are straightforward, either a no-slip condition for solid objects or fixed velocity at inlets is specified. Outlet boundary conditions can be made with the assumption of a zero gradient for velocity, a Neumann condition. The zero gradient assumption on velocity at outflow requires the computational domain or grid to be constructed to match this imposed boundary condition. This statement can be relaxed with the use of the viscous boundary condition (Gresho, 1985), described as follows.

9.5.1.1 *Viscous boundary condition for velocity*

Another boundary condition exists when a weak statement is created. Weakening the second derivative viscous term results in

$$\int_{\Gamma}\{N_i\}\{\mu+\mu_t\}\left[\frac{\partial N_j}{\partial x_i}\right]\{V_i\}\,\mathbf{n}_i d\Gamma . \qquad (9.117)$$

This equation is zero for walls where no-slip conditions apply and is zero where an inlet or outlet velocity is normal to the boundary. Otherwise these components of the boundary integral are evaluated and used to relax the requirement of zero gradient of velocity at an outflow, when combined with the calculated pressure at the outflow.

9.5.2 Boundary conditions for pressure and velocity correction

Because the equation for pressure is elliptic, boundary conditions must be imposed at all surfaces of the computational domain. The Neumann boundary condition for pressure is simply

$$n \bullet \nabla P = \frac{\partial P}{\partial n} = -\mathbf{n} \bullet \frac{\partial \mathbf{V}}{\partial t} . \qquad (9.118)$$

This boundary condition, when combined with a Dirichlet condition at some reference point to eliminate the singularity in the equation set, is sufficient to determine the pressure up to an arbitrary constant (Gresho and Sani, 1987).

The second-half step is the step related to inviscid flow, the portion of the decomposition without curl since it relies on the determination of the proper scalar gradient to make the decomposition true. Therefore, the proper boundary condition would be related to the normal component of penetration through the boundary, that is,

$$\mathbf{n} \bullet \mathbf{V}^{n+1}|_{\Gamma} = \mathbf{n} \bullet \mathbf{f}^{n+1} , \qquad (9.119)$$

where f^{n+1} could be the prescribed boundary conditions or be evaluated from viscous terms at the boundary given by

$$\int_{\Gamma} \mu\{N_i\}\left[\frac{\partial N_j}{\partial x_i}\right]\{\mathbf{V}_i\}\mathbf{n}_i d\Gamma .$$

If a Lagrangian multiplier is substituted for pressure, the boundary conditions for the projection equation (the Euler–Lagrange variational statement) are found in the same manner as the pressure Poisson equation. The boundary condition for this multiplier is derived from

$$\nabla P = \frac{\mathbf{V}-\mathbf{V}^*}{dt}, \tag{9.120}$$

and combined with the boundary conditions for pressure given by Eq. 9.118. The resulting traction equation is

$$n \bullet \nabla\lambda = \frac{\partial \lambda}{\partial n} = n \bullet \left[\mathbf{V}^* - \mathbf{V}\right]. \tag{9.121}$$

Clearly, continuity applies to $\mathbf{n} \bullet \mathbf{V}$ by definition of **V**, so the boundary condition for λ is (Ramaswamy, 1990)

$$\frac{\partial \lambda}{\partial n} = \mathbf{n} \bullet \mathbf{V}^* \quad \text{on } \Gamma . \tag{9.122}$$

9.5.3 Boundary conditions for turbulent kinetic energy and specific dissipation rate

Boundary layer flow in the presence of turbulence is thought to consist of a defect and inner region. The defect region can made to include a buffer zone between the defect layer and the viscous sublayer that is next to the wall. Where these meet is a buffer zone if considering a three equation inner region, otherwise the viscous sublayer and defect region are blended by a single logarithmic equation and only the sublayer and the defect layer exist. The various regions are shown in Fig. 9.5.

A log layer melds the defect layer with the buffer zone with the outer region or it melds the defect layer with the outer region, depending on whether the model is a two or three equation wall region. The outer portion of the defect region is a fully turbulent region, hence the idea of requiring a thrid buffer region to distinguish the fully turbulent portion of the defect layer.

Traditionally turbulence is modeled only to some point near the wall and not through the inner layer. Using non-dimensional terms, u+ and y+ we can define equations to represent the regions shown in Fig. 9.5. With $y^+ = y\rho u^* / \mu$ and $u^+ = u / u^*$ where u^* is the friction velocity given by

$$u^* = \sqrt{\tau_w / \rho}, \tag{9.123}$$

and $\tau_w = \mu du / dy$ is the wall shear stress, the viscous sublayer (Wilcox, 2006) is

$$u^+ = y^+ = y\rho\sqrt{\tau_w / \rho} / \mu . \tag{9.124}$$

This sublayer equation is valid for $y^+ \leq 2 \approx 8$. Above this range for incompressible flow over a smooth surface in the absence of a pressure gradient, the buffer zone will be in the range $8 < y^+ \leq 50$ and the wall law equation Hoffman and Chiang (2000), is

$$u^+ = 5\ln(y^+) - 3.05. \tag{9.125}$$

At larger distances, up to $50 < y^+ \leq 200 \approx 400$, u^+ is

$$u^+ = 2.5\ln(y^+) + 5.0. \tag{9.126}$$

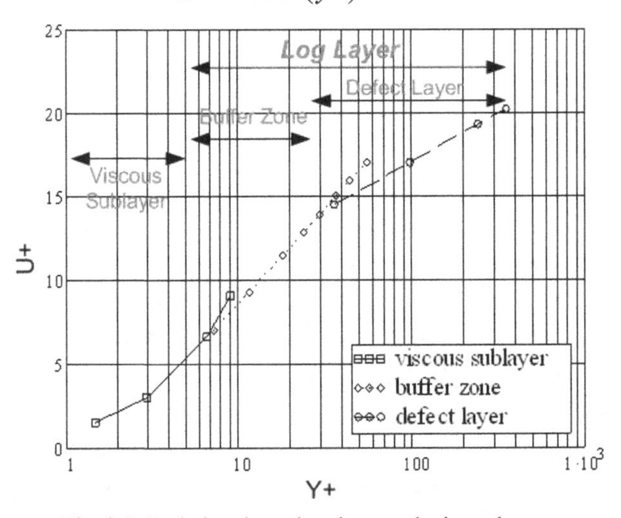

Fig. 9.5 Turbulent boundary layer – the inner layer.

The above equations form the definition of the boundary conditions at points within the inner layer, depending on the location of that point. Again, these conditions hold for smooth walls when the flow has a positive pressure gradient. The solution to the wall condition progresses by iteration since it is transcendental relation. Wilcox (2006) demonstrates the appropriate equation to solve at some point a distance 'y_p' from the wall in the presence of an adverse pressure gradient is

$$u_p = u^* \left[\frac{1}{k_{vk}} \ln\left(\frac{\rho u^* y_p}{\mu} \right) + B - 1.13 \frac{y_p}{\rho \left(u^*\right)^2} \frac{dP}{dx} \right], \quad (9.127)$$

where u_p is the tangential velocity at that grid point, **B** is a surface roughness factor, and $k_{kv} = 0.41$ is the von Karman constant. The equations for turbulent kinetic energy, k and ϖ in the inner layer are

$$k_p = \frac{\left(u^*\right)^2}{\sqrt{\beta^*}} \left[1 + 1.16 \frac{\rho u^* y_p}{\mu} P^+ \right], \quad (9.128)$$

$$\varpi_p = \frac{u^*}{k_{vk} y \sqrt{\beta^*}} \left[1 - 0.30 \frac{\rho u^* y_p}{\mu} P^+ \right]. \quad (9.129)$$

Surface roughness can be understood in terms of sand grain size, as found experimentally by Nikuradse (see Schlichting, 1979),

$$k_s^+ = \rho u^* k_s / \mu, \quad (9.130)$$

where k_s is the average grain height. A surface roughness parameter, $S_R = f(k_s)$ is incorporated into the roughness factor B,

$$B = 8.4 + \frac{1}{k_{vk}} \ln\left(\frac{S_R}{100} \right). \quad (9.131)$$

Wilcox (2006) suggests using the following correlations between S_R and k_s^+

$$S_R = \begin{Bmatrix} \left(50 / k_s^+\right)^2, & k_s^+ < 25 \\ 100 / k_s^+ & , k_s^+ \geq 25 \end{Bmatrix}. \quad (9.132)$$

A wall boundary condition for the specific dissipation rate can be stated as a function of average sand grain height (Wilcox, 2006), i.e., ϖ is a function of surface roughness,

$$\varpi = \frac{2500\mu / \rho}{k_s^2}, \tag{9.133}$$

to be applied at the wall. Integration to the wall can proceed by using Eq. 9.133 and specifying turbulent kinetic energy equal to zero at the wall and requiring that the gradient of the specific dissipation rate be zero. At solid wall boundaries, we set the turbulent eddy viscosity equal to zero because there is now flow at that point and not turbulent kinetic energy.

9.5.4 *Boundary conditions for thermal and species transport*

Thermal and species transport equations have either specified flux (Neumann), or fixed (Dirichlet) conditions. As noted earlier, with the FEM method, a zero flux is automatically applied if no other boundary condition exists. For solid walls, species concentration is fixed at zero. When a molecule of any species attaches itself to the wall, i.e., is deposited to a wall, it is no longer part of the transported material. In order to count the amount of species deposited to a wall, the flux can be calculated from the gradient and the diffusivity in the boundary layer. This is performed in FEM similarly to that of heat flux and is presented later where we discuss finding a local Nusselt number with specified wall temperature.

The integral of thermal flux is calculated for the energy transport equation by

$$\int_{\Gamma} \{N_i\} \left[N_j \right] \{q_i\} \, d\Gamma, \tag{9.134}$$

remembering that the shape functions are now for 1-D line elements or 2-D surfaces elements depending on whether the problem is 2-D or 3-D, respectively. By integrating momentum and energy to the boundaries with values determined for the turbulent viscosity $\kappa + \mu_t / \Pr_t$ in the wall function (or by integration of the closure model to the boundaries), Eq. 9.134 is applied in the transport equations.

No special treatment for temperature at a point within the Log Layer, i.e., the regime determined by the law of the wall, is required. Species flux from a surface, that is, mass injection would also be calculated in an identical fashion.

9.5.5 Thermal and species flux calculation in the presence of Dirichlet boundaries

Often it is necessary to calculate a flux at a boundary given the solution to the flow field in the presence of Dirichlet boundary temperatures. For example, determining the rate of mass and heat flux is of engineering interest. This quantity is not known from the model equations when Dirichlet boundaries are applied. At any point in time the governing equation can be solved for the flux given the current state of the dependent variables using the thermal energy equation

$$\begin{aligned}
&\left(\int_{\Gamma}\{N_i\}\left[N_j\right]\{q_i\}d\Gamma\right) = \\
&-\left(\rho c_p \int_{\Omega}\{N_i\}(N_k V_k)\left[\frac{\partial N_j}{\partial x_i}\right]d\Omega\right)\{T_i\} \\
&+\left(\int_{\Omega}\{N_i\}\left[\frac{\partial N_j}{\partial x_i}\right]\left\{\frac{\mu_t}{Pr_t}\right\}\left[\frac{\partial N_j}{\partial x_i}\right]d\Omega\right) \\
&-\left(\int_{\Omega}\left\{\kappa + \frac{\mu_t}{Pr_t}\right\}\left\{\frac{\partial N_j}{\partial x_i}\right\}\left[\frac{\partial N_j}{\partial x_i}\right]d\Omega\right)\{T_i\}.
\end{aligned} \tag{9.135}$$

It is a simple matter at this point to calculate local Nusselt numbers from a known local conductance κ_L, and calculated local heat flux, q_L,

$$Nu_L = \frac{h_L \Delta y}{\kappa_L} = \frac{q_L \Delta y}{\kappa_L (T_w - T_L)}, \tag{9.136}$$

where h_L is the local convectivity coefficient, Δy is the normal distance from where the wall temperature, T_w, is determined to the point in the domain where T_L, the local fluid temperatures, is calculated. These local quantities are averaged over the cell's area since the flux is the cell's surface flux. Nusselt values and mass transfer coefficients, flux, for species from a surface (mass injection), would also be calculated in an identical fashion, with Pr_t replaced with the turbulent Schmidt number Sc_t.

9.6 Validation of Turbulence Models

Algorithms and computer codes are verified and then validated with experimental data. When numerical solutions of benchmark problems have already been established these solutions also can be used to validate new algorithms and software. Verification involves the process of understanding the model equations and their implementation. Verification answers the question of, "Are the model equations represented correctly." Validating a particular implementation is the process of determining whether the model equations and their numerical representation are capable of solving the modeled phenomena correctly.

Here we present two problems that will gage the accuracy of the model's implementation, (1) flow in a duct, (2) flow over the a 2-D backward-facing step. Figure 9.6 shows the geometric configuration for flow in a 2-D duct.

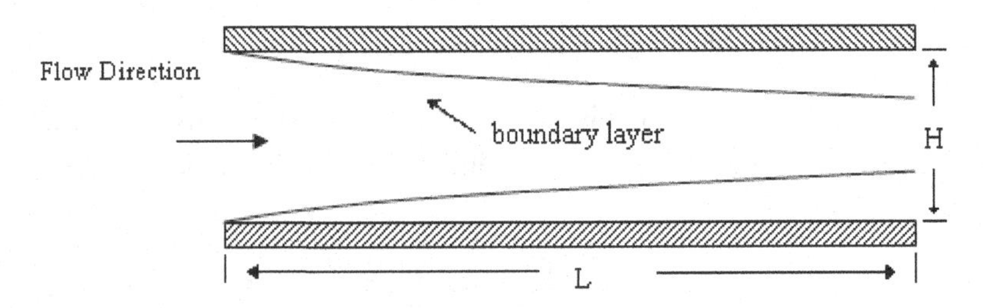

Fig. 9.6 Schematic of the two-dimensional duct.

Flow and heat transfer over a backward facing step (Fig. 9.7) is more challenging. This problem is of interest to researchers in the areas of combustion, solidification, environmental flow and compact heat exchanger design, etc. Study of flow and heat transfer over a backward-facing step has been a good benchmark problem due to its simplistic geometry, richness of flow physics, and availability of experimental data. Some of these include Schram, *et al.* (2004) and Kostas, *et al.* (2002) backward-facing step flow investigating with particle velocimetry. Armaly, *et al.* (1983) provided both experimental and theoretical investigation to the backward-facing step flow. Gartling (1990) proposed using the backward-facing step geometry as a test problem for outflow boundary conditions.

Flow over a backward-facing step includes interesting physical phenomena such as unsteady behavior, separation, recirculation, reattachment, and three-dimensionality. In combustion applications chemical reaction and radiation play a dominant role. Depending on the Reynolds number, various recirculation zones set up at the center plane as shown in Fig. 9.7.

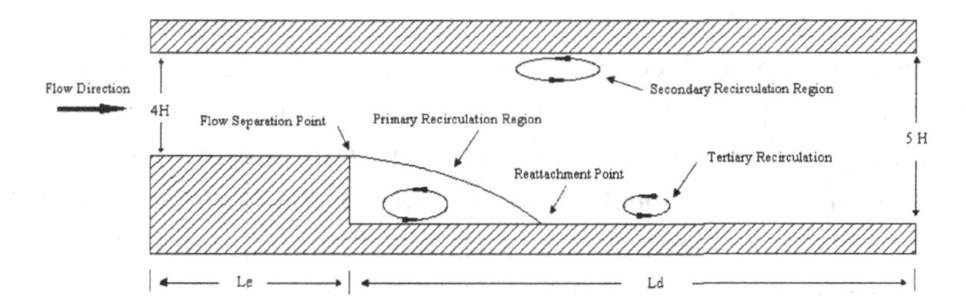

Fig. 9.7 Schematic of the two-dimensional backward-facing step showing recirculation areas for unsteady flow.

At low Reynolds numbers (*Re*) flow is laminar and steady. When *Re* increases, a recirculation zone downstream of the back step forms and flow becomes unsteady. The loci of reattachment points (reattachment line) are known to oscillate back and forth. In transition regime in addition to the primary recirculation pocket downstream of the back step two additional recirculation pockets are formed. One is on the top of the duct and one is downstream of the primary recirculation pocket as shown in Fig. 9.7. As *Re* increases further these two additional recirculation pockets disappear and flow becomes turbulent.

Investigating flow over the 2-D backward-facing step using a k-ω two-equation turbulent closure modeled by FEM and using an h-adaptive grid demonstrates that the model appears to be well suited for the transition to turbulent regimes. The solution method incorporates uncoupled enthalpy transport. Eventually chemical reactions and radiation phenomena will be added to the model.

The physics of flow over backward-facing step in the laminar, transitional and low Reynolds number turbulent regimes was studied experimentally by Armaly *et al.* (1983).

The 2-D backward-facing step is merely a tool for benchmarking. Assuming flow in a very wide duct achieves two-dimensionality or is somehow symmetric along the center plane is not precisely accurate. So, investigators have been developing solutions to flow over a 3-D backward-facing step, where the symmetry assumptions may not be applicable in the unsteady and higher flow regimes.

Williams and Baker (1997) performed numerical investigations of laminar flow over a three-dimensional backward-facing step. Williams and Baker employ the continuity constraint method (CCM) in conjunction with the Galerkin finite element technique to solve the unsteady three-dimensional Navier–Stokes equations. An implicit scheme was used to march in time. They found agreement of their results with the experimental data of Armaly *et al.* (1983).

Pepper and Carrington (1997) introduced forced convective heat transfer using a finite element method and the pressure projection method for this problem. They have reported good agreement of their results with the experimental data in the low transition regime at Re = 800 (Carrington and Pepper, 2002).

Chiang and Sheu (1999) also simulated the three-dimensional laminar flow over a backward-facing step. Euler implicit scheme was used for the time derivatives. Good agreement with the experimental data was found.

Example 9.6.1 Solution for flow in a 2-D duct: Solutions for flow in a 2-D duct are presented in Figs. 9.8 through 9.15. The flow is at a Re = 13,750 as determined by a density of 1.1774 kg/m, an inlet velocity of 4.312m/s, a dynamic viscosity $\mu = 1.846\text{x}10^{-5}$ (N sec/m^2), and an inlet diameter (hydraulic diameter) of 0.1 meters. The turbulent Re_t number = 440.0 as determined by

$$Re_\tau = \frac{\delta}{\nu / u^*} = \frac{H}{2\nu\sqrt{\rho / \tau_{wall}}} = \frac{.05}{1.135e-04} = 440. \qquad (9.137)$$

The inlet turbulent kinetic energy is k = 0.025 and specific dissipation rate at the inlet is ϖ = 68.0 for an inlet turbulent viscosity of $3.7\text{x}10^{-4}$ and a initial value of $1.0\text{x}10^{-5}$. The grid consists of 12,544 elements, and 13,021 nodes.

The state of the flow as shown in the figures is at steady state, that is, when the sup ||residual|| norm of the dependent variables is to be less than 1e-05. The thermal properties are a conductance of κ=0.02624 (W/m-K°), specific heat at constant pressure c_p = 1.057 x10^3 (J/kg-K°) and having a turbulent Prandlt number Pr_t = 1.0. At two levels of adaptation the final grid density was 66,429 nodes and 63,070 elements.

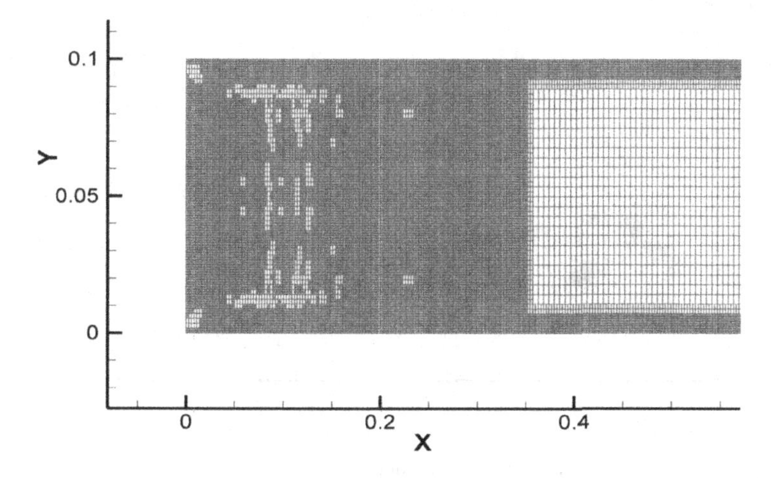

Fig. 9.8 Adapted grid for flow in 2-D duct – two levels of adaptation used at Re=13,750.

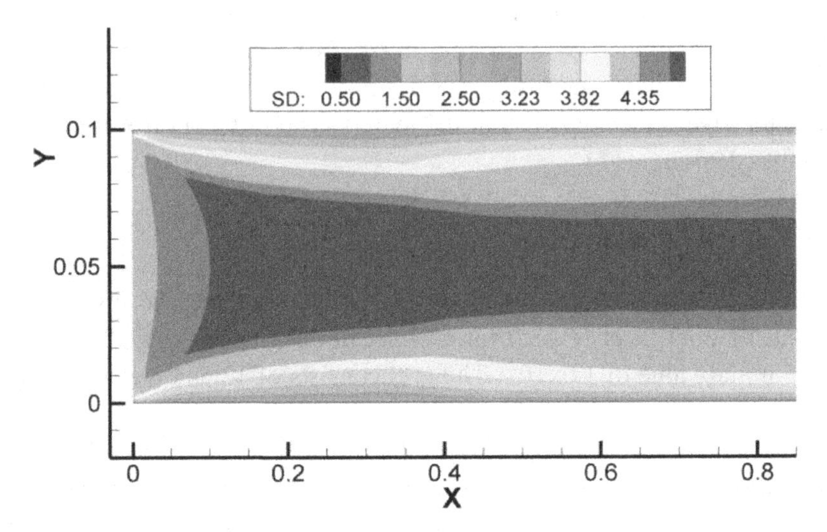

Fig. 9.9 Two-dimensional duct turbulent flow – speed contours.

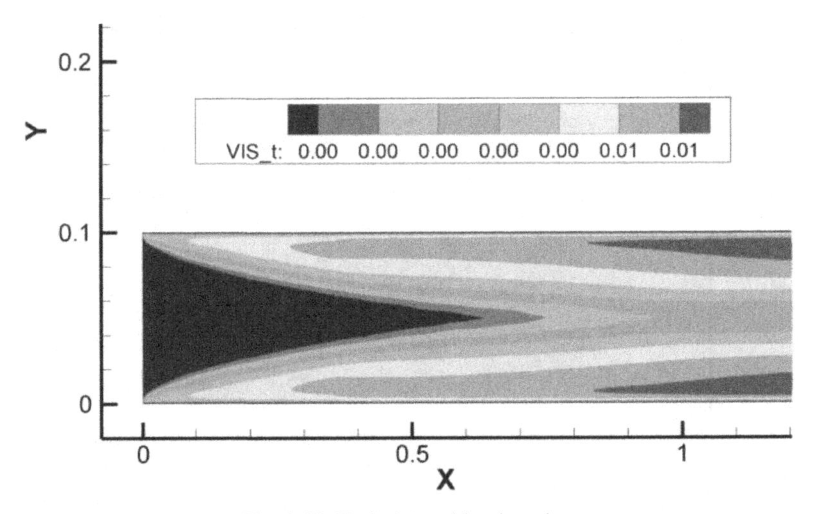

Fig. 9.10 Turbulent eddy viscosity.

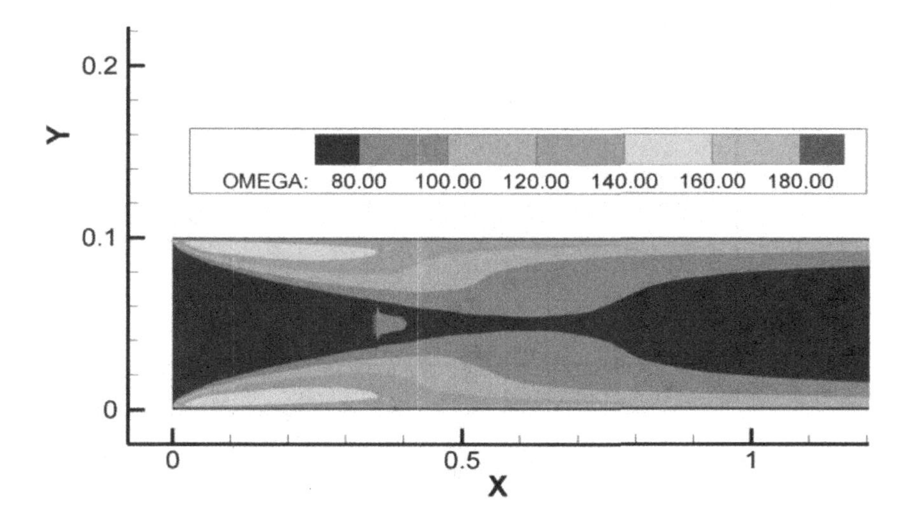

Fig. 9.11 Isopleths of specific dissipation rate.

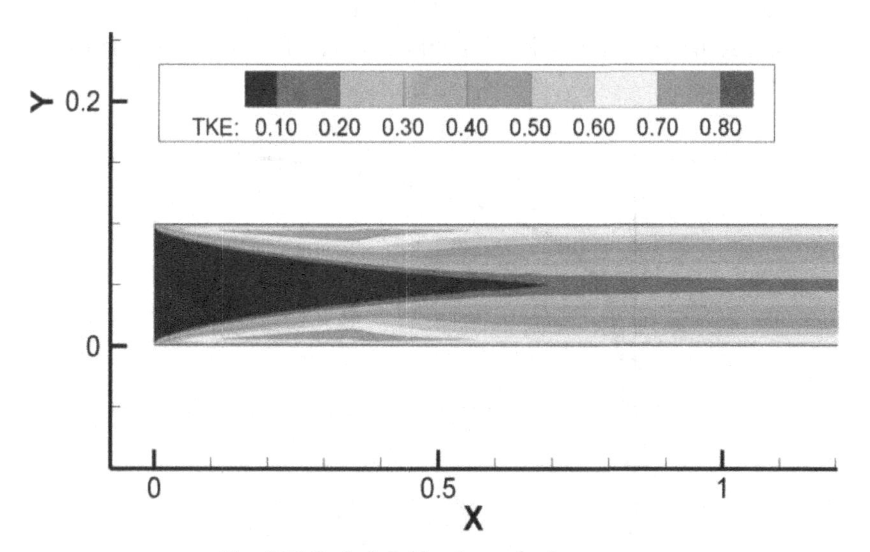

Fig. 9.12 Turbulent kinetic energy in contour.

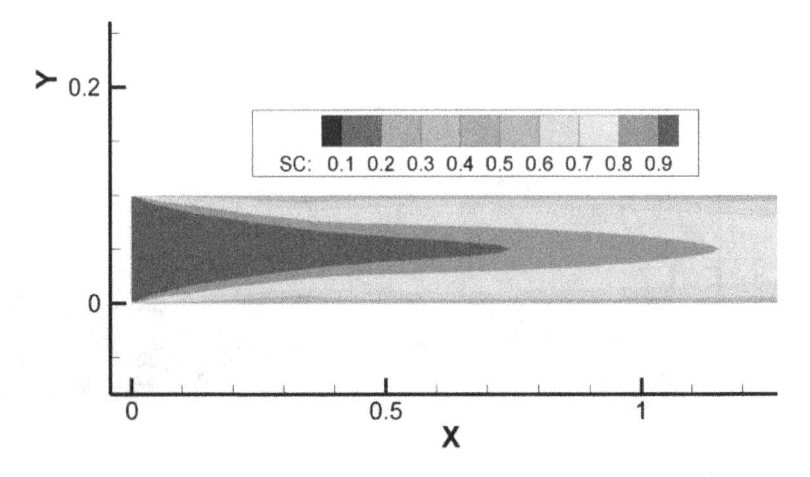

Fig. 9.13 Species concentration.

The velocity profile matches well with direct numerical simulation as shown in Fig. 9.14. The boundary layer, wall layer, as a function of y+ is compared to that of Wilcox's model in Fig. 9.15, with a very good resemblance using two levels of adaptation. The first grid point is shown in the figure.

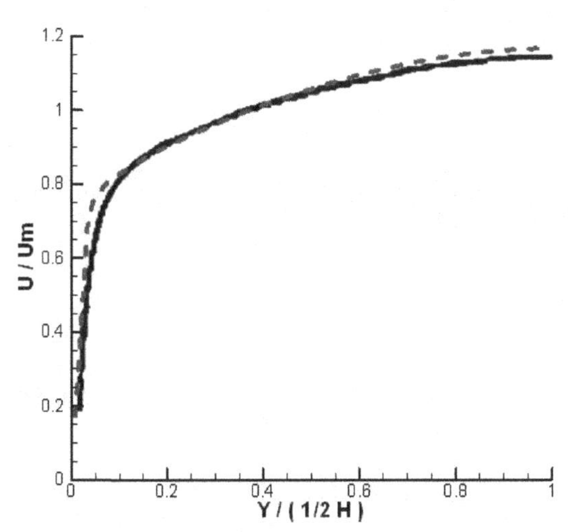

Fig. 9.14 Boundary Layer Profile at 22 hydraulic diameters from inlet. Dashed line FEM k-w using strain rate limiting, solid line from Kim, *et al.* (2000).

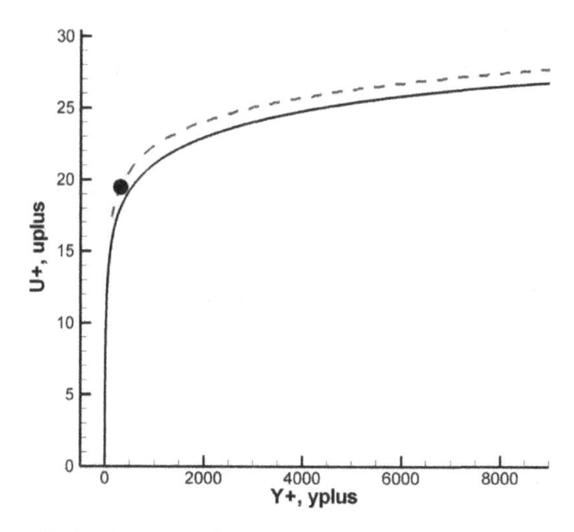

Fig. 9.15 U+ vs. Y+ for developed flow. Solid line from Wilcox's Pipe Flow Program (2006) k-w solution for duct flow vs. FEM predictions, Re = 13,750.

Example 9.6.2 Flow over 2-D backward-facing step: Solutions for flow over a 2-D backward-facing step are presented in Figs. 9.16 through 9.20. The flow Re=42,000, scaled by the step height. The Re is determined by a density of 1.1774 kg/m, an inlet velocity of 13.2 m/sec., a dynamic viscosity $\mu = 1.846\text{x}10^{-5}$ (N sec/m^2), and a step height of 1/3th the overall height, or a hydraulic diameter of 0.05 meters. The outlet is 0.15 meters high, and inlet of 0.1 meters. The inlet turbulent kinetic energy is k = 0.28 and specific dissipation rate at the inlet is ϖ = 770, resulting in an inlet turbulent viscosity of 3.7 x10^{-3}. The grid consists of 11,128 elements, and 11,385 nodes. The inlet turbulent Re_t number is determined by

$$Re_{\tau} = \frac{\rho k}{\varpi \nu} = \frac{1.1774 \times 0.28}{770 \times 1.846\text{e-}05} = 24. \tag{9.138}$$

The upper and lower walls have a heat flux, q_{wall} applied equal to 0.04 (W/m^2). The heat flux is into the domain from both the lower and upper walls. The final grid consists of 35,128 elements, and 35,595 nodes at one-level adaptation. The state of the flow is shown in the figures at steady state as determined by the L-infinity norm of the dependent variables to be less than 1e-06. The thermal properties are conductance of κ=0.02624 (W/m-K°), specific heat at constant pressure c_p = 1.057 x10^3 (J/kg-K°) and turbulent Prandlt number Pr_t=1.0.

Figure 9.16 shows the velocity vectors, at every 9th node point (and one level of adaptation). The results shown in Figs. 9.16 through 9.18 compare favorably to other's (Nallasamy, 1987). In particular, the recirculation length, or region matches experimental results, at 7.1 h, where h is the step height. This length is the standard gauge of the benchmark. Fig. 9.20 shows the predicted species concentration given an constant inlet of 1.0 g/cm^2.

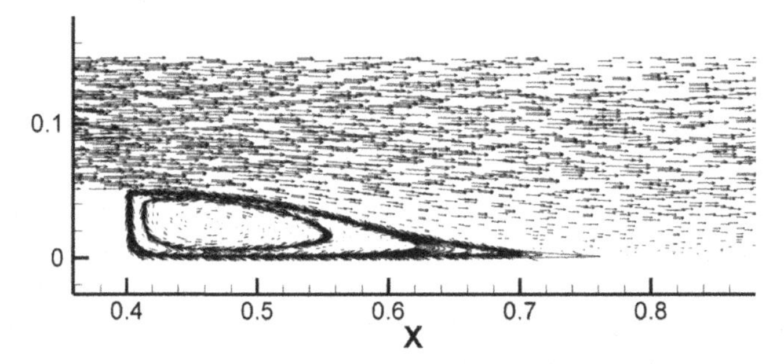

Fig. 9.16 Vectors on refined grid for flow over a 2-D backward-facing step at Re = 42,000.

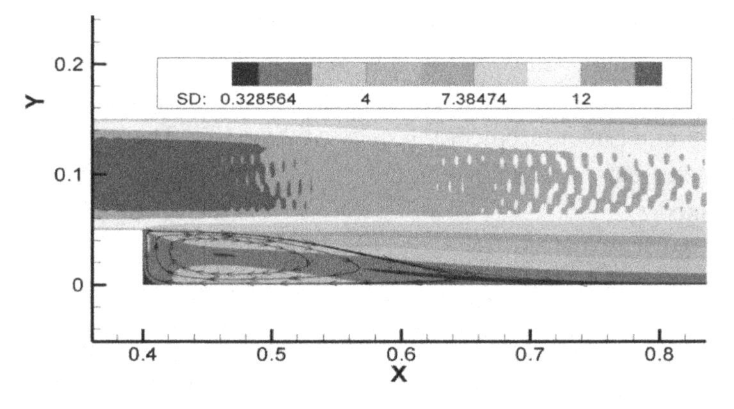

Fig. 9.17 Re = 42,000: Speeds contour, and streamlines showing the recirculation zone out to 7.1 step heights (7.1 h).

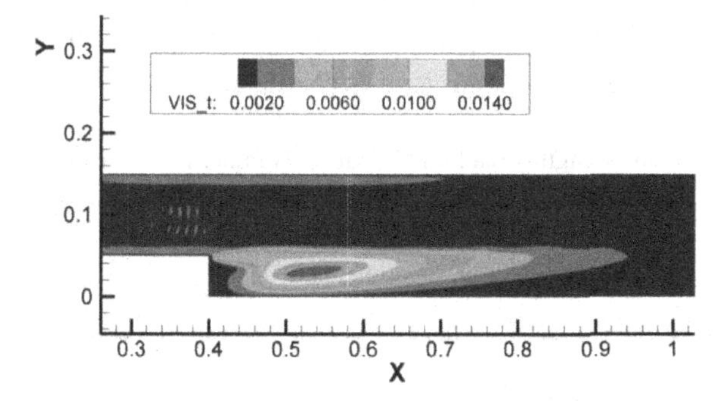

Fig. 9.18 Re = 42,000: Effective viscosity distribution behind step.

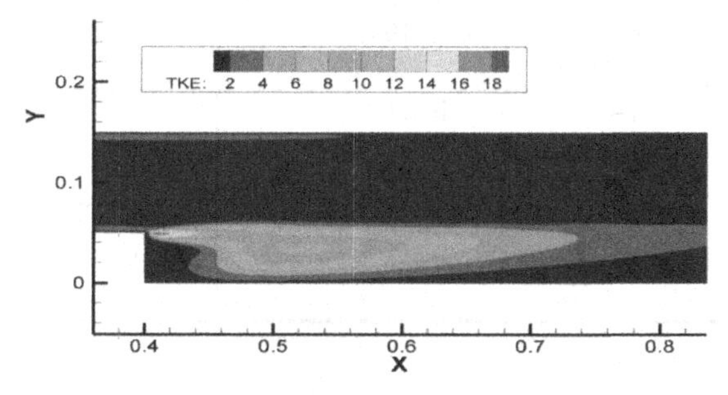

Fig. 9.19 Re = 42,000: Turbulent kinetic energy distribution behind step.

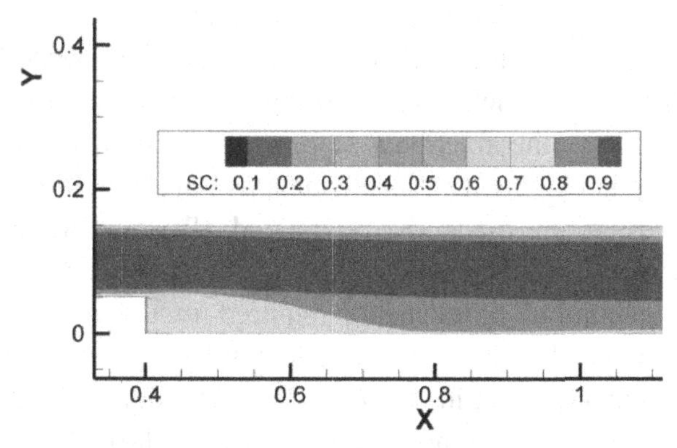

Fig. 9.20 Re = 42,000: Two-dimensional turbulent flow over a backward-facing step. Species concentration at steady state.

Table 9.1 shows results of the recirculation zone length from various researchers. In general the predicted reattachment point has a wide spread, as does the experimental results (Nallasamy, 1987).

Table 9.1 Recirculation Lengths as a Function of Re Number from Various Investigators.

Re = 47,625	6.51 h – *k-w* FEM Steady State Ilinca *et al.*, 1998)	Experimental Values 7.0±0.5 Nallasamy, 1987)
47,625	6.9 h – Algebraic Stress Model (Launder *et al.*, 1981)	
47,625	7.2 h – *k-w* Finite Volume Steady State Ilegbusi and Spalding, 1983)	
42,000	7.1 h – 1st-order-in-time *k-w* FEM Carrington, 2007)	
3,025	4.5 h – *k*- ε Steady State Taylor *et al.* 1981)	6.0 (Denham *et al.*, 1975)

9.7 Comments

The effective viscosity formulation, as demonstrated with the two-equation closure models in this chapter, is reasonably accurate for many engineering type flows, including modeling of indoor air pollution. For incompressible flow, the fractional step, or projection method, works effectively in conjunction with a locally adaptive grid scheme.

Examples using an h-adaptive stabilized (Petrov–Galerkin) finite element framework show the ability to handle complex geometries while minimizing the number of elements required in the model. Element enrichment also works well, while automatically selecting cells for adaptation. An enhancement in the adaptation procedure can be achieved by including p-refinement along with the h, i.e., an h-p adaptation. This procedure leads to exponential convergence – which is sought when attempting to reach a specific error level in the solutions.

The ability to pick regions/cells is useful, although certainly not the most efficient of methods. The mesh adaptation process works better using an *a posterior* error method. This can be driven by solution from course and finer grids using either a remeshing technique (Pelletier and Llinca, 1994) or using an enrichment process (Wang and Pepper, 2007).

The boundary layer has great agreement with existing *k-w* models and with a DNS formulation for flow in a 2-D duct. The model precisely predicts recirculation for the backward-facing step benchmark. This is a detachment-reattachment problem with adverse pressure gradient. However, the recirculation zone and reattachment lengths are known to be somewhat a function of the arbitrary inlet conditions set for turbulent kinetic energy and dissipation rate and this parameter needs to be set with consideration. Most models do fall short of matching experimental data, and the ranges are widely varying as reported by Nallasamy (1987).

The second moment methods or algebraic stress models are better than the two-equation model presented here, particularly for flow where the Boussinesq approximation is no longer valid. Wilcox (2006) provides an excellent chapter on methods for use in this regime. Turbulence modeling with LES methods is thought to be a direction of the future, but in the near term, perhaps some combination of k-ω and LES is more practical from the point of view of computational requirements.

Chapter 10

Homeland Security Issues

The threat of chemical and biological agents being dispersed within a building has become a reality, and is an important cornerstone of the homeland security initiative and formal Department of Homeland Security recently instigated by the US government. This issue became evident when a letter contaminated with anthrax was sent to former Senator Tom Daschle's office in October 2001. Senator Daschle's office resided in the Hart Senate Office Building, which is a nine-story complex located near the Capital Building in Washington, DC. Fumigation and cleanup of the building took approximately three months and cost about $14M. Traces of anthrax were found in other rooms; however, it is unknown exactly how the aerosolized spores dispersed from the envelope to other parts of the building. In 2004, ricin was discovered in the Dickson Senate Office Building, and another letter containing anthrax was intercepted.

In this closing chapter, we briefly discuss some of the issues regarding dispersion of hazardous pollutants into an indoor environment, and describe the use of several models for simulating the spread of such agents.

10.1 Introduction

One gram of anthrax contains about 100 billion spores. Only about 10,000 spores are needed to generate a lethal dose attributed to inhalation. Anthrax ranges in size from 2–4 microns – a fairly large aerosol – which makes the spores susceptible to gravitational settling.

This means that anthrax can quickly settle onto carpets, tabletops, desktops, keyboards, etc. Generally smaller aerosols tend to adhere to ceilings and walls. Assuming someone inadvertently opens an envelope containing one gram of anthrax, some of the anthrax will remain in the envelope, some will settle onto the floor, some will become dispersed into the air, and some may likely be deposited onto and into the person opening the letter. Since people move from room to room, the anthrax will likely be tracked, resuspended, and redeposited within the facility. In addition, the air circulation and HVAC will aid in the redistribution of the spores through the vents and ducts. Interestingly, most airflow models do not account for the movement of people. Efforts to develop a more sophisticated model that would include many of these factors are under development at the Lawrence Berkeley Laboratory's Indoor Environment Department (Dan Krotz, dakrotz@lbl.gov).

10.2 Potential Hazards

Delivery methods vary for introducing chemical and biological agents, e.g., non-exploding means such as open gas cylinders, open containers of liquid agents left to evaporate, aerosol generators, spray tanks, and dry powder. Explosive means vary from gigantic eruptions to small explosive charges.

Chemical agents tend to degrade or disperse in a few hours to a few weeks when exposed to the elements. Biological agents are either in viral or bacterial forms. Viruses range from 0.01–0.30 microns while bacteria range from 0.3–35 microns in diameter. Typical examples of biological agents are anthrax, botulism, plague, smallpox, tularemia, and hemorrhagic fever (for which there are over 12 types of viruses).

While biological agents typically will not cause immediate symptoms, chemical agents almost always cause instant symptoms. An example of the immediate effects of chemical weapons was evident during World War I when mustard gas was released and dispersed into the trenches, affecting thousands of soldiers. Similar effects were felt by US troops during the Vietnam War in the 1960s when agent orange was dropped as a defoliating agent. More recently, Saddam Hussain used chemical agents in Iraq following the Gulf War of 1991 in a genocide that killed thousands of Kurds. The invasion of Iraq in 2003 by the US was based on the premise that Iraq was stockpiling weapons of mass destruction including biological and chemical agents.

In 2002, approximately 50 Chechen terrorists entered and overpowered a Moscow theatre containing nearly 700 people. The siege continued for a number of days and eventually Russian commandos released a fantanyl-based gas to overpower the terrorists. While most of the terrorists died, unfortunately about 100 of the innocent hostages also died. Total fatalities were on the order of 17% of the occupants of the hall. While the intent was noble, the deleterious effects upon the hostages and the high loss ratio of innocent life are clearly unacceptable. Had the commandos been able to utilize proper modeling and risk assessment techniques, the loss of life could certainly have been reduced. For example, the fantanyl could have been introduced in a more controlled manner, preferentially at strategic locations to expose just the terrorists to higher doses, and lower or none discharged to the hostages.

Terrorist seek to promote fear and confusion by selecting targets of opportunity where people generally feel safe, e.g., shopping malls, sporting events, churches, and major performances. The September 11, 2001 World Trade Center disaster clearly illustrated the risks and consequences of terrorist attacks on buildings that can hold over 50,000 people. Dealing with terrorist attacks require new and innovative ways to counter such events as well as preventive measures to deter their thrusts. Terrorists incidents have numbered in the thousands in some countries. In fact, the bulk of terrorist events worldwide have generally been very specific and isolated over the past century, and have not focused on social issues or money. However, recent incidents indicate a dramatic shift of terrorist ideology towards mass destruction and global visibility.

A large toxic release dispersed outdoors tends to affect people severely when they become exposed to the substance. For example, birds will fall from trees, people will collapse, and animals will lie down. For example, a large container of H_2S (~100 tons) began to leak at a major nuclear facility located in the southern US many years ago. This leak went unnoticed until some of the workers began to collapse. Since around 10 ppm is all that is required to cause someone to pass out, many people began falling to the ground. When the accident was finally recognized, efforts were immediately undertaken to evacuate the area and transfer the H_2S to another container. However, when H_2S becomes exposed to air, the result can be an explosion which produces H_2SO_4, whereby only 8 ppm is needed to kill people. Transfer of the toxic material required over 36 hours utilizing intensive and delicate processes. Needless to say, the event was quite stressful on the workers and administration during the incident.

An external release will usually affect a wide range of people within a building since the ventilation constantly receives outdoor makeup air. An indoor release tends to be more isolated, affecting some areas of a building more quickly and severely than other parts of the building. If there are no visible signs of an external release, and if some areas of a building appear to be more severely affected than others, an indoor release should be assumed.

A list of dangerous agents is shown in Table 10.1 through 10.6. These agents are the most common set of hazardous materials that have either been used or considered as terrorist weapons. Some of the biological and chemical substances were employed in World War I and World War II, and later stockpiled during the Cold War.

Table 10.1 List of Radiological Agents

RADIOACTIVE AGENTS	Form
Americum-241	powder
Californium-252	powder
Cesium-137	powder
Iridium-192	powder
Plutonium-239	powder
Strontium-90	powder
Uranium-235	powder

Table 10.2 List of Biological Agents

BIOLOGICAL AGENTS	FORM
Category A	
Anthrax	spore
Botulism	spore
Plague	spore
Smallpox	spore
Viral hemorrhagic fevers	spore
Category B	
Brucellosis	virus
Epsilon toxin of Clostridium perfringens	virus
Food safety threats	various
Glanders	virus
Melioidosis	virus
Psittacosis	virus
Q fever	virus
Ricin toxin	spore
Staphylococcal enterotoxin B	virus
Typhus fever	bacillus
Viral encephalitis	virus
Water safety threats	liquid
Category C	
Emerging infectious diseases (Nipah virus; hantavirus)	virus

Table 10.3 List of Biotoxic Chemical Agents

AGENT	Form
Abrin	bacillus
Ricin	bacillus
Strychnine	bacillus
Blister Agents/Vesicants	
Mustards	
Distilled mustard (HD)	particulate
Mustard gas (H) (sulfur mustard)	gas
Mustard/lewisite (HL)	particulate
Mustard/T	particulate
Nitrogen mustard	gas
Sesqui mustard	gas
Sulfur mustard	gas
Lewisites/chloroarsine agents	
Lewisite	particulate
Mustard/lewisite (HL)	particulate
Phosgene oxime	particulate

Table 10.4 List of Blood Type Chemical Agents

AGENT	Form
Cyanide	
Cyanogen chloride (CK)	gas
Hydrogen cyanide (AC)	gas
Potassium cyanide (KCN)	gas
Sodium cyanide (NaCN)	gas
Caustics (Acids)	
Hydrofluoric acid (hydrogen fluoride)	liquid
Hydrogen fluoride (hydrofluoric acid)	liquid
Choking/Lung/Pulmonary Agents	
Ammonia	gas
Chlorine (CL)	gas
Hydrogen chloride	gas
Phosgene	
Diphosgene (DP)	gas
Phosgene (CG)	gas
Phosphine	gas
Phosphorus, elemental, white or yellow	particulate

Table 10.5 List of Incapacitating Chemical Agents

AGENT	Form
Fentanyls and other opioids	liquid
Long-Acting Anticoagulants	
Super warfarin	liquid
Metals	
Arsenic	powder
Mercury	liquid
Thallium	powder
Nerve Agents	
G agents	
Sarin (GB)	gas
Soman (GD)	gas
Tabun (GA)	gas
V agents	
VX	gas

Table 10.6 List of Organic Solvent Agents

AGENT	Form
Riot Control Agents/Tear Gas	
Various agents and combinations	
Bromobenzylcyanide (CA)	gas
Chloroacetophenone (CN)	gas
Chlorobenzylidenemalononitrile (CS)	gas
Chloropicrin (PS)	gas
Dibenzoxazepine (CR)	gas
Toxic Alcohols	
Ethylene glycol	liquid
Vomiting Agents	
Adamsite (DM)	powder

10.2.1 Prevention and protection

The following guidelines should be considered in the event of a toxic release (from Miller, *The Military Engineer*, June 2003):

External Release:

1. Shut off supply fans that are not equipped with proper filtration systems
2. Keep supply fans running equipped with high-performance filtration
3. Shut off all exhaust fans
4. Close fresh air intakes and building openings
5. Recall all elevators to the lobby
6. Lock all doors, including dock and garage
7. Have all occupants remain indoors near the core of the building

Indoor Release:

1. Execute prearranged tenant communications
2. Close duct dampers to isolate zones of release
3. Activate stairwell pressurization system (100% outside air)
4. Evacuate occupants to a predestinated location
5. Segregate people during a biological release to avoid contamination

When dealing with chemical releases, the concentration exposure to people must be minimized. Areas surrounding the floor and release sites should be flooded with 100% outdoor air. One should provide 100% exhaust only to the floor or area of release. If this is not possible, then the entire area should be filled with 100% outdoor air and 100% exhaust mode. If this scenario is not possible, then all supply and return fans should be shut down until the type of hazard and dispersion pattern can be determined. Once this information is obtained, the HVAC system can be considered for reactivation. The normal operation of most HVAC systems in buildings provides some outdoor air and exhausts some indoor air, which helps to dilute some of the hazardous material.

When encountering biological releases, the number of people exposed must be minimized. All supply, return, and exhaust fans should be shut down. Such systems should not be restarted until sufficient information has been obtained, and competent authority authorizes startup of the HVAC network.

A building environment should maintain integrity, i.e., protect the interior from contaminated outdoor air. Air tightness is a term used to quantify how well the exterior of the building serves as a barrier. As a general rule of thumb, exterior air intakes should be positioned atop the building or at least four storeys above grade. As an alternative, solid walls can be used to surround each intake with sloped screen tops. Non-public building areas should be air balanced in order to provide a positive pressure to prevent infiltration from the outside. Quick-closing dampers should be utilized in supply and return ducts, along with dedicated exhaust systems in each public area.

HEPA filters should be installed in ductwork to provide emergency ventilation in stairways with intakes at low levels. Generally high-grade infiltration is suggested due to the critical nature of stairs and concentration of people in the event of an emergency evacuation. Dealing with stairways can be difficult, and can require weeks to adequately clean during a shutdown.

ASHRAE Standard 52.2–1999 is the current standard for determining performance of particulate filters. The efficiency to capture particulates is ranked by a Minimum Efficiency Reporting Value (MERV) that varies from 1 to 20. The higher the MERV rating, the better the capture efficiency (especially when trying to capture small micron size particulates). Air pressure, fan-motor horsepower, and energy costs decrease as filtration efficiency increases.

The most commonly used filter is an upgraded 2-inch panel filter with a MERV rating of 5 to 8. However, for chemical and biological agents, higher MERVs will be required. Such filters include High Efficiency Particulate Arrestanc (HEPA) filters, 90–95 percent high efficiency filters (MERV 14), 95 percent (MERV 15), 95 percent DOP (MERV 16), and activated carbon filters. Table 10.7 lists the MERV rating and level of filtering.

Table 10.7 Particulate Filter Levels (from Miller, *The Military Engineer*, June, 2003)

MERV LEVEL	Dust Spot	Particulate Filter Type	% 0.3–1 μm	% 1–3 μm	% 3–10 μm
1	NA	Low efficiency fiberglass and	Too low efficiency to be		
2	NA	Synthetic media disposable	applicable to ASHRAE 52.2		
3	NA	Panels, cleanable filters, and			
4	NA	Electrostatic charged panels			
5	NA	Pleated filters, cartridge/cube			20–35
6	NA	Filters, and disposable multi-			36–50
7	25–30%	Density synthetic link panels			50–70
8	30–35%				>70
9	40–45%	Enhanced media pleated filters,		>50	>85
10	50–55%	bag filters of either fiberglass or		50–65	>85
11	60–65%	synthetic media, rigid box filters		65–80	>85
12	70–75%	using lofted or paper media		>80	>90
13	80–85%	Bag filters, rigid box filters,	>75	>90	>90
14	90–95%	minipleat cartridge filters	75–85	>90	>90
15	>95%		85–95	>90	>90
16	98%		>95	>95	>95
17	NA	HEPA/ULPA filters using IEST	99.97% IEST Type A		
18	NA	MoT. Types A through D yield	99.99% IEST Type C		
19	NA	Efficiencies @ 0.3 μm and Type	99.999% IEST Type D		
20	NA	F@0.1 μm	>99.999% IEST Type F		

Over the past few years, Penn State University has been developing a terrorist-resistant air conditioning concept that is cheaper to operate, costs less to install, and is more energy efficient that conventional industrial standards. If a biological or chemical contaminant is released within an office, standard forced-air cooling can transport the agent to other locations within the building.

The Penn State system de-couples the process of supplying fresh air to the occupants within the building from the heating and cooling functions of the HVAC system. The system, called Dedicated Outdoor Air System (DOAS), couples an independent fresh air supply with radiant cooling panels. The radiant panels, which utilize cool circulating water and can be integrated into the building's fire sprinkler system, have been employed in Europe for over 15 years.

The DOAS/radiant approach does not use recirculated air. Hazardous agents released in the interior of the building are not transported to other parts of the building by the HVAC system but are diluted and exhausted from each particular space. Since the fresh air supply is independent from the HVAC function, less air is required and can be treated and de-humidified at less cost, including running the exiting air through an energy recovery system.

10.3 A Simple Model

Classrooms, auditoriums, and public buildings tend to have transient or variable occupancy. Ventilation rates in these enclosures are typically varied to maintain acceptable contaminant concentrations at all times. A pollutant can be indoors before the start of occupancy, produced by people, processes or materials placed within the building, or supplied from the outside through exterior ventilation.

To determine the variation of a concentration level over time, the mass balance equation of the pollutant over the entire enclosure must be solved. Assuming a time step, dt, we can denote the change in the concentration of indoor air as dc. This change represents the quantity of pollutant generated within the interior plus the quantity dispersed by the ventilation into the enclosure minus the quantity leaving the enclosure, i.e., returning to Eq. 5.3,

$$\forall dC = (S + QC_{air} - QC)dt, \tag{10.1}$$

or

$$\forall \frac{dC}{dt} = S + Q(C_{air} - C), \tag{10.2}$$

where

$\forall$ = effective volume of the enclosure (m^3)
Q = exterior air supply rate (m^3/s)
C_{air} = ambient concentration of pollutant (gm)
C = concentration of pollutant at time t (gm)
S = pollutant source (gm-m^3/s)

For simplicity, if we assume perfect mixing and no density changes within the enclosure, integrating Eq. 10.2 gives the indoor concentration at time t:

$$C = \frac{QC_{air} + S}{Q}\left(1 - e^{\frac{-Qt}{\forall}}\right) + C_o e^{\frac{-Qt}{\forall}}. \tag{10.3}$$

where C_o is the indoor concentration at time t = 0. We can simplify the form of this general equation further based on various practical conditions.

1. If we assume that the initial concentration within the room is zero ($C_o = 0$), then Eq. 10.3 can be simplified to the following form:

$$C = \frac{QC_{air} + S}{Q}\left(1 - e^{\frac{-Qt}{\forall}}\right). \tag{10.4}$$

2. If the pollutant in outdoor air is zero ($C_{air} = 0$) and the initial concentration is zero ($C_o = 0$), the concentration equation can be rearranged and simplified to:

$$C = \frac{S}{Q}\left(1 - e^{\frac{-Qt}{\forall}}\right). \tag{10.5}$$

Solutions to Eq. 10.5 can be graphically displayed as a family of curves representing the ratio Q/S. Assuming values for Q, $\forall$, S, and time t are known, the concentration indoors can be obtained directly from Fig. 10.1.

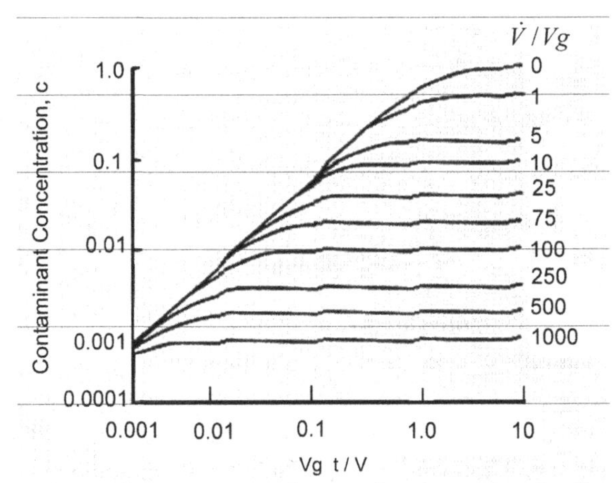

Fig. 10.1 Indoor pollutant concentration as a function of time for $C_o = C_{air} = 0$.

If the pollutant in the outdoor air is zero ($C_e = 0$) and there is no indoor contaminant generation (S = 0), Eq. 10.2 simplifies to the form:

$$C = C_o e^{-Nt}, \tag{10.6}$$

where N is the air change rate per second if t is in seconds or hours if t is in hours. This equation denotes simple decay commonly used in measuring ventilation rates within a building using tracer gases.

Assuming steady-state conditions ($t \rightarrow \infty$), equilibrium indoor concentration levels are reached as $t \rightarrow \infty$, giving the final concentration as:

$$C_\infty = \frac{QC_{air} + S}{Q}, \tag{10.7}$$

where one can see that the final concentration c_∞ is independent of the interior volume, V. Note that the value of V affects the rate at which C $\rightarrow C_\infty$. Likewise, the initial concentration, C_o, has no influence on the final concentration.

It is assumed that perfect mixing of the room air and the supply air occurs within the enclosure, resulting in perfect dilution of the indoor contaminant. This rarely happens, if ever. In actuality, the supply air does not mix perfectly with the indoor air in the occupied zone, with the end result of the outdoor air being exhausted before it has a chance to adequately absorb some of the indoor contaminant. Thus, different concentration rates will exist in the occupied zone. This leads to the need for larger air supply rates in order to achieve the threshold limit.

Ventilation efficiency or ventilation effectiveness is used to describe the degree of mixing of supply air with room air. There are two main categories used to define ventilation efficiencies for steady-state conditions (see Sandberg, 1981).

Relative ventilation efficiency: this describes the variability of a system's ventilation abilities among different parts of a room. It is expressed as either an average or overall relative efficiency for the entire occupied zone or as a local relative efficiency. The local value is expressed as

$$E_r = \frac{C_x - C_{air}}{C - C_{air}}, \tag{10.8}$$

and the average relative ventilation efficiency is written as

$$\bar{E}_r = \frac{C_x - C_{air}}{\bar{C} - C_{air}}, \tag{10.9}$$

where

C = contaminant concentration at a point, ppm
$\bar{C}$ = mean concentration in the occupied zone, ppm
C_{air} = contaminant concentration in the outdoor air supply, ppm
C_x = contaminant concentration in the exhaust air, ppm

The absolute ventilation efficiency: this term relates the ability of the ventilation system to reduce the pollution concentration relative to a theoretical maximum. The relation is normally expressed as:

$$E_a = \frac{C_o - C}{C_o - C_{air}}, \tag{10.10}$$

where
C_o = initial concentration at a point, ppm
C = concentration at the same point after time t, ppm

The relative ventilation efficiency is a measure of pollutant dispersion and doesn't take either the absolute concentration levels or changes in concentration from initial values into account. The value of E is always positive and can either be less than, equal to, or greater than 1, depending on the location in the room and air distribution method. The absolute ventilation efficiency represents the change in concentration as a result of change in the ventilation rate and is always less than 1.

In order to overcome the effects of imperfect dilution of indoor contaminant by the outdoor air, air supply rates greater than that provided in the above equations is necessary. This is expressed quantitatively by replacing V in these equations with E_rV. The value of E_r is dependent on the type of air distribution system used to supply and extract air to the room. The types of ventilation systems include local exhaust ventilation, piston ventilation, displacement ventilation, and mixing ventilation. These various ventilation schemes are described in more detail in the ASHRAE guidelines and ventilation textbooks.

10.3.1 Example – analytical model of anthrax dispersion: To illustrate the application of the above set of simple relations, assume that a room contains a package (e.g., a letter, box, etc.) that has just been opened containing anthrax. We assume the anthrax is dispersing at S = 0.001 gm-m^3/s. The room is 4 m x 4 m x 3 m, giving a volume of 48 m^3. The office complex is shown in Fig. 10.2. This is the same problem as shown in Example 5.1.1. What is the concentration within the office after 10 minutes?

There are two rooms in the office complex; one is for the secretary and the other is for the manager. Several tables are laid out in the office complex. The terrorist puts the anthrax in the secretary's room. Due to the door being opened and ventilation flowing into the office at U = 1.0

m/s, the powder disperses within the room. The boundary conditions and problem definition are shown in Fig. 10.3.

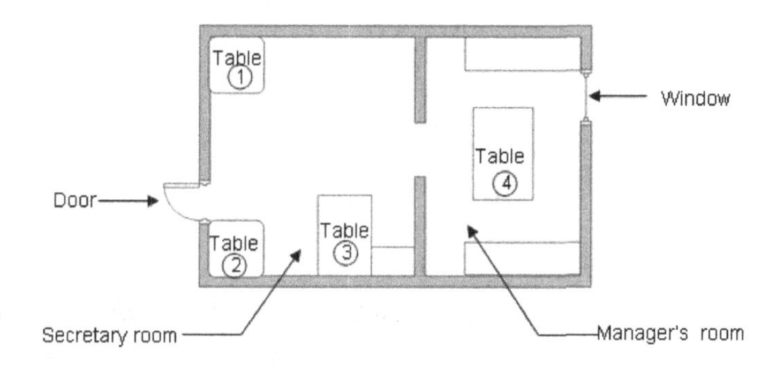

Fig. 10.2 Office complex layout.

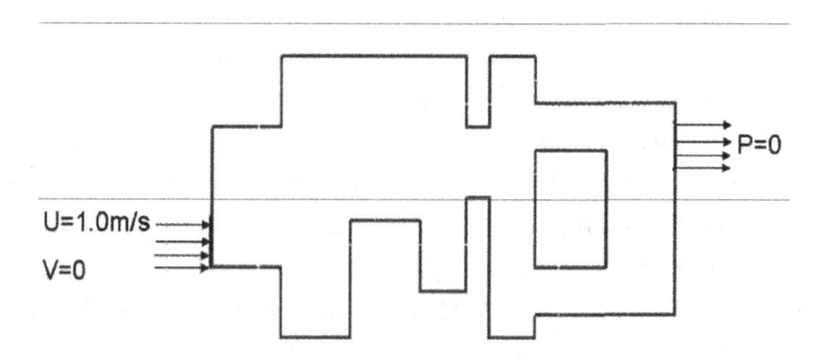

Fig. 10.3 Boundary condition setting.

We assume the door has an opening of 3 m x 1 m; thus Q = AU = 3 m^3/s. We use the following input values:

$\forall$ = 48 m^3
Q = 3 m^3/s
C_{air} = 0 gm
C = 0 gm
S = 0.001 gm-m^3/s
t = 600 s

We use Eq. 10.4 to solve for the dispersion within the room since the initial concentration is zero. After 10 minutes, the concentration within the room would be

$$C = \frac{QC_{air} + S}{Q}\left(1 - e^{\frac{-Qt}{V}}\right),$$

$$C = \frac{3 \cdot 0 + 0.001}{3}\left(1 - e^{\frac{-3 \cdot 600}{48}}\right) = 0.000333(1 - e^{-37.5}),$$

$$C = 0.000333 \text{ gm}$$

The value only tells us the ensemble concentration within the room. The solution of Eq. 10.4 doesn't tell us about the spread of the anthrax within the office, or the possible dispersion pathway.

10.3.2 Example – numerical model of anthrax dispersion: We repeat the example anthrax dispersion problem of 10.3.1, but this time we use COMSOL, which is an FEM commercial code that permits mesh adaptation. Here we can refine the location of the source and assume there could be three different point locations. The added detail inherent within a numerical model permits us to be more specific with regards to the location of the source, with interesting outcomes based on the source location. The specific steps and procedures involved in setting up the problem to be solved by COMSOL can be found on the website: www.iaqcodes.edu (see Pepper and Wang, 2006).

The initial computational mesh has 161 quadrilateral elements and 213 nodes, which is shown in Fig. 10.4.

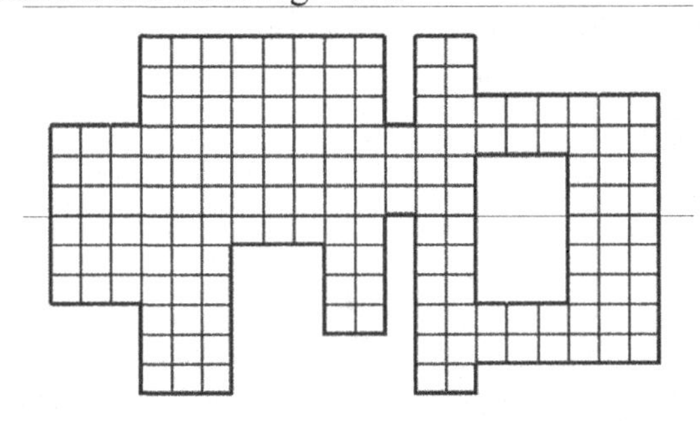

Fig. 10.4 Initial coarse mesh.

Simulated air distribution patterns and pathways of anthrax dispersing within the office complex are shown in the following series of pictures. Both the door and the windows are open, and the contaminant powder spreads into the inner office. The sequence of pictures shows a plan view of the flow of air and velocity vectors in Fig. 10.5, velocity contour in Fig. 10.6, and a 2-D plan view of the streamlines in Fig. 10.7. Differences in trajectories are observed when the pollutant source is placed at different locations inside the secretary's room. In one case the pollutant source is placed on the top right corner of table 2, while in the other case it is placed between table 1 and table 2 in the outer secretarial room.

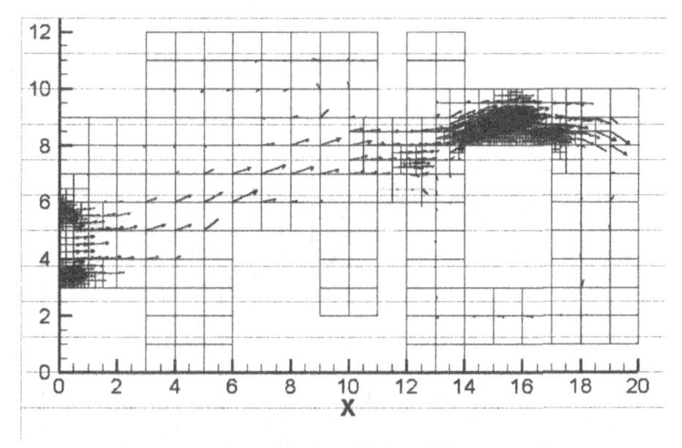

Fig. 10.5 Flow of air within office complex.

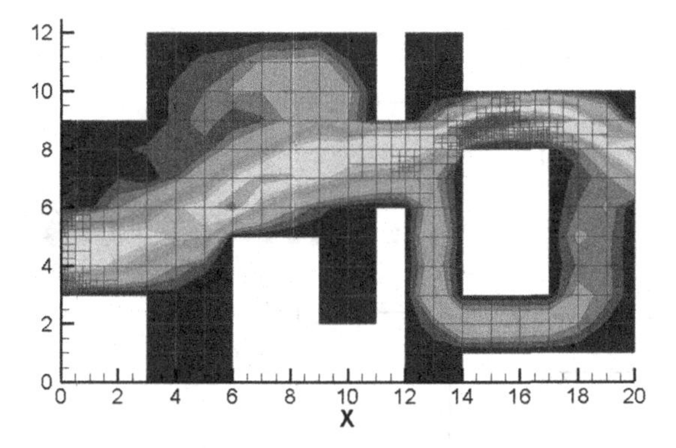

Fig. 10.6 Velocity contours.

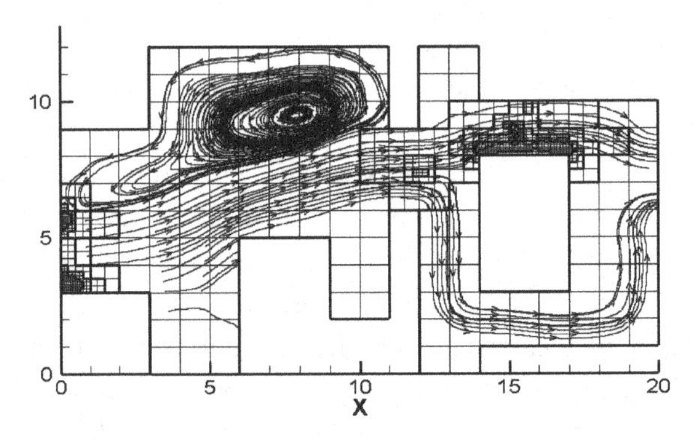

Fig. 10.7 Streamlines.

Particle dispersion pattern (with large dot denoting contaminant source) is shown in Figs. 10.8, 10.9, and 10.10. The contaminant source has been placed on table 1, center of the room and table 2, respectively.

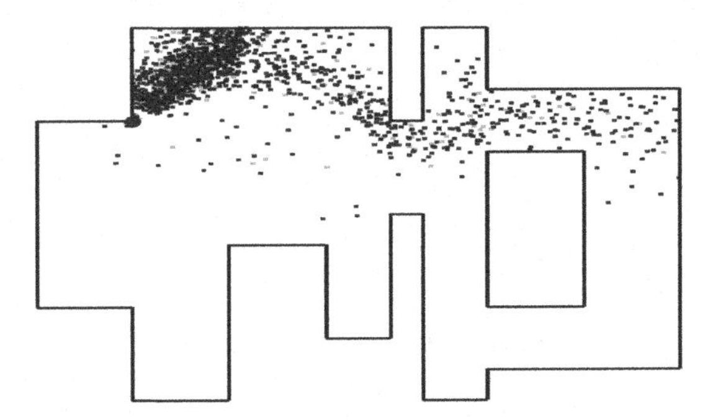

Fig. 10.8 Particle dispersion pattern case 1.

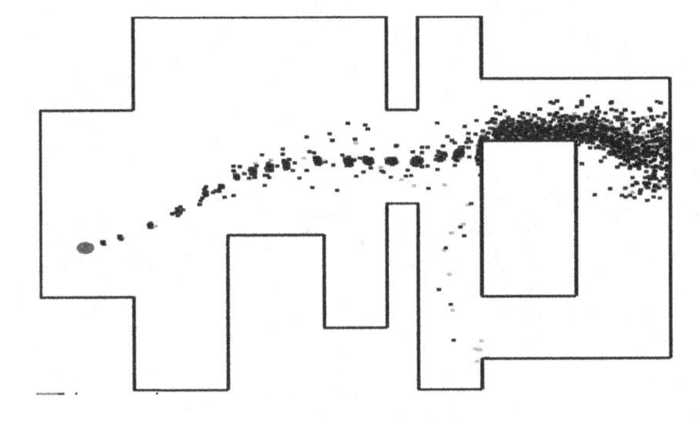

Fig. 10.9 Particle dispersion pattern case 2.

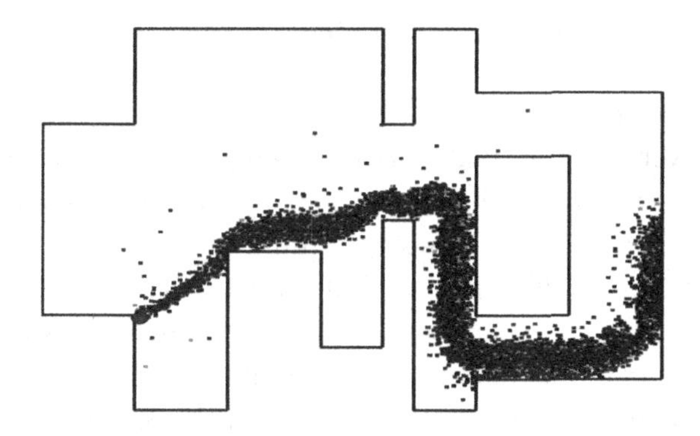

Fig. 10.10 Particle dispersion pattern case 3.

As can be seen in the particle dispersion patterns, the pollutant is transported and diffused by the ventilation pattern that affects the office complex. It is easy to see that the anthrax essentially remains trapped within the large recirculation zone (Fig. 10.7) near the upper wall in the outer room, as seen in Fig. 10.8. If the terrorist decides to place the source in different locations, the anthrax dispersion pattern becomes quite different, as we see in Figs. 10.9 and 10.10. In this instance, we see the added value of performing a more detailed simulation using a CFD model – the added physics allows us to more accurately plot the trajectories, especially as they progress into the inner office.

When first responders arrive at an incident location, it is important that they be aware of the trajectory of the spreading contaminant. It is also critically important that inhabitants be aware of the contaminant pathway, and take evasive action. For example, the manager in the inner office could move to the upper left corner of his room until reached by a rescue team, as seen in Fig. 10.10. Likewise, the secretary would be better off waiting at her desk instead of walking into the plume of particles, based on the pattern shown in Fig. 10.8.

10.4 Other Indoor Air Quality Models

There are several nice computer codes that can be downloaded for free from the web. These codes have been developed at government laboratories and are available for use by researchers and academic institutions in the US Although not quite as robust as some of the commercial CFD codes, they are still effective and you can't beat the price. There are several commercial CFD codes that are particularly well suited for indoor air pollution simulation; while they can be very expensive, they are very convenient and can be learned fairly quickly. FLUENT, STAR-CD, ANSWER, and CFX are general CFD codes that can be configured for practically any flow-related problem; COMSOL allows one to input multiphysics, including equations and MATLAB script, in the overall program procedure. FLOVENT is a FVM model, similar to FLUENT, that is aimed primarily at the HVAC and IAQ industry. This code permits one to readily input boundary conditions and source locations associated with IAQ problems using a simple CAD interface.

10.4.1 CONTAM 2.4 (NIST)

CONTAM 2.4 is an indoor dispersion model that was initially developed at the National Institute of Standards and Technology (NIST) in Gaithersburg, MD, some years ago and has been revised and updated (see Walton and Dols, 2006). The model is rather sophisticated, with lots of input required from the user.

The most current version, CONTAM 2.4, can be accessed from the web: http://www.bfrl.nist.gov/IAQanalysis/CONTAM/overview/3.htm. The model allows the user to output numerous data, and even includes a sketchpad capability for configuring floor plans and building configurations, along with weather input for outdoor conditions. Figure 10.11 shows an example of a building that is first converted into a series of blocks that ultimately are refined into floor plans that can then be modeled.

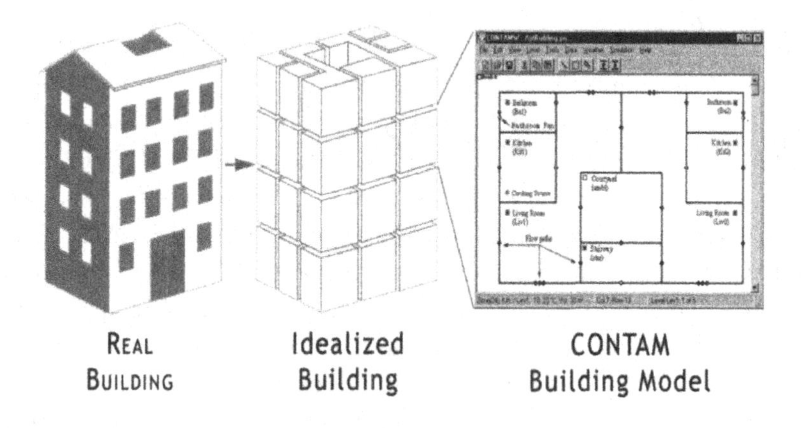

Fig. 10.11 CONTAM 2.4 program.

CONTAM 2.4 has the ability to calculate building airflows and relative pressures among zones within a building – which can be valuable in determining ventilation rates. The model has been used to help in the design and analysis of smoke management, and in the design decisions related to ventilation systems and material selections. Deposition sink models, a 1-D advection/diffusion contaminant model for ducts and diffuers, and contaminant filter models are also included. A nice feature of the model is the prediction of exposure of building occupants to airborne contaminants for risk assessment.

10.4.2 I-BEAM (EPA)

I-Beam is a computer model developed by the EPA. This program permits one to implement an energy audit of a building or series of rooms, as well as conduct an IAQ assessment (I-BEAM, 2002). This code can be downloaded from the web: http://www.epa.gov/iaq/largebldgs/i-beam/index.html. I-BEAM stands for IAQ **B**uilding **E**ducation and **A**ssessment **M**odel. The model is quite comprehensive, and permits the user to vary numerous parameters regarding boundary conditions and initial values. The user's manual is extensive, and provides a very good background on IAQ fundamentals and related energy issues. Figure 10.12 shows a screen image of the code that comes up on a PC when the code is accessed by the user.

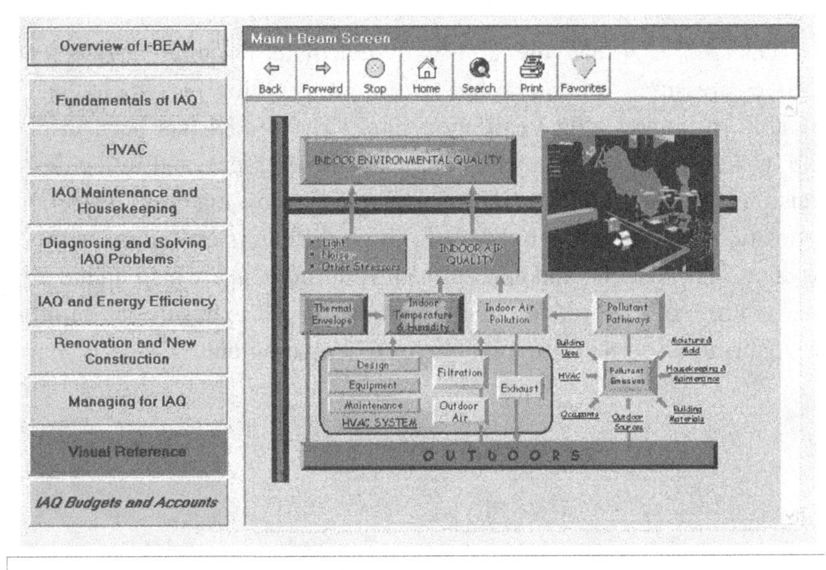

Fig. 10.12 I-BEAM program.

A particularly nice feature of the model is the ability to access active modules from the I-BEAM website, i.e., a series of downloads that enable the user to activate and animate various processes. Clicking on one of the modules in the screen image above immediately sends the user to another set of modules related to the parent topic. The model is well designed, but does take some time to become acquainted with the loading of boundary conditions and the display of results.

10.4.3 COMIS-MIAQ (APTG-LBNL)

An anthrax model was developed by the Airflow and Pollutant Transport Group at Lawrence Berkeley National Laboratory in 2002. The basis of the LBNL model is COMIS, with much of its context coming from MIAQ (Multi-Chamber Indoor Air Quality Modeling Package) originally developed as a Ph.D. thesis at Cal Tech by Nazaroff (1988). The model can be used to predict the transport of anthrax spores within a building, including what fraction settles on floors, in carpets and on walls, potential for resuspension, and amounts that could be caught in ducts and air filtration units as it leaves the building through cracks, doors, and windows.

COMIS/MIAQ is very detailed and permits many variations on contaminant dispersion within an indoor environment. This model can be downloaded from the web: http://eetd.lbl.gov/ie/APT/APT.html. Figure 10.13 shows a building floor plan and dispersion within a three-story facility generated by the model.

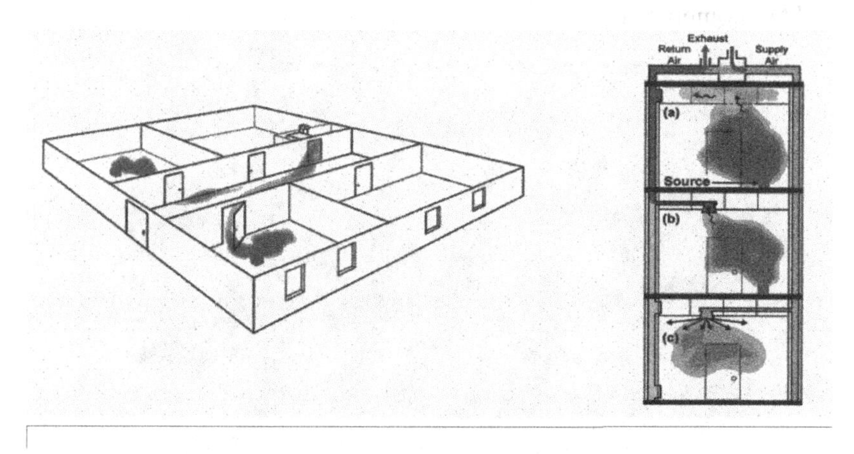

Fig. 10.13 COMIS/MIAQ LBNL model.

The ability of COMIS/MIAQ to display 2 and 3-D dispersion patterns throughout a building is a plus, and allows one to incorporate complex flows and ventilation settings within an office complex or building. The computer code requires some time to become familiar with loading and displaying data.

10.4.4 FLOVENT (Flomerics, Inc.)

A very nice indoor air pollution model was developed by Flomerics, Inc., London, UK, several years ago known as FLOVENT. This model is a commercial CFD code based on the FVM, much like FLUENT and other similar FVM commercial CFD models. FLOVENT permits an easy I/O interface for conducting indoor ventilation and dispersion simulations, and uses a CAD interface for creating a floor or building layout. An example of the flow and dispersion within a hospital floor is shown in Fig. 10.14.

FLOVENT is a very robust and easy to use program aimed at the IAQ and HVAC group. The model permits the user to develop a problem configuration using a SolidWorks protocol, and then helps the user establish flow and source boundary conditions in units typical of HVAC and IAQ nomenclature.

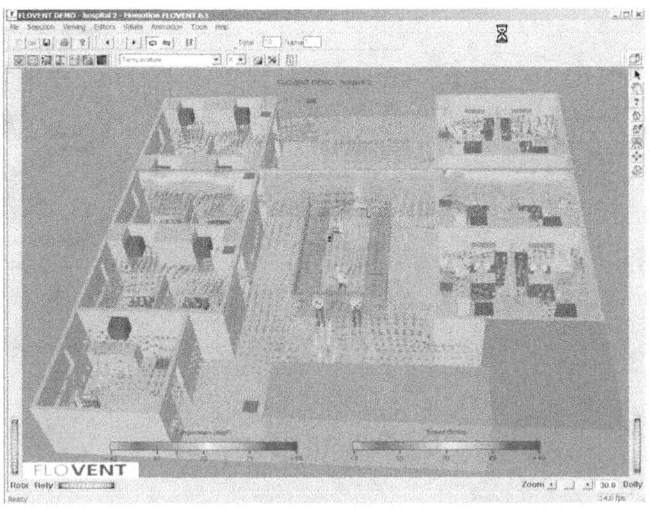

Fig. 10.14 FLOVENT model output – hospital complex.

10.5 Comments

Issues involving homeland security will be with us for many years. Terrorist activity will dictate that the US and other countries must be prepared for attempts of local as well as mass destruction on innocent people for the unforeseeable future. The recent creation of the US Department of Homeland Security is one means of being prepared and trying to stay ahead of such terrorist activities. Considerable research and work has begun throughout the US on ways to counter terrorism and assess the consequences of attacks on the populace. Models and sensing equipment will continue to evolve as new approaches to inflicting mass destruction arise throughout the world.

The models presented in this chapter are fairly mature and provide reliable results when understood and mastered by the user. However, one must remember that these models only serve as a family of numerical tools that can be used effectively for certain classes of problems. Locking into using only one specific numerical technique or commercial code can become very limiting. On the other hand, finding a good, reliable technique or code that handles a wide range of problems can be very beneficial – as long as one does not try to force fit the model into problems for which it would be an overkill or grossly inaccurate.

When an emergency occurs, there is no time to sit and ponder about the best numerical model to use, or wait on a lengthy calculation to conclude while everyone is dying from exposure. On the other hand, it can be just as dangerous to make grossly inaccurate guesses or rely on assumptions that could be the opposite of the best scenario or evasion tactic. A quick, reasonable estimate is all that is needed in the first moments of an emergency – follow up analysis and evaluation can always come later when people are safe and out of harm’s way.

Appendix A Diffusion Coefficients in Gas

(Experimental values of diffusion coefficients in gases at 1 atm (Cussler, 1997)

Gas Pair	Temperature (^oK)	Diffusion Coefficient (cm^2/s)
Air-CH_4	273.0	0.196
Air-C_2H_5OH	273.0	0.102
Air-CO_2	276.2	0.142
	317.2	0.177
Air-H_2	273.0	0.611
Air-D_2	296.8	0.565
Air-H_2O	289.1	0.282
	298.2	0.260
	312.6	0.277
	333.2	0.3050
Air-He	276.2	0.6242
Air-O_2	273.0	0.1775
Air-n-hexane	294.0	0.080
Air-n-heptane	294.0	0.071
Air-bezene	298.2	0.096
Air-toluene	299.1	0.086
Air-chlorobenzene	299.1	0.074
Air-aniline	299.1	0.074
Air-nitrobenzene	298.2	0.0855
Air-2-propanol	299.1	0.099
Air-butanol	299.1	0.087
Air-2-butanol	299.1	0.089
Air-2-pentanol	299.1	0.071
Air-ethylacetate	299.1	0.087
CH_4-Ar	298.0	0.202
CH_4-He	298.0	0.675
CH_4-H_2	298.0	0.726
CH_4-H_2O	307.7	0.292
CO-N_2	295.8	0.212
^{12}CO-^{14}CO	373.0	0.323
CO-H_2	295.6	0.743
CO-D_2	295.7	0.549
CO-He	295.6	0.702
CO-Ar	295.7	0.188

Gas Pair	Temperature (°K)	Diffusion Coefficient (cm^2/s)
CO_2-H_2	298.0	0.646
CO_2-N_2	298.2	0.165
CO_2-O_2	293.2	0.160
CO_2-He	298.0	0.612
CO_2-Ar	276.2	0.1326
CO_2-CO	296.1	0.152
CO_2-H_2O	307.5	0.202
CO_2-N_2O	298.0	0.117
CO_2-SO_2	263.0	0.064
$^{12}CO_2$-$^{14}CO_2$	312.8	0.125
CO_2-propane	298.0	0.0863
CO_2-ethyleneoxide	298.0	0.0914
H_2-N_2	297.2	0.779
H_2-0_3	273.2	0.697
H_2-D_2	288.2	1.24
H_2-He	298.2	1.132
H_2-Ar	287.9	0.828
H_2-Xe	341.2	0.751
H_2-SO_2	285.5	0.525
H_2-H_20	307.1	0.915
H_2-NH_3	298.0	0.783
H_2-acetone	296.0	0.424
H_2-ethane	298.0	0.537
H_2-n-butane	287.9	0.361
H_2-n-hcxane	288.7	0.290
H_2-cyclohexane	288.6	0.319
H_2-benzene	311.3	0.404
H_2-SF_4	286.2	0.396
H_2-n-heptane	303.2	0.283
H_2-n-decane	364.1	0.306
N_2-0_2	273.2	0.181
	293.2	0.22
N_2-He	298.0	0.687
N_2-Ar	293.0	0.194
N_2—NH_3	298.0	0.230
N_2-H_20	307.5	0.256
N_2-$S0_2$	263.0	0.104
N_2-ethylene	298.0	0.163
N_2-ethane	298.0	0.148
N_2-n-butane	298.0	0.096
N_2-isobutane	298.0	0.0905
N_2-n-hexane	288.6	0.076

Gas Pair	Temperature (°K)	Diffusion Coefficient (cm^2/s)
N_2-n-octane	303.1	0.073
N_2-2,2,4-trimethy1pentane	303.3	0.071
N_2-n-decane	363.6	0.084
N_2-benzene	311.3	0.102
O_2-He (He trace)	298.2	0.737
(0_2 trace)	298.2	0.718
O_2-He	298.0	0.729
O_2- H_20	308.1	0.282
O_2-CCl_4	296.0	0.075
O_2-benzene	311.3	0.101
O_2-cyclohexane	288.6	0.075
O_2-n~hexane	288.6	0.075
O_2-n-octane	303.1	0.071
O_2-2,2,4-trimethylpentane	303.0	0.071
He-D_2	295.1	1.250
He-Ar	298.0	0.742
He-H_2O	298.2	0.908
He-NH_3	297.1	0.842
He-n-hexane	417.0	0.1571
He-benzene	298.2	0.384
He-Ne	341.2	1.405
He-methanol	423.2	1.032
He-ethanol	298.2	0.494
He-propanol	423.2	0.676
He-hexanol	423.2	0.469
Ar-Ne	303	0.327
Ar-Kr	303	0.140
Ar-Xe	329.9	0.137
Ar-NH_3	295.1	0.232
Ar-SO_2	263	0.077
Ar-n-hexane	288.6	0.066
Ne-Kr	273	0.223
Ethylene-H_2O	307.8	0.204
Ethane-n-hexane	294	0.0375
N_20-propane	298	0.0860
N_20-ethyleneoxide	298	0.0914
NH_3-SF_6	296.6	0.1090
Freon-12-H_2O	298.2	0.1050
Freon-12-benzene	298.2	0.0385
Freon-12-ethanol	298.2	0.0475

Source: Data from Hirschfelder *et al.* (1954) and Reid *et al.* (1977).

The most popular formula for determining diffusion coefficients in gases is the Chapman–Enskog equation, which is normally written in the form

$$D_{12}(\text{cm}^2/\text{s}) = \frac{\beta T^{3/2}\left(\frac{1}{M_1}+\frac{1}{M_2}\right)^{1/2}}{P d_{12}^2 \Omega},$$

where Ω is a collision integral obtained from look-up tables, M_2 is the molecular weight of the gas that compound 1 is diffusing through, and $\beta = 1.83 \times 10^{-3}$, P is the pressure in atmospheres, T is the absolute temperature in °K, and $d_{12} = 1/2(d_1 + d_2)$ where d_1 and d_2 are the equivalent spherical diameters of the molecules (in Angstroms). A common adjustment to the Chapman Enskog equation is the Wilke–Lee value for β,

$$\beta = 2.17\text{x}10^{-3} - 5\text{x}10^{-4}\left(\frac{1}{M_1}+\frac{1}{M_2}\right)^{1/2}.$$

Example: Determine the diffusivity of water vapor in air. Assume a temperature ot 25°C and a pressure of 1 atm. Using the Chapman–Enskog equation and denoting water vapor as subscript 1 and air as subscript 2, we first find the values for the variables from a chemistry handbook (e.g., CRC Handbook of Chemistry and Physics, or look them up on the web),

$T = 298°\text{K}$
$M_1 = 18$ g/mole
$M_2 = 29$ g/mole
$P = 1$ atm
$d_1 = 2.64$ Å
$d_2 = 0.8(3.8) + 0.2(3.5) = 3.74$ Å
$d_{12} = ½(d_1 + d_2) = 3.17$ Å
$\varepsilon_1 / k = 809°\text{K}$
$\varepsilon_2 / k = 0.8(71.4)+0.2(106.7) = 78.5°\text{K}$
$\varepsilon_{12} / k = (e_1/k\ e_2/k)^{1/2} = 252°\text{K}$
$KT/\varepsilon = 298°\text{K}/252°\text{K} = 1.18$
$\Omega = 1.33$

Thus,

$$D_{12} = 0.21 \text{ cm}^2/\text{s} = 2 \text{ x } 10^{-5} \text{ m}^2/\text{s}.$$

The Wilke–Lee adjustment gives the value

$$D_{12} = 0.23 \text{ cm}^2/\text{s} = 2.3 \text{ x } 10^{-5} \text{ m}^2/\text{s}.$$

The experimental value obtained from Table 1 is

$$D_{12} = 0.26 \text{ cm}^2/\text{s} = 2.6 \text{ x } 10^{-5} \text{ m}^2/\text{s}.$$

APPENDIX B 2-D Office Simulations: COMSOL and ANSWER Software

B.1 COMSOL Model – Report Output

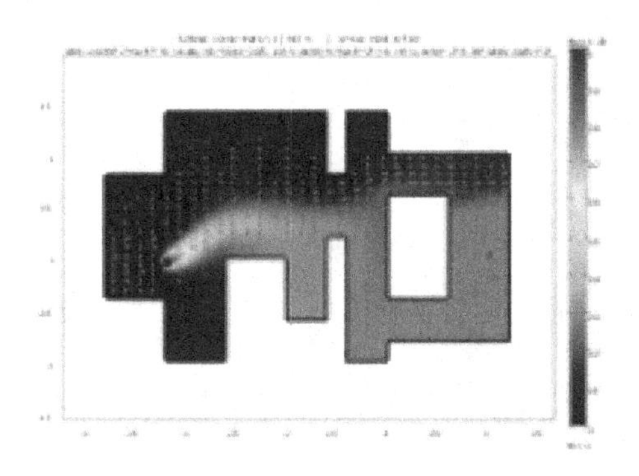

B.1.1 Model properties

Property	Value
Model name	
Author	
Company	
Department	
Reference	
URL	
Saved date	Apr 7, 2008 3:45:57 PM
Creation date	Mar 10, 2008 2:34:11 PM
COMSOL version	COMSOL 3.4.0.248

Application modes and modules used in this model:
Geom1 (2D)
Incompressible Navier–Stokes
Convection and Diffusion

B.1.2 Geometry

Number of geometries: 1

B.1.2.1 Geom1

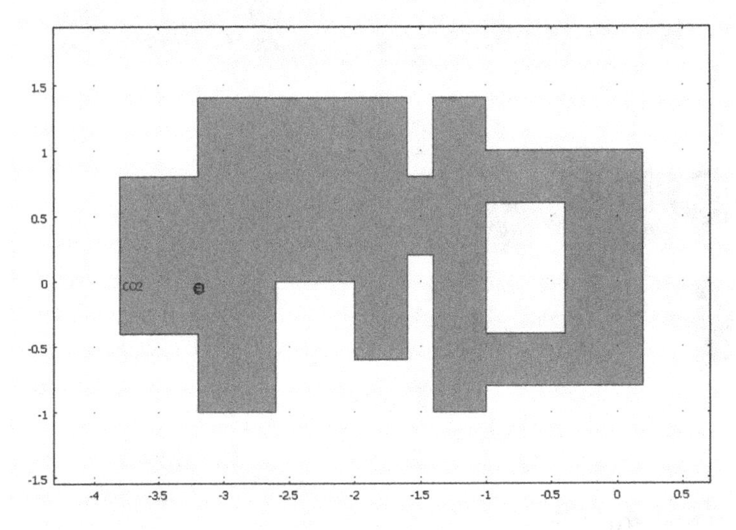

Fig. B.1.1 Domain of 2-D office configuration for COMSOL MultiPhysics software.

B.1.2.2 Point mode

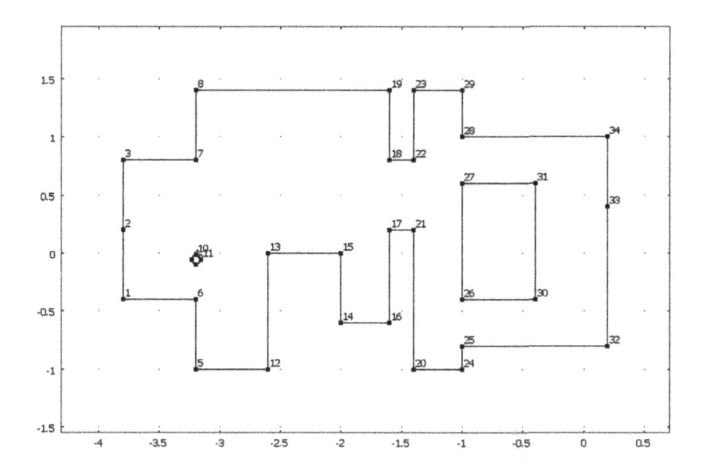

Fig. B.1.2 Grid vertices for 2-D office configuration for COMSOL MultiPhysics software.

B.1.2.3 Boundary mode

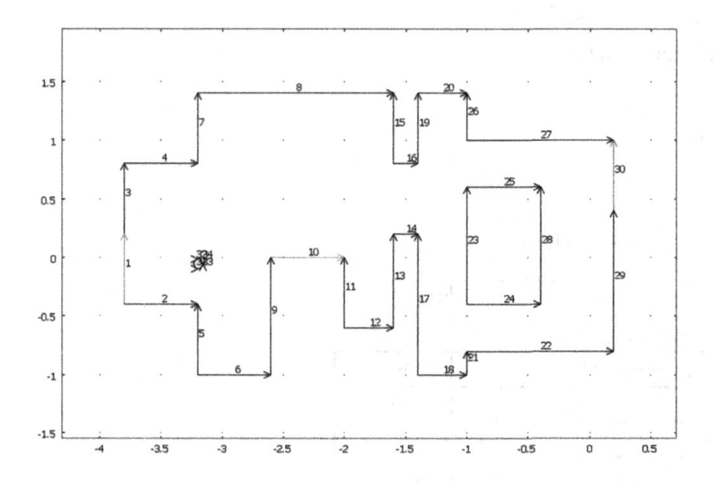

Fig. B.1.3 Boundary edges of 2-D office configuration for COMSOL MultiPhysics software.

B.1.2.4 Subdomain mode

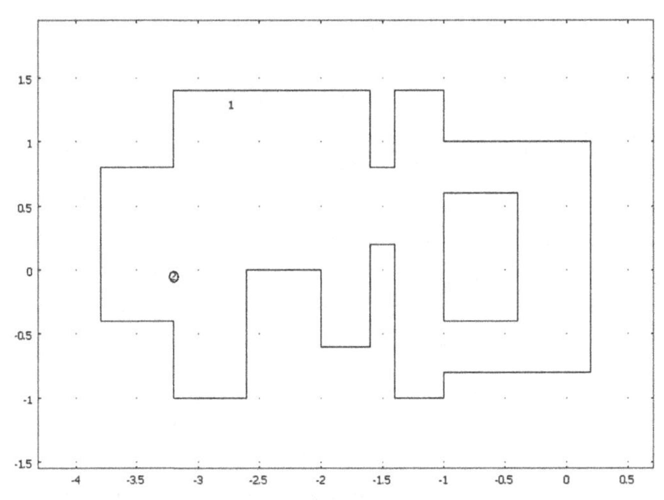

Fig. B.1.4 Grid subdomains of 2-D office configuration for COMSOL MultiPhysics software.

B.1.3 Geom 1

Space dimensions: 2-D
Independent variables: x, y, z

B.1.4 Mesh

B.1.4.1 Mesh statistics

Number of degrees of freedom	81186
Number of mesh points	6357
Number of elements	12229
Triangular	12229
Quadrilateral	0
Number of boundary elements	509
Number of vertex elements	34
Minimum element quality	0.665
Element area ratio	0.007

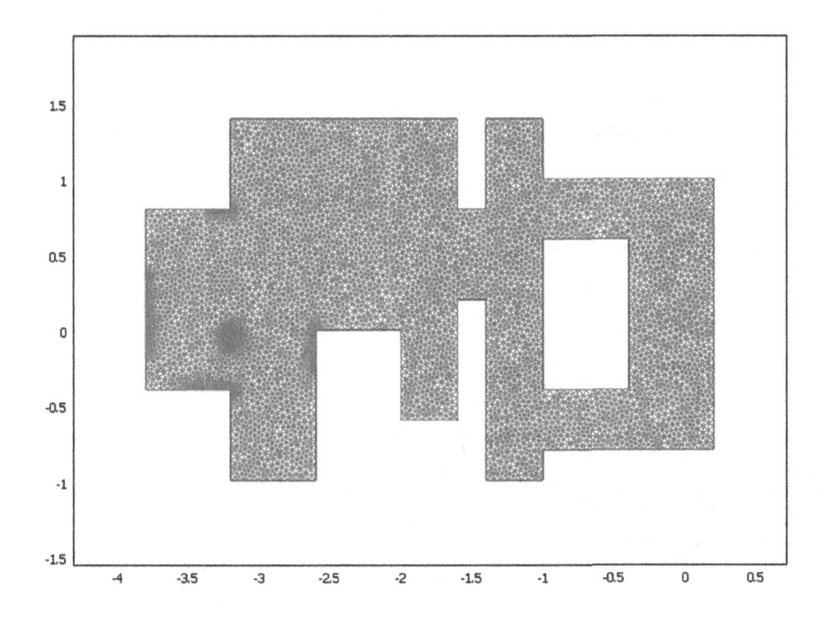

Fig. B.1.5 2-D grid (triangles) of office configuration generated by COMSOL MultiPhysics software.

B.1.5 Application mode: Incompressible Navier–Stokes

Application mode type: Incompressible Navier–Stokes
Application mode name: ns

B.1.5.1 Application mode properties

Property	Value
Default element type	Lagrange - P_2 P_1
Analysis type	Stationary
Corner smoothing	Off
Frame	Frame (ref)
Weak constraints	Off
Constraint type	deal

B.1.5.2 Variables

Dependent variables: u, v, p, nxw, nyw
Shape functions: shlag(2,'u'), shlag(2,'v'), shlag(1,'p')
Interior boundaries active
Locked Points: 6
Locked Boundaries: 1

B.1.5.3 Boundary settings

Boundary		1	2–22, 26–27, 29	23–25, 28
Type		Inlet	Wall	Wall
Inttype		cont	cont	uv

Boundary		30	31–34
Type		Outlet	Interior boundary
Inttype		cont	cont

B.1.5.4 Subdomain settings

Subdomain		1–2
Integration order (gporder)		4 4 2
Constraint order (cporder)		2 2 1

B.1.6 Application mode: Convection and diffusion

Application mode type: Convection and diffusion
Application mode name: cd

B.1.6.1 Application mode properties

Property	Value
Default element type	Lagrange – Quadratic
Analysis type	Stationary
Equation form	Non–conservative
Frame	Frame (ref)
Weak constraints	Off
Constraint type	Ideal

B.1.6.2 Variables

Dependent variables: c
Shape functions: shlag(2,'c')
Interior boundaries active
Locked Boundaries: 1

B.1.6.3 Boundary settings

Boundary		1, 10	2–9, 11–29	30
Type		Insulation/ Symmetry	Insulation/ Symmetry	Convective flux
Concentration (c0)	mol/m^3	1	0	0

Boundary		31–34
Type		Concentration
Concentration (c0)	mol/m^3	1

B.1.6.4 Subdomain settings

Subdomain		1–2
Diffusion coefficient (D)	m^2/s	0.01
x-velocity (u)	m/s	u
y-velocity (v)	m/s	v

B.1.7 Solver settings

Solve using a script: off

Analysis type	Stationary
Auto select solver	On
Solver	Stationary
Solution form	Automatic
Symmetric	auto
Adaption	Off

B.1.7.1 Direct (PARDISO)

Solver type: Linear system solver

Parameter	Value
Preordering algorithm	Nested dissection
Row preordering	On
Pivoting perturbation	1.0E-8
Relative tolerance	1.0E-6
Factor in error estimate	400.0
Check tolerances	On

B.1.7.2 Stationary

Parameter	Value
Linearity	Automatic
Relative tolerance	5.0E-4
Maximum number of iterations	25
Manual tuning of damping parameters	Off
Highly nonlinear problem	On
Initial damping factor	1.0
Minimum damping factor	1.0E-4
Restriction for step size update	10.0

B.1.7.3 Advanced

Parameter	Value
Constraint handling method	Elimination
Null-space function	Automatic
Assembly block size	1000
Use Hermitian transpose of constraint matrix and in symmetry detection	Off
Use complex functions with real input	Off
Stop if error due to undefined operation	On
Store solution on file	Off
Type of scaling	None
Manual scaling	
Row equilibration	On
Manual control of reassembly	Off
Load constant	On
Constraint constant	On
Mass constant	On
Damping (mass) constant	On
Jacobian constant	On
Constraint Jacobian constant	On

B.1.8 Postprocessing

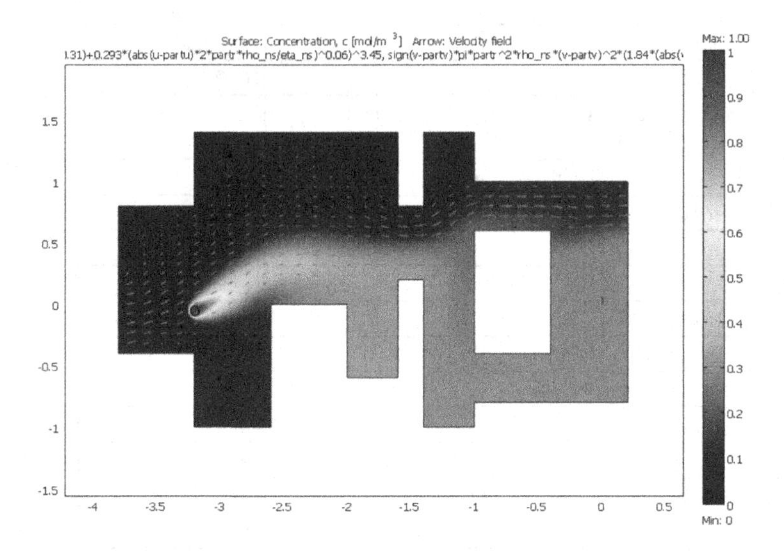

Fig. B.1.6 Concentration isopleths in 2-D office simulated by COMSOL MultiPhysics software.

B.2 ANSWER Model

B.2.1 Answer input deck

```
/====   PROBLEM IDENTIFICATION              ====/
TITLe Indoor_Air_Office_2d
USER Dave Carrington
/====   GEOMETRY SPECIFICATIONS              ====/
COORdinates X MIN -10 MAX +10
COORdinates Y MIN -6  MAX +6
GRID which 42 by 26
/// define objects
LOCAte ID=LLC    COOR (-10, -6) to ( -7, -3)
LOCAte ID=ULC    COOR (-10, +3) to (-7, +6)
LOCAte ID=MB1    COOR ( -4, -6) to (-1, -1)
LOCAte ID=MB2    COOR ( -1, -6) to (+1, -4)
LOCAte ID=MB3    COOR ( +1, -6) to (+2,  0)
LOCAte ID=MU1    COOR ( +1, +3) to (+2, +6)
LOCAte ID=MBG    COOR ( +4, -3) to (+7, +2)
LOCAte ID=LRC    COOR ( +4, -6) to (10, -5)
LOCAte ID=URC    COOR ( +4, +4) to (10, +6)
LOCAte ID=INLET  COOR (-10, -2) to (-9, -1)
LOCAte ID=OUTLET COOR ( +9, +2) to (10, +3)
LOCAte ID=SOURCE COOR LIST ( -7, -2)
/====   INITIAL & BOUNDARY CONDITIONS        ====/
BLOCK  ID=LLC
BLOCK  ID=ULC
BLOCK  ID=MB1
BLOCK  ID=MB2
BLOCK  ID=MB3
BLOCK  ID=MU1
BLOCK  ID=MBG
BLOCK  ID=LRC
```

```
BLOCK  ID=URC
INLEt  ID=INLET X-
OUTLet at X+ for SubRegion ID=OUTLET ;1.0
BOUNdary AT ID=INLET X- U=1.
WALL at everywhere unless otherwise specified
/====   FLUID PROPERTIES & CONSTANTS        ====/
DENSity 1.00
VIScosity 0.01
SCHMIDT NUMBER FOR C=1
/====   NATURE OF FLOW                 ====/
TURBulent model HRe Ke
/====   SOURCE & SINK SPECIFICATIONS       ====/
SOURce C ID=SOURCE CONSTANT 1.
/====   OUTPUT CONTROL                 ====/
SAVE U V SPEED C 'Indoor_Air_Office_2d.sav' FORMatted
CONVergence C 1.0E-6 MAX 5 ITERATIONS
/====   OPERATIONAL CONTROL              ====/
SOLVe  STEADy-state Nmax=500
END
```

B.2.2 Answer solutions

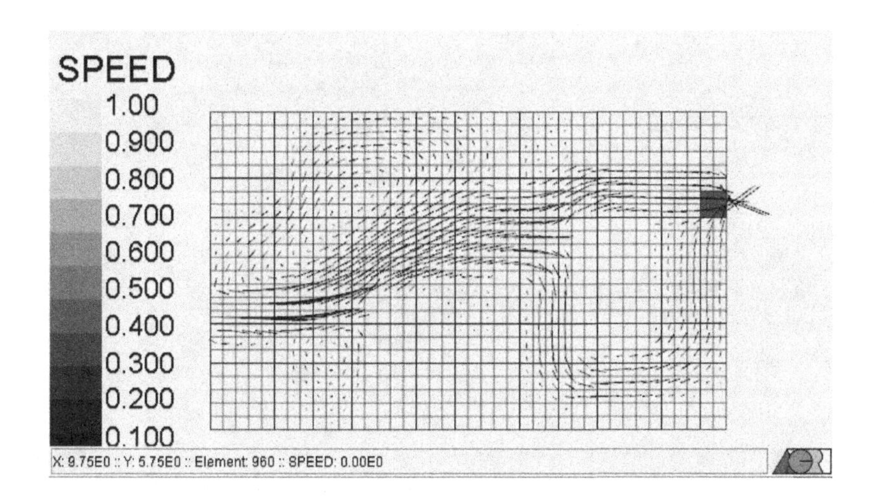

Fig. B.2.1 Predicted velocity vectors and grid for 2-D office configuration simulated by ANSWER CFD software.

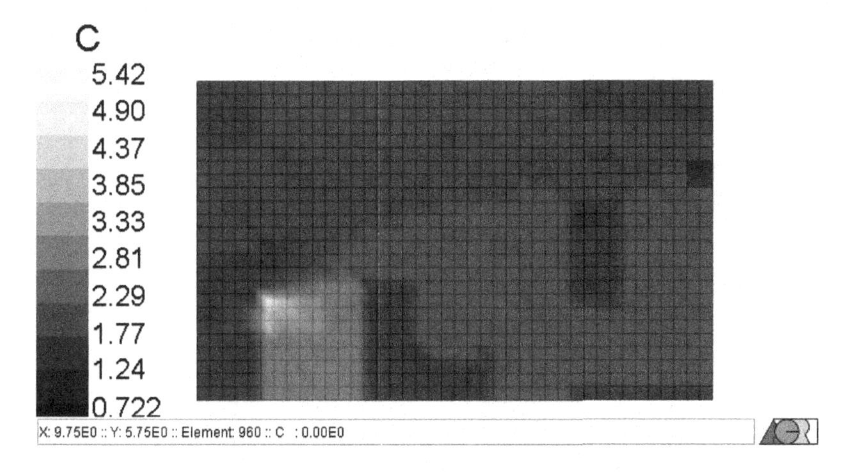

Fig. B.2.2 Predicted concentration and grid for 2-D office configuration simulated by ANSWER CFD software.

Bibliography

American Society of Heating, Refrigerating and Air-Conditioning Engineers (1981). *ASHRAE Handbook 1981 Fundamentals Volume*, Chpt. 22, Atlanta, GA, pp. 22.1–22.20.

Ames, B. N., Magaw, R., and Gold, L. S. (1987). Ranking possible carcinogenic hazards, *Science*, 236, 17 April, pp. 271–280.

Amsden, A. A. (1966). The particle-in-cell method for calculation of the dynamics of compressible fluids, Los Alamos Scientific Laboratory report LA-3466, Los Alamos, NM.

Anderson, J. D. (2001). *Fundamentals of Aerodynamics*, 3rd Ed., McGraw-Hill, NY.

Anderson, D. A, Tannehill, J. C., and Pletcher, R. H. (1997). *Computational Fluid Mechanics and Heat Transfer*, 2nd Ed., Taylor and Francis, Wash., DC.

Armaly, B. F., Durst, F., Pereira, J. C. F., and Schonung, B., (1983). Experimental and theoretical investigation of backward-facing step flow, *J. Fluid Mech.*, 127, pp. 473–496.

Atluri S. N. and Shen S. (2002). *The Meshless Local Petrov–Galerkin (MLPG) Method*, Tech Science Press Encino, Calif.

Atluri S. N. and Zhu, T. (1999). A meshless numerical method based on the local boundary integral equation (LBIE) to solve linear and non-linear boundary value problems, *Engr, Anal. Bndry. Elem.*, 23, no. 5–6, pp.375–389.

Awbi, H. B. (1989). Application of computational fluid dynamics in room ventilation, *Build. Environ.*, 24, pp. 73–84.

Awbi, H. B. (1991). *Ventilation of Buildings*, Chapman and Hall, London, England.

Babuška, I., Chandra J., and Flaherty, J. E. (1983). Adaptive computational methods for partial differential equations, SIAM, Philadelphia.

Beer, G. and Watson J. O. (1992). *Introduction to Finite and Boundary Elements for Engineers*, John Wiley & Sons, NY.

Beer, G. (2001). *Programming the Boundary Element Method,* John Wiley & Sons, Sussex, England.

Bejan, A. (1984). *Convection Heat Transfer*, John Wiley & Sons, NY.

Belytschko T., Lu, Y. Y., and Gu, L. (1994). Element-free Galerkin methods, *Int, J. Num. Meth. in Engr.*, 37, pp. 229–256.

Bochev, P. B. and Gunzburger, M. D. (1998). Finite element methods of least-squares type, *SIAM Review*, 40, no. 4, pp. 789–837.

Boussinesq, J. (1877). Essai sur la theorie des eaux courantes., *Mem. Pres. Acad. Sci. XXIII*, 46, Paris.

Brain, J. D. and Beck, B. D. (1985). Bioassays for mineral dusts and other particulates, in vitro effects of mineral dusts, E. B. Beck and J. Bignon (Eds.), NATO ASI Series Vol. G3, Springer Verlag, Berlin, pp. 323–335.

Brebbia, C. A. and Dominguez, J. (1989). *Boundary Elements, An Introductory Course*, Comp. Mech. Pub., Southampton, England.

Brooks, A. N. and Hughes, T. J. R. (1982). Streamline upwind/Petrov–Galerkin formulations for convective dominated flows with particular emphasis on the incompressible Navier–Stokes equations, *Comp. Meth. App. Mech. and Engr.*, 32, North–Holland, pp. 199–259.

Brueckner, F. and Heinrich, J. C. (1991). Petrov–Galerkin finite element model for compressible flows, *Int. J. Num.. Meth. in Engr.*, 32, pp. 255–274.

Burgraf, O. R. (1966). Analytical and Numerical Studies of the Structure of Steady Separated Flows, *J. Fluid Mechanics*, 24, Part 1, pp. 113–152.

Burnett, D. S. (1987). *Finite Element Analysis*, Addison–Wesley, Reading, MA.

Callen, H.B. (1985). *Thermodynamics and an Introduction to Thermostatistics*, John Wiley & Sons, NY.

Carrington D. B. and Pepper, D. W. (1998). Prediction of species transport in urban canyons using an h-adaptive finite element approach, *Development and Application of Computer Techniques to Environmental Studies*, Pepper, Brebbia, Zannetti (eds.), *Comp. Mech. Pub.*, WIT Press, pp. 53–64.

Carrington D. B. and Pepper, D. W. (1999). A boundary element method for indoor air quality simulation, *Boundary Element Technology XIII*, Chen, Brebbia, Pepper (eds.), *Comp. Mech. Pub.*, WIT Press, pp. 53–61.

Carrington, D. B. (2000). Finite elements with h-adaptation for momentum, heat, and mass transport with application to environmental flow, Ph.D. Thesis, University of Nevada Las Vegas.

Carrington, D. B., and Pepper, D. W. (2002). Convective heat transfer downstream of a 3-D backward-facing Step, *Num. Heat Transfer, Part A*, 41, pp. 555–578.

Carrington, D. B. and Pepper, D. W. (2002). Predicting winds and air quality within the Las Vegas Valley, *Comm. Num. Meth. Engr.*, Vol. 18, Issue 3, pp. 195–201, John Wiley & Sons, Ltd., March 2002.

Carrington, D. B.,(2007). An h-adaptive k-w finite element model for turbulent thermal flow, Los Alamos National Laboratory, report. LA-UR-06-8432, Los Alamos, NM.

Carati, D. and Vanden Eijnden, E. (1997). On the self-similarity assumption in dynamic models for large eddy simulations, *Phys. of Fluids,* 9, no.7, pp. 2165–7.

Carey, G. (1997). *Computational Grids: Generation, Adaptation, and Solution Strategies*, Taylor & Francis, Washington, DC.

Carslaw, H. S. and Jaeger, J. C. (1947). *Conduction of Heat in Solids*, Oxford University Press, London, UK.

Cermak, J. E. (1976). Aerodynamics of buildings, *Annu. Rev. Fluid Mech.*, 8, pp. 75–106.

Chapman, S. and Cowling, T. G. (1970). *The Mathematical Theory of Non-uniform Gases; An Account of the Kinetic Theory of Viscosity, Thermal Conduction, and Diffusion in Gases*, Cambridge University Press, Cambridge, UK.

Chiang T. P. and Sheu, T. W. H. (1999). Time evolution of laminar flow over a three-dimensional backward-facing step, *Int. J. Num. Meth. Fluids*, 31, pp. 721–745.

Chorin, A. J. (1968). Numerical solution of the Navier–Stokes equations, *Math. of Comp.*, Vol. 22, American Mathematical Society, Providence, RI, pp. 745–762.

Chorin, A. J. and Manderson, J. E. (1993). *A Mathematical Introduction to Fluid Mechanics*, 3rd edition, Springer–Verlag, N.Y., N.Y., pp. 36–39.

Chow, C. Y. (1983). *An Introduction to Computational Fluid Mechanics*, Seminole Pub. Co., Boulder, CO.

Chung, T. J. (2002). *Computational Fluid Dynamics*, Cambridge University Press, Cambridge, UK.

Cooper, D. C. and Alley, F. C. (1994). *Air Pollution Control, A Design Approach*, 2nd Ed., Waveland Press, Inc., Prospect Heights, Illinois

Cussler, E. L. (1997). *Diffusion: Mass Transfer in Fluid Systems*, Cambridge University Press, NY.

Davidson, L. (1989). Numerical simulation of turbulent flow in ventilated rooms, Ph.D. Thesis, Chalmers Univ. of Tech., Gotenborg.

Davies, C. N. (1966). Chapter XII Deposition from Moving Aerosols, C. N. Davies (ed.), *Aerosol Science*, New York, Academic Press, pp. 409–428.

Deardorff, J. W. (1970). A numerical study of three-dimensional channel flow at large Reynolds numbers, *J. Fluid Mech.*, 41, pp. 453–480.

Deardorff, J. W. (1973). The use of subgrid transport equations in a three-dimensional model of atmospheric turbulence, *Proc. Applied Mechanics and Fluids Engineering,* ASME.

Denham, M. K., Briard, P. and Patrick, M. A. (1975). A directionally sensitive laser anemometer for velocity measurements in highly turbulent flow, *J. Physic E: Scientific Instruments*, 8, pp. 681–683.

Driscoll, A. (1996). A MATLAB toolbox for Schwarz–Christoffel mapping, *ACM Trans. Math. Soft.*, 22, pp. 168–186.

Duarte, C. A., and Oden, J. T. (1996). An h-p adaptive method using clouds, *Comp. Meth. App. Mech. and Engr.*, 139, no. 1–4, pp. 237–262.

Eguchi, Y., Yagawa, G., and Fuchs, L. (1988). A conjugate-residual-FEM for incompressible viscous flow analysis, *Comp. Mech.*, 3, no. 1, pp. 59–72.

Ferziger, J. H. (1977). Large eddy Numerical Simulations of Turbulent Flows, *AIAA Journal*; 15, no.9, pp.1261–7.

Feynman, R. P., Leighton, R. B., and Sands, M. (1963). *The Feynman Lectures on Physics,* Addison–Wesley, Reading, UK.

Fletcher, C. A. J. (1991). *Computational Techniques for Fluid Dynamics: Volume 1 – Fundamentals and General Techniques*, 2nd Ed., Springer–Verlag, NY.

Fletcher, C. A. J. (1991). *Computational Techniques for Fluid Dynamics – Volume II,* 2nd Ed., Springer–Verlag, NY, pp. 47–78.

Fogelson, A. L. (1992) Particle-method solution of two-dimensional convection-diffusion equations, *J. Comp. Phys.*, 100, no. 1, pp. 1–16.

Franke, R. (1982). Scattered data interpolation: tests of some methods, *Math. of Comp.*, 38, no. 157, pp. 181–200.

Fuchs, N. A. (1964). *The Mechanics of Aerosols*, Pergamon Press, NY.

Fureby, C. (1999). Large-eddy simulation of rearward-facing step flow, *AIAA J.*, 37, no. 11.

Fureby, C. and Grinstein, F. F. (1999). Monotonically integrated large eddy simulation of free shear flows, *AIAA J.*, 37, no. 5, pp. 544–556.

Gartling, D. K., (1990). A test problem for outflow boundary conditions—flow over a backward-facing step, *Int. J. Num. Meth. Fluids*, 11, pp. 953.

George, P. (1991). *Automatic Mesh Generation: Application to Finite Element Methods*, John Wiley & Sons, Ltd., Paris.

Gobeau, N. and Zhou, X. X. (2003). Constructive comments, *Fire Prevention*, 368, p.55–57.

Gosman, A. D., Pun, W. M., Runchal, A. K., Spalding, D. B., and Wolfshtein, M. (1969). *Heat and Mass Transfer in Recirculating Flows*, Academic Press, London.

Gosman, A. D., Nielsen, P. V., Restivo, A., and Whitelaw, J. H. (1980). The flow properties of rooms with small ventilation openings, *J. Fluid Engr.*, 102, pp. 316–323.

Greenspan, D. (2005). Molecular modeling of flow past an inclined flat plate, *Math. Comp. Modeling*, 42, no. 13, pp. 1491–1504.

Gresho, P. M., Chan, S. T., Lee, R. L., and Upson, C. D. (1984). Modified finite element method for solving the time-dependent incompressible Navier–Stokes equations, Part 1, Theory, *Int. J. Num. Meth. Fluids*, 4, pp. 557–598.

Gresho, P. M. (1985). A modified finite element method for solving the incompressible Navier–Stokes equations, *Lectures in Applied Mathematics*, 22, Part 1, eds. Engquist B. E., Osher, S., and Somerville, R. C. J., American Mathematical Society, Providence, RI, pp. 193–240.

Gresho, P. M. and Sani, R. (1987). On pressure boundary conditions for the incompressible Navier–Stokes equations, *Int. J. Num. Meth. Fluids*, 7, John Wiley & Sons, Ltd., pp. 1116–1145.

Gresho, P. M. and Chan, S. T. (1990). On the theory of semi-implicit projection methods for viscous incompressible flow and its implementation via a finite element method that also introduces a nearly consistent mass matrix. Part 2: Implementation, *Int. .J. Num. Meth. Fluids*, 11, John Wiley & Sons, Ltd., pp. 621–659.

Grinstein, F. F. and Fureby, C. (2002). Recent progress on MILES for high Reynolds number flows, *J. Fluids Engr.*, 124, pp. 848–861.

Gulliver, J. S. (2007). Introduction to chemical transport in the environment, Cambridge University Press, NY.

Gunzburger, M. (1989). *Finite Element Methods for Viscous Incompressible Flows – A Guide to Theory, Practice, and Algorithms*, Academic Press, Inc., San Diego, CA, p. 60.

Hackbusch, W. and Trottenberg, U. (1982). Multi-grid Methods, Proceedings, Koln–Porz, Nov. 1981, *Lecture Notes in Mathematics 960*, Springer, Berlin.

Hadjisophocleous, G. V., Benichou, N. (1999). Performance criteria used in fire safety design, *Automation in Construction,* 8, no. 4, 1999, pp. 489-501

Hansen, G. A. (1996). Scalability of preconditioners as a strategy for parallel computation of compressible fluid flow, Ph.D. Dissertation, University of Idaho, Idaho Falls, ID.

Harlow, F. H. and Amsden, A. A. (1970). The SMAC method a numerical technique for calculating incompressible fluid flows, Los Alamos Scientific Laboratory report LA-4370, Los Alamos, NM.

Harlow, F. H. and Welch, J. E. (1965). Numerical calculation of time-dependent viscous incompressible flow of fluid with free surfaces, *Phys. Fluids*, 8, pp. 2182–2189.

Harlow, F. H. and Hirt, C. W. (1969). Generalized transport theory of anisotropic turbulence (Generalized transport theory of anisotropic turbulence), Los Alamos Scientific Laboratory report LA-4086, Los Alamos, NM.

Hayes, S. M., Gobbell, R. V., and Ganick, N. R. (1995). *Indoor Air Quality: Solutions and Strategies*, McGraw–Hill, Inc., NY.

Heinrich, J. C. and Yu, C. C. (1988). Finite elements simulation of buoyancy-driven flows with emphasis on natural convection in a horizontal circular cylinder, *Comp. Meth. Apps. in Mech. Engr.*, 69, pp. 1–27.

Heinrich, J. C. and Pepper, D. W. (1999). *Intermediate Finite Element Method – Fluid Flow and Heat Transfer Applications,* Taylor & Frances, Philadelphia, PA.

Heinsohn, R. J. (1991). *Industrial Ventilation: Engineering Principles*, John Wiley & Sons, NY.

Hindmarsh, A. C., Gresho, P. M., and Griffiths, D. F. (1984). The stability of explicit Euler time-integration for certain finite difference approximations of the multi-dimensional advection–diffusion equation, *Int. J. Num. Meth. Fluids*, 4, John Wiley & Sons, Ltd., pp. 853–897.

Hjertager, B. H. and Magnussen, B. F. (1977). Numerical prediction of three-dimensional turbulent buoyant flow in a ventilated room, *Heat Transfer and Turbulent Buoyant Convection*, D. B. Spalding and N. Afgan (eds.), Vol. II, Hemisphere, Washington, DC, pp. 429–441.

Hinds, W. C. (1982). *Aerosol Technology – Properties, Behavior, and Measurement of Airborne Particles*, John Wiley & Sons Inc., NY, pp. 155–156.

Hiorns, N. and Sinai, Y. (1999). Comfort and safety in the Millennium dome, *CFX Update*, 20, Spring.

Hoffman, K. A., and Chiang, S. A. (1993). *Computational Fluid Dynamics for Engineers, Vol. 1, Engineering Education System™*, Wichita, Kansas, US.

Hoffman, K. A., and Chiang, S. A. (2000). *Computational Fluid Dynamics for Engineers, Vol. 3, Engineering Education System™*, Wichita, Kansas, US.

Hoppe, P. and Martinac, I. (1998). Indoor climate and air quality, *Int. J. Biometeorol.*, 42, pp. 1–7.

Hughes, T. J. R. (1987). Recent progress in the development and understanding of SUPG methods with special reference to the compressible Euler and Navier–Stokes equations, *Int. J. Num. Meth. Engr.*, 7, John Wiley & Sons, Ltd., pp. 1261–1275.

Huang, H. C. and Usmani, A. S. (1994) *Finite Element Analysis for Heat Transfer, Theory and Software*, Springer–Verlag, London, UK.

Hung L., Parviz, M., and Kim, J. (1997). Direct numerical simulation of turbulent flow over a backward-facing step, *J. Fluid Mech.*, 330, pp. 349–374.

Hunt, J. C., Abell, C. J., Peterka, J. A., and Woo, H. (1978). Kinematical studies of the flows around free or surface mounted obstacles; applying topology to flow visualization, *J. Fluid Mech.*, 86, 1, pp.179–200.

Ilegbusi, J. O. (1983). A revised two-equation model of turbulence, Ph.D. Thesis, Imperial College of Science and Technology, London, UK.

Ilegbusi, J. O. and Spalding, D. B. (1983). Turbulent flow downstream of a backward facing step, *Proc. 4th Int. Symp. on Turbulent Shear Flows*, Karlsruhe, West Germany.

Incropera, F. P. and DeWitt, D. P. (2002). *Fundamentals of heat and mass transfer*, John Wiley & Sons, Inc., NY.

Ilinca, F., Hetu, J. F., and Pelletier, D. (1998). A unified finite element algorithm for the two-equation models of turbulence, *Computers & Fluids*, 27, no. 3, Elsevier Science, Ltd., pp. 291–310.

Irons, B. (1968). Curved, isoparametric, quadrilateral elements for finite element analysis, *Int. J. of Solids and Structures*, 4, no. 1, pp. 31–42.

Issa, R. I. (1986). Solution of the implicitly discretised fluid flow equations by operator-splitting, *J. Comp. Phys.*, 62, pp. 40–65.

Ivanov, V. I. and Trubetskov, M. K. (1994). *Handbook of Conformal Mapping with Computer-Aided Visualization*, CRC Press, Boca Raton, FL.

Jiang, B. N. and Lin, T. L. (1993). Least-squares finite element solutions for three-dimensional backward-facing step flow, *Proc. 5th Int. Symp. Comp. Fluid Dynamics*, Sendai, Japan.

Jones, W. P. and Launder, B. E. (1972). The prediction of laminarization with a two-equation model of turbulence, *Int. J. Heat Mass Transfer*, 15, pp. 301–314.

Jovic. S., and Driver, D. M. (1994). Backward-facing step measurement at low Reynolds number, Re = 5000, NASA Tech. Memo. 108807.

Kalla, N. D. and D. W. Pepper (2008), A meshless radial basis function method for fluid flow with heat transfer, *Proc. ICCES '08*, Honolulu, HI, March 17-22, 2008.

Kalla, N. D. and D. W. Pepper (2008), A meshless radial basis function method for fluid flow with heat transfer, *ICCES*, Vol. 142, No. 1, pp. 1-6.

Kane, J. H. (1994). Boundary element analysis on vector and parallel computers, *Comp. Sys. Engr.*, 5, no. 3, pp. 239–252.

Kansa, E. J. (1990). Multiquadrics – A scattered data approximation scheme with applications to computational fluid dynamics – II solutions to parabolic, hyperbolic and elliptic partial differential equations, *Comp. Math. with Applications,* 19, no. 8–9, pp. 147–161.

Karlsson, B. and Quintiere, J. G. (2000). *Enclosure Fire Dynamics*, CRC Press, Boca Raton, FL.

Kelly, D. W., Nakazawa, S., Zienkiewicz, O. C., and Heinrich, J. C. (1980). A note on upwinding and anisotropic balancing dissipation in the finite element approximations to convective diffusion problems, *Int. J. Num. Meth. Engr.*, 15, John Wiley & Sons, pp. 1705–1711.

Kim, J., Monin, P., and Moser, R. (1987). Turbulence statistics in flully developed channel flow at low Reynolds numbers, *J. Fluid Mech.*, 177, pp. 133–166.

Klote, J. H. (1999). CFD simulations of the effects of HVAC induced flows on smoke detector response. *Proc. ASHRAE Winter Meeting*, 105, no. 1, ASHRAE Transactions, pp. 395–409.

Kolmogorov, A. N. (1942). Equations of Turbulent Motion of an Incompressible Fluid, *Izvestia Academy of Sciences, USSR: Physics*, 6, 1–2, pp. 56–58.

Kostas, J., Soria, J., and Chong, M. S. (2002). Particle image velocimetry measurements of a Backward-Facing Step Flow, *Exper. in Fluids*, 33, pp. 838–853.

Koster, K. T. and Dickerson, M. H. (1990). An updated summary of MATHEW/ADPIC model evaluation studies, UCRL-JC-104134, CONF-9008157-1, May 1, 1990.

Koutmos, P., and Mavridis, C. (1997). A computational investigation of unsteady separated flows, *Int. J. Heat Fluid Flow*, 18, pp. 297–306.

Krotz, D. (2002). Berkeley-Trent: A BETR model of toxic transport, http://uclbl.org/Science-Articles/Archive/EETD-BETR-Krotz.html, Feb. 15, 2002.

Kuehn, T. H. and Goldstein, R. J. (1976). An experimental and theoretical study of natural convection in the annulus between horizontal concentric cylinders, *J. Fluid Mech.*, 74, pp. 695–719.

Lange, R. (1973). ADPIC: A three-dimensional computer code for the study of pollutant dispersal and deposition under complex conditions, Lawrence Livermore Laboratory report, No. UCRL—51462, Livermore, CA.

Launder, B. E. and Spalding, D. B. (1972). The numerical computation of turbulent flows, *Comp. Meth. in Appl. Mech. and Engr.*, 3, no. 2, pp. 269–289.

Launder, B. E. and Sharma, B. I. (1974). Application of the energy dissipation model of turbulence to the calculation of flow near a spinning disc, *Letters in Heat and Mass Transfer*, 1, 2, pp. 131–138.

Leonard, B. P. (1979). The QUICK finite difference method for the convection-diffusion equation (Quadratic Upstream Interpolation for Convective Kinematics), *Proc. Third Int. Symp. for advances in computer methods for partial differential equations*, Bethlehem, Pa.

Levinson, N. and Redheffer, R. M. (1970). *Complex Variables*, Holden–Day, Inc., San Francisco, CA.

Lewis, R. W., Morgan, K., Thomas, H. R., and Seetharamu, K. N. (1996). *The Finite Element Method in Heat Transfer Analysis*, John Wiley & Sons, Chichester, UK.

Li, J., Pepper, D. W., and Chen, Y. (2003), RBF based meshless methods for hyperbolic equations, *Nat. Heat Transfer Conf. (invited talk)*, Rio Hotel and Casino, Las Vegas, NV, July 20–23.

Liu, G. R and Gu, Y. T. (2001). A local radial point interpolation method (LRPIM) for free vibration analyses of 2-D solids, *J. Sound and Vibration,* 246, no. 1, pp. 29–46.

Liu, G. R., Yan, L., Wang, J. G. and Gu., Y. T. (2002). Point interpolation method based on local residual formulation using radial basis functions, *Structural Engr. & Mech.*, 14, no. 6, pp. 713 –732.

Liu, G. R. and Gu, Y. T. (2002). Comparison of two meshfree local point interpolation methods for structural analyses, *Comp. Mech.*, 29, no. 2, pp.107 –121.

Liu, W. K and Jun, S. (1998). Multiple-scale Reproducing Kernel Particle Methods for large deformation problems, *Int. J. Num. Meth. in Engr.*, 41, no. 7, pp. 1339 – 1362.

Löhner, R. (1990). A fast finite element solver for incompressible flows, *Proc. 28th AIAA Aerospace Sciences Meeting,* AIAA-90-0398.

Lougheed, G. D. and Hadjisophocleous, G. V. (2001). The smoke hazard from a fire in high spaces, *Proc. 2001 Winter Meeting Conference,* 107, no. 1, ASHRAE Transactions, pp. 720–729.

Lucy, L. (1977). A numerical approach to testing the fission hypothesis, *A. J.,* 82, pp. 1013–1024.

Maple ®, http://www.maplesoft.com.

Marchioro, C. and Pulvirenti, M. (1994). *Mathematical Theory of Incompressible Nonviscous Fluids*, Springer–Verlag, New York, NY, pp. 49–50.

Markatos, M. C. and Cox, G. (1984). Hydrodynamics and heat transfer in enclosures containing a fire source, *Phys. Chem. Hydrodyn.*, 5, pp. 53–66.

Marro, L. (1980). *Méthodes de Réduction de la largeur de bande et du profil efficace des matrices creuses.* Ph.D. thesis, Université de Nice.

Martin, R. A., Tang, P. K., Harper, A. P., Novat, J. D., and Gregory, W. S. (1983). Material transport analysis for accident-induced flow in nuclear facilities, Los Alamos National Laboratory report NUREG/CR-3527, Los Alamos, NM.

Mathematica, http://www.wolfram.com/company/background.html.

Matlab ®, http://www.mathworks.com.

Melenk, J. M. and Babuska, I. (1996). Partition of unity finite element method: basic theory and applications, *Comp. Meth. in Appl. Mech. And Engr.*, 139, no.1–4, pp.289–314.

Mills, F. A. (2001). Case study of a fire engineering approach to a large, unsprinklered, naturally ventilated atrium building, *Proc. of ASHRAE 2001 Winter Meeting*, Atalanta, GA, ASHRAE Transactions, 107, no. 1, pp. 744–752.

Miller, J. D. (2003). Guarding Against "ChemBio" Attack, *The Military Engineer*, 95, 623, May–June, pp. 37–40.

Moult, A. and Dean, R. B. (1980). CAFE - A computer program to calculate the flow environment, CAD 80, *Proc. 4th Int. Conf. on Computers in Design Engineering*, Brighton, UK.

Murakami, S., Tanaka, T., and Kato, S. (1983). Numerical simulation of air flow and gas diffusion in room model – correspondence between numerical simulation and model experiment, *Proc. 4th CIB Int. Symp. on the Use of Computers for Environmental Engineering Related to Buildings*, Tokyo, pp. 90–95.

Murakami, S., Kato, S., and Suyama, Y. (1987). Three-dimensional numerical simulation of turbulent airflow in a ventilated room by means of a two-equation model, *ASHRAE Transactions*, 93, no. 2, pp. 621–642.

Nagda, N. L. (1993). Modeling of Indoor Air Quality and Exposure, ASTM STP 1205.

Nallasmy, M. (1987). Turbulence models and their applications to the prediction of internal flows: a review, *Computers and Fluids*, 15, no. 2, pp. 151–194.

Nayroles, B., Touzot, G. and Villon, P. (1992). Generalizing the finite element method: Diffuse approximation and diffuse elements, *Comp. Mech.*, 10, no.5, pp.307–318.

Nazaroff, W. W. and Cass, G. R. (1989). Mathematical modeling of indoor aerosol dynamics, *Environ. Sci. Tech.*, 23, no. 2, pp. 157 - 166

Nielsen, P. V. (1974). Flow in air conditioned rooms, Ph.D. Thesis, Tech. Univ. of Denmark.

Nielsen, P. V. (1981). Contamination distribution in industrial areas with forced ventilation and two-dimensional flow, *Proc. Joint Meeting E1*, Essen, W. Germany, Int. Inst. of Refrigeration, pp. 223–230.

Oden, J. T., Strouboulis, T., and Devloo, P. (1986). Adaptive finite element methods for the analysis of inviscid compressible flow. Fast refinement/unrefinement and moving mesh methods for unstructured meshes, *Comp. Meth. App. Mech. Engr.*, 59, no. 3, pp. 327–62.

Osher, S. (1984). Shock capturing algorithms for equations of mixed type, *Num. Meth. for Partial Differential Equations*, Longman Sci. Tech., Harlow. UK, pp. 103—139.

Pasquill, F. and Smith, F. B. (1985). *Atmospheric Diffusion*, 3rd ed., Ellis Horwood Ltd., Chichester, UK.

Patankar, S. V. (1980). *Numerical Heat Transfer and Fluid Flow*, Hemisphere Publishing, NY.

Patankar, S. V. and Spalding, D. B. (1972). A calculation procedure for heat, mass, and momentum transfer in three-dimensional parabolic flows, *Int. J. Heat Mass Transfer*, 15, pp. 147–163.

Pelletier, D. and Hetu, J. F. (1992). Fast, adaptive finite element scheme for viscous incompressible flows, *AIAA Journal*, 30, no.11, pp. 2677–82.

Pelletier, D. and Ilinca, F. (1994). An adaptive finite element method for mixed convection heat transfer, *Proc. 32nd AIAA Aerospace Sci. Meeting*, AIAA-94-0347.

Pepper, D. W. (1981). Results from the savannah river laboratory model validation workshop, Nov. 19–21, *Proc. 5th Symp. on Turbulence, Diffusion, and Air Pollution.*

Pepper, D. W. (1987). Modeling of natural convection in three dimensions using a time-split finite element method, *Num. Heat Transfer*, 11, pp. 31–55.

Pepper, D. W. and Baker, A. J. (1980). A simple one-dimensional finite-element algorithm with multidimensional capabilities, *Num. Heat Transfer*, 2, pp. 81 –95.

Pepper, D. W. and Carrington, D. B. (1997), Convective Heat Transfer over a 3-D Backward Facing Step, *ICHMT Int. Symp. on Advances in Computational Heat Transfer*, May 26–30, Cesme, Turkey.

Pepper, D. W., Chen, C. S., and Li, J. (2002). Modeling heat transfer using adaptive finite elements, boundary elements, and meshless methods, *Proc. 7th International Conference on Advanced Computational Methods in Heat Transfer*, Kassandra, Greece, pp. 349–360.

Pepper, D. W. and Emery, A. F. (1994) Atmospheric transport prediction using an adaptive finite element method, *Proc. 5th Int. High Level Rad. Waste Manag. Conf.*, pp. 1946–1952.

Pepper, D. W. and Harris, S. D. (1977). Fully implicit algorithms for solving partial differential equations, *J. of Fluids Engr.*, 99, pp. 781–783.

Pepper, D. W. and Heinrich, J. C. (1992). *The Finite Element Method, Basic Concepts and Application*, Taylor & Francis Publishing, Hemisphere Publishing Co.

Pepper, D. W., Sarler, B., and Strojniski, V. (2005). Application of meshless methods for thermal analysis, 51, no. 7–8, *Proc. ASME / ZSIS International Thermal Science Seminar*, Bled, Slovenia, pp. 476–483.

Pepper, D. W. and Singer, A. P. (1990). A modified finite element method for the personal computer, *Num. Heat Transfer*, 11, pp. 31–55.

Pepper, D. W. and Stephenson D. E. (1995). An adaptive finite-element model for calculating subsurface transport of contaminant, *Ground Water*, 33, no.3, pp.486-496

Pepper, D. W. and Wang, X. (2006). Modeling Indoor Air Pollution, COMSOL Users Conference 2006, Las Vegas, NV, Oct. 26–28.

Pepper, D. W., Wilcox, T., and Gewali, L. (2000). HTADAPT: An h-adaptive finite element model for heat transfer, *Proc. Sixth International Conference on Advanced Computational Methods in Heat Transfer, Computational Studies*, 3, pp. 395–405.

Persson, L., Fureby, C., and Svanstedt, N. (2002). On homogenization-based methods for large-eddy simulation, *J. Fluids Engr.*, 124, pp. 892–903.

Pitz, R. W., and Daily, J. W., (1981). *Experimental study of combustion: the turbulent structure of a reacting shear layer formed at a rearward-facing step*, NASA CR Rep. 165427.

Plog, B. A. (1988). *Fundamentals of industrial hygiene* Plog B. A. (ed.), Chicago, Ill, National Safety Council.

Pope. S. B. (2000). *Turbulent Flows*, Cambridge University Press, Cambridge, UK.

Powell, M. J. D. (1992). The theory of radial basis function approximation in 1990. *Advances in numerical analysis, Vol. II*, Oxford Univ. Press, NY, pp.105-210.

Pozrikidis, C. (1999). *Little Book of Streamlines*, Academic Press, San Diego, CA.

Pozrikidis, C. (2002). Dynamical simulation of the flow of suspensions: wall-bounded and pressure-driven channel flow, *Industrial Engr. Chem. Research*, 41, no. 25, pp. 6312–6322.

Prandtl, L. (1925). Über die ausgebildete Turbulenz, *ZAMM*, 5, pp. 136–139.

Prandtl, L. (1945). Über ein neues Formelsystem für die ausgebildete Turbulenz, *Nacr. Akad. Wiss. Göttingen, Math-Phys. Kl. 1945*, pp. 6–19.

Pruppacher, H. R. and Klett, J. D. (1978). *Microphysics of Clouds and Precipitation*, D. Reidel Pub., Dordrecht, Holland.

Rai, M. M. and Moin, P. (1991). Direct simulation of turbulent flow using finite-difference schemes, *J. Comp. Phys.*, 96, pp. 15–33.

Ramachandran, P. A. (1994). Boundary Element Methods in Transport Phenomena, *Comp. Mech. Pub.*, Southampton, UK, co-published with Elsevier Applied Science, London.

Ramakrishnam, R., Bey, K. S. and Thornton, E. A. (1990). Adaptive Quadrilateral and Triangular Finite Element Scheme for Compressible flows, *AIAA J.*, 28, 1, pp. 51–59.

Ramaswamy, B. (1990). Efficient finite element method for two-dimensional fluid flow and heat transfer problems, *Num. Heat Transfer*, 17, Hemisphere Publishing, Inc., pp. 123–154.

Ramaswamy, B., Jue, T. C., and Akin, J. E. (1992). Semi-implicit and explicit finite element schemes for coupled fluid/thermal problems, *Int. J. Num. Meth. Engr.*, 34, John Wiley & Sons, pp. 675–696.

Reist, P. C. (1993). *Aerosol Science and Technology*, 2nd Ed., McGraw–Hill, NY.

Repace, J. L. and Lowery, A. H. (1980). Indoor air pollution, tobacco smoke, and public health, *Science*, 208, May, pp. 464–472.

Reynolds, O. (1895). On the dynamical, theory of incompressible viscous fluids and the determination of the criterion, *Phil. Trans. Roy Soc.*, 186, pp. 123–164.

Roache, P. J. (1972). *Computational Fluid Dynamics*, Hermosa, Albuquerque, NM.

Roache, P. J. (1998). *Verification and Validation in Computational Science and Engineering*, Hermosa, Albuquerque, NM.

Runchal, A. K. (1980). A random walk atmospheric dispersion model for domplex terrain and meteorological conditions, *Proc. 2nd AMS Joint Conf. on Appl. of Air Pollution Meteor.*, New Orleans, LA.

Saffman, P. G. (1970). A model for inhomogeneous turbulent flow, *Proc. Roy. Soc. Lond.*, A317, pp. 417–433.

Sakamoto, Y. and Matsuo, Y. (1980). Numerical predictions of three-dimensional flow in a ventilated room using turbulence models, *Appl. Math. Modeling*, 4, pp. 67–71.

Sandberg, M. (1981). What is ventilation efficiency?, *Building and Environment,* 16, 2, pp. 123–135.

Sarler, B., Perko, J., and Chen, C. (2002). Radial basis function collocation method solution of natural convection in porous media, I*nt. J. Num. Meth. for Heat & Fluid Flow*, 14, no. 2, pp. 187–212.

Schlichting, H. (1979). *Boundary-Layer Theory*, 7th Ed., McGraw–Hill, NY.

Schram, C., Rambaud, P., and Riethmuller, M. L. (2004). Wavelet based eddy structure education from a backward-facing step flow investigated using particle image velocimetry, *Expr. in Fluids*, 36, pp. 233–245.

Settles, G. S. (2006). Fluid Mechanics and Homeland Security, *Annu. Rev. Fluid Mech.*, 38, pp. 87–110.

Sinclair, R. (2001). CFD simulation in atrium smoke management system design, *Proc. ASHRAE 2001 Winter Meeting*, Atlanta, GA, pp. 843–850.

Sklarew R. J., Fabrick, A. J., and Prager, J. E. (1971). A particle-in-cell method for numerical solution of the atmospheric diffusion equation, and applications to air pollution problems, volume 1 (PICK method of particle-in-cell techniques for numerical solution of atmospheric diffusion equation applied to air pollution problems, Systems Science and Software Rep., La Jolla, CA, no. PB-209290, REPT-3SR-844-VOL-1, APTD-0952.

Smagorinsky, J. (1963). General circulation experiments with the primitive equations, *Mon. Weather Rev.,* 91, no. 3, Washington, DC.

Shapiro, R. A. and Murman, E. M. (1988). Adaptive finite element methods for the Euler equations, *Proc. AIAA 26th Aerospace Sciences Meeting,* Reno, NV.

Spaulding, M. (1976). Numerical modeling of pollutant transport using a Lagrangian marker particle technique, NASA-TM-X-73930.

Taylor, C., Thomas, C. E., and Morgan, K. (1981). Modeling flow over a backward-facing step using the FEM and the two-equation model of turbulence, *Int. J. Num. Meth. Fluids*, 1, pp. 295–304.

Tennekes, H., and Lumley, J. L. (1972). *A First Course in Turbulence*, MIT Press, Cambridge, MA.

Thomas, L. C. (1999). *Heat Transfer – Professional Version*, 2nd Ed., Capstone Publishing Co., Tulsa, OK.

Thompson, J. F., Warsi, Z. U. A., and Martin, C. W. (1985). *Numerical Grid Generation,* North–Holland, NY.

Van Doormaal, J. P., Raithby G. D., and McDonald, B. H. (1986). The segregated approach to predicting viscous compressible fluid flows, *Proc. ASME, 31st International Gas Turbine Conference and Exhibit,* Duesseldorf, West Germany.

Vargaftik, N. B. (1975). *Tables on the Thermophysical Properties of Liquids and Gases,* 2nd Ed., Hemisphere Pub. Corp., Washington, DC.

Versteeg, H. K. and Malalasekera, W. (2007). *An Introduction to Computational Fluid Dynamics: The Finite Volume Method,* 2nd Ed., Pearson Prentice–Hall, London.

Walton G. N. and Dols, W. S. (2006). NISTIR 7251 – CONTAM 2.4 User manual, Multizone airflow and contaminant transport analysis software. Tech. Rpt. NISTIR 6921, National Institute of Standards and Technology, October.

Wang, X. and Pepper, D. W. (2007). Application of an hp-Adaptive FEM for solving thermal flow problems, *AIAA J. Thermophysics. and Heat Trans.*, 21, 1, pp.190–198,

Warsi, Z. U. (1999) *Fluid Dynamics: Theoretical and Computational Approaches*, CRC Press, Boca Raton, FL.

Wilcox, D. C. (2006). *Turbulence Modeling for CFD*, 3rd Ed., DWC Industries, La Canada, CA.

Wilke, C. R., Lee, C. Y. (1955). Estimation of diffusion coefficients for gases and vapors, *Indus. Engr. Chem.*, 47, no. 6, pp. 1253-1257.

Williams, P. T. and Baker, A. J. (1997). Numerical simulations of laminar flow over a 3–D backward-facing step, *Int. J. Num. Meth. Fluids*, 24, pp. 1159–1183.

Woo, T. C. and Hwang, C. C. (2000). *Classic Analytical Problems in Mechanical Engineering*, R. T. Edwards, Inc., Flourtown, PA.

Wrobel, L. C. (2002). *The Boundary Element Method, Vol. 1: Application in Thermo-Fluids and Acoustics.*, John Wiley & Sons, Sussex, UK.

Wu, Y. L and Liu, G. R. (2003) A meshfree formulation of local radial point interpolation method (LRPIM) for incompressible flow simulation, *Comp. Mech.*, 30, no. 6–5, pp. 355–365.

Yakhot, V., Orszag, S. A., Thangam, S., Gatski, T. B. and Speziale, C. G. (1992). Development of Turbulence Models for shear flows by a double expansion technique, *Physics of Fluids A*, 4, pp. 1510–1520.

Yu, C. C. and Heinrich, J. C. (1986). Petrov–Galerkin methods for the time-dependent convective transport equation, *Int. J. Num. Meth. Engr.*, 23, p. 883–890.

Zerroukat, M., Power, H. and Chen, C. S. (1998). Numerical method for heat transfer problems using collocation and radial basis functions, *Int. J. Num. Meth. Engr.*, 42, no. 7, pp. 1263–78.

Zhai, Z., Srebric, J. and Chen, Q. (2003). Application of CFD to Predict and Control Chemical and Biological Agent Dispersion in Buildings, *Int. J. of Ventilation*, 2, 3, pp. 251–264.

Zienkiewicz, O. C. (1977). *The Finite Element Method*, McGraw–Hill, London.

Zienkiewicz, O. C. and Taylor R. L. (2000). The finite element method, Butterworth-Heinemann, Boston, Vol 1, 5th Ed. Chpt. 15, pp. 401.

Zlamal, M. (1973). The finite element method in domains with curved boundaries, *Int. J. Num. Meth.Engr.*, 5, no. 3, pp. 367–373.

Index